Connections

Connections

The Geometric Bridge between Art and Science

Jay Kappraff
Department of Mathematics
New Jersey Institute of Technology

McGraw-Hill, Inc.

New York   St. Louis   San Francisco   Auckland   Bogotá
Caracas   Hamburg   Lisbon   London   Madrid
Mexico   Milan   Montreal   New Delhi   Paris
San Juan   São Paulo   Singapore
Sydney   Tokyo   Toronto

Library of Congress Cataloging-in-Publication Data

Kappraff, Jay.
    Connections: the geometric bridge between art and science/
    Jay Kappraff.
        p.   cm.
    1. Geometry. 2. Mathematics. 3. Design. I. Title.
    QA447.K37 1990   516—dc20   89-78206
    ISBN 0-07-034250-4   CIP
    ISBN 0-07-034251-2 (pbk.)

1 2 3 4 5 6 7 8 9 0   DOC/DOC 9 5 4 3 2 1 0

ISBN 0-07-034250-4

ISBN 0-07-034251-2   {PBK.}

The sponsoring editor for this book was Joel Stein, the editing
supervisor was Nancy Young, the designer was Naomi Auerbach, and
the production supervisor was Suzanne W. Babeuf. This book was set
in Century Schoolbook. It was composed by McGraw-Hill's
Professional & Reference Division composition unit.

Printed and bound by R. R. Donnelley & Sons Company.

For more information about other McGraw-Hill materials,
call 1-800-2-MCGRAW in the United States. In other countries,
call your nearest McGraw-Hill office.

Contents

Preface

The writing of this book has been a personal exploration for me in the widest sense of the word. Its origins can be traced to my friendship with Mary Blade, an engineer, artist, and descriptive geometer who developed a project-oriented course on the relationship between mathematics and design and taught it for many years at the Cooper Union. I am a mathematician and 10 years ago I presented some of Professor Blade's ideas to a number of colleagues from the Mathematics and Computer Science Departments and the School of Architecture at the New Jersey Institute of Technology. These discussions led to the offering of a course for students from the School of Architecture in the Mathematics of Design. Over the past 10 years, I have had the pleasure of observing beautiful works of art and designs created by my students, based on the mathematical ideas that I have presented to them. It was only years after I started that I learned that I was rediscovering a well-established field of inquiry known to some as design science. This book is meant to be an introduction to this field. I have attempted to make it as comprehensive a survey of the field as space and my own involvement in it permits.

What is design science? It is a subject that has advanced from the twin perspectives of the designer and the scientist sometimes in concert with each other and sometimes on their own, and may be considered to be a geometric bridge between art and science. Design science owes its beginnings to the architect, designer, and inventor Buckminster Fuller. In a meeting with Nehru in India in 1958, Fuller said

> The problem of a comprehensive design science is to isolate specific instances of the pattern of a general, cosmic energy system and turn these to human use.

The chemical physicist Arthur Loeb is one of the individuals most responsible for recognizing design science as an independent discipline.

He considers it to be the grammar of space and describes it as follows:

> Just as the grammar of music consists of harmony, counterpoint, and form
> which describes the structure of a composition, so spatial structures,
> whether crystalline, architectural, or choreographic, have their grammar
> which consists of such parameters as symmetry, proportion, connectivity,
> stability, etc. Space is not a passive vacuum; it has properties which
> constrain as well as enhance the structures which inhabit it.

This book is an exploration of this grammar of space, with the objective
to show, by way of demonstration, that this grammar can be the basis
of a common language that spans the subjects of art, architecture,
chemistry, biology, engineering, computer graphics, and mathematics.
Perhaps design science's greatest value lies in its potential to reverse
the trend toward fragmentation resulting from the overspecialization
of our scientific and artistic worlds and to alleviate some of the isolation
of discipline from discipline that has been the result of that overspecialization.

Design science is an interdisciplinary endeavor based on the work
of mathematicians, scientists, artists, architects, and designers. The
early pioneers, some of whom have been influential in its development
in varying degrees, include the inventor Alexander Graham Bell, the
biologist D'Arcy Thompson, R. Buckminster Fuller, the structural inventor Robert Le Ricolais, Arthur Loeb, the recreational mathematician Martin Gardner, the artist and designer Gyorgy Kepes, the artist
M. C. Escher, and several architectural designers who have contributed
continually to the field. These include David Emmerich, Stuart Dun-
can, Janos Baracs, Anne Tyng, Steve Baer, Michael Burt, Peter Pearce,
Keith Critchlow, and Haresh Lalvani. Reference to these people and
others is found throughout the chapters and in the bibliography to this
book.

Mathematics serves as the foundation of design science, and the
mathematicians who have had the most profound influence on my own
thinking on this subject are H. S. M. Coxeter, Branko Grünbaum, and
Benoit Mandelbrot. Special mention must also be made of the work
gathering and disseminating ideas on the part of the structural topology group at the University of Montreal under the leadership of Janos
Baracs and the mathematician Henry Crapo. In addition, the chemist
Istvan Hargittai has done enormously valuable work editing two large
volumes on symmetry as a unifying force behind science and art and
starting a new journal entitled *Symmetry*. In addition, I would like to
acknowledge another journal, *Space Structures,* which is devoted primarily to structures from an architectural and engineering standpoint.

The unsung heroes of design science also deserve a large share of
the credit for its development. These are people who, for a variety of

reasons have labored, often on a single idea, in their studios, labora-
tories, or studies to discover parts of the thread which binds this discipline together. Today, mathematicians have, for the most part, given
up the study of the roots of their subject in two- and three-dimensional
geometry in order to delve into greater and greater realms of abstraction. As Branko Grünbaum (1981) has lamented:

> It is a rather unfortunate fact (for mathematics) that much of the creative
> introduction of new geometric ideas is done by nonmathematicians, who
> encounter geometric problems in the course of their professional activities.
> Not finding the solution in the mathematical literature, and often not
> finding even a sympathetic ear among mathematicians, they proceed to
> develop their solutions as best they can and publish their results in the
> journals of their own disciplines.

At the same time computer scientists have added their own form of
abstraction to the study of geometry by replacing the constructive
aspects of this subject with two-dimensional pictures on a computer
screen. It is into this dearth of geometrical thinking that artists, architects, designers, crystallographers, chemists, structural biologists,
and individuals from other disciplines have come with their extraordinary constructions and discoveries. A large part of this book is devoted to bringing their ideas to light.

A book such as this must have boundaries and so certain topics were
regrettably omitted. For example, Chaps. 7 through 10, devoted to
polyhedra, leave off where B. M. Stewart's fascinating toroidal polyhedra begin (Stewart, 1980). Also, most of the topics of this book relate
to euclidean geometry, yet projective geometry is a far richer system
of geometry as shown in the work of Janos Baracs and Henry Crapo
and the many books and monographs on the synthetic approach to
projective geometry published by the Rudolf Steiner Institute (Crapo,
1978) (Edwards).

It was only at the conclusion of my work on this book that I discovered
what it was about. On one level, this book is a collection of special
topics in ancient and modern geometry. On another it introduces the
reader to many of the ways that geometry underlies the creation of
beautiful designs and structures. At a deeper level, this book shows
how geometry serves as an intermediary between the unity and harmony of the natural world and the capability of humans to perceive
this order. Le Corbusier has expressed this role of mathematics eloquently (Le Corbusier, 1968b):

> The flower, the plant, the tree, the mountain . . . if the true greatness of
> their aspect draws attention to itself, it is because they seem contained
> in themselves, yet producing resonances all around. We stop short, con-
> scious of so much natural harmony; and we look, moved by so much unity
> commanding so much space; and then we measure what we see.

In this book we shall measure and study the consequences of these measurements but try not to lose sight of the spiritual elements which give meaning and life to the study of design science.

The book is written so that the theory is illustrated at each step by either a design or an application. However, no attempt has been made to be exhaustive in either theory or practice. Each chapter of the book is written so that it can be read separately. However, as is characteristic of design science, each chapter is also tightly interwoven with each of the others. As a result, the reader can choose a variety of paths through the book. Design science is a dynamic discipline. It is forever changing as each practitioner brings his or her new perspective to bear on the subject. In this spirit, the reader is invited to actively participate in the discovery of design science by carrying out some of the constructions, experiments, and problems suggested throughout the book and to think about how the ideas arise in the reader's own discipline.

Although this book was not written as a textbook, if supplemented by a manual of additional exercises, problems, projects, and a guide to instructors, it can be used to teach a course like the one I teach at New Jersey Institute of Technology. McGraw-Hill is considering publishing such a supplementary manual.

Jay Kappraff

Acknowledgments

I would like to acknowledge support that The Graham Foundation offered to make the writing of this book possible. In addition to the people already mentioned, I would like to acknowledge the invaluable help of Alan Stewart, who taught the mathematics of design with me for several years and made many contributions to its development, and to Denis Blackmore, Bill Strauss, and Steve Zdepski, who also worked with me on the early development of the ideas found in this book. A special thanks goes to the generations of students who have taken my course and who, through their creations, have inspired me to develop the ideas found in this book. I wish to acknowledge the help of Branko Grünbaum and Denis Blackmore who read and commented on the manuscript in its early stages and for the help and encouragement of Istvan Hargittai. I am indebted to Haresh Lalvani who made the results of his advanced research in design science generously available to me. He helped me to see how the many parts of this subject fit together, and you will see much of his work displayed throughout this book. N. Rivier and Janos Baracs were also generous in sharing the results of their work with me. I am also grateful for the help of Eytan Carmel, Hyung Lee, and David Henig-Elona, who created many of the drawings, and Richard McNally, Rebeca Daniels, and Vedder Wright, who contributed their comments, ideas, and encouragement. A special thanks goes to Bruce Brattstrom who played a major role in creating drawings and models and in offering a calming influence as final deadlines approached. My patient family also deserves thanks since without their encouragement the completion of this task would have been more difficult and less enjoyable. Finally, McGraw-Hill has been an ideal partner in the creation of this manuscript. I have enormous appreciation for their venturesome spirit in the production of this unusual book. I could not have had two finer editors to work with than Joel Stein and Nancy Young. I, of course, take full responsibility for any errors of content found within these covers.

Credits

Figure 3.21(c) *Photo by Michael Ziegler.*

Figure 3.22 *From Stevens, 1974. © 1974 by Peter H. Stevens. By permission of Little, Brown and Company.*

Figures 3.25, 26, 28, and 29 *From Lendvai, 1966.*

Figure 4.2 *From Williams, 1972.*

Figures 4.9 and 4.10 *From Trudeau, 1976.*

Figure 4.12 *From Beck, et al., 1969.*

Figures 4.15, 4.19, 4.21, and 4.22 *From Baglivo and Graver, 1983. © Cambridge Univ. Press.*

Figure 4.25 *From Trudeau, 1976.*

Figure 4.28 *From Baglivo and Graver, 1983. © Cambridge Univ. Press.*

Figure 4.29 *From Tietze, 1965.*

Figures 4.33, 4.34, and 4.39. *From Firby and Gardiner, 1982. © Wiley.*

Figure 4.41 *From Struble, 1971.*

Figure 4.43 *From Beck et al., 1969.*

Figures 4.45 through 4.48 *From Szilassi, 1986.*

Figure 4.50 *From Baglivo and Graver, 1983. © Cambridge Univ. Press.*

Figure 4.51 *From March and Steadman, 1974. © Methuen, Ltd.*

Figure 4.52 *From Rowe, 1976.*

Figures 4.53 through 4.60 *From Baglivo and Graver, 1983. © Cambridge Univ. Press.*

Figure 4.61 *From Euler, 1979.*

Figures 4.63 and 4.64 *From Baglivo and Graver, 1983. © Cambridge Univ. Press.*

Figure 4.66(a) *From Beck, et al., 1969.*

Figure 4.66(b) and 4.68 *From Coxeter, 1955.*

Figure 5.1 *From* Geometry and Visualization. *© Creative Publications.*

Figures 5.4 and 5.5 *From Davis and Chinn, 1969.*

Figure 5.7 *From Zimmer, H., Kunstf und Yoga im Indischen Kultbild, Berlin, 1920.*

Figure 5.8 *From Michell, 1988.*

Figure 5.9 *From Loeb, 1976.*

Figure 5.10 *From Grünbaum, 1977.*

Figure 5.12 *From Edmondson, 1987.*

Figure 5.14 *From Grünbaum, 1977.*

Figure 5.15 *Drawing by Hyung Lee.*

Figure 5.16 *From Loeb, 1976.*

Figure 5.18 *By Kathleen Slevin-Buchanan.*

Figures 5.19 and 5.20 *From Grünbaum, 1977.*

Figures 5.21 and 5.22 *Courtesy, William Varney.*

Figures 5.23, 5.24, and 5.25. *Courtesy, Janusz Kapusta.*

Figures 5.27, 5.28, and 5.29 *From Williams, 1972.*

Figure 5.31 Consternation *from the Basic Design Studio of William S. Huff; by Scott Grady.*

Figure 5.32(b) *By Dan Wall.*

Figure 5.34 *By Edward Godek.*

Figure 5.35 *From* The Mathematical Tourist *by Ivars Peterson. Copyright © 1988 by I. Peterson. Reprinted by permission by W. H. Freeman and Co.*

Figures 5.36 and 5.37 *Courtesy, Peter Engel.*

Figure 5.38 *By Peter Engel. Photo by Quesada/Burke.*

Figure 5.39 *From Burckhardt, 1976. By permission of World of Islam Publishers.*

Figure 5.40 *From Bourgoin, 1973.*

Figures 5.41 and 5.43(a) *From Critchlow, 1984.*

Figure 5.44 *From Chorbachi, 1988.*

Figure 6.1, 6.2, 6.3, 6.5, and 6.7 *From Stevens, 1974. © 1974. Peter S. Stevens. By permission of Little, Brown and Company.*

Figure 6.8 *From Dormer (1980).*

Figure 6.11 *Construction by Bici Pettit in the Teaching Collection of the Carpenter Center for the Visual Arts at Harvard University. Reproduced with the permission of the Curator.*

Figure 6.15 *From Loeb, 1976.*

Figures 6.16 through 6.23 *From Ash, 1988. Reproduced with the permission of Walter Whiteley.*

Figures 6.29 through 6.32 *From Gilbert, 1983.*

Figure 6.33(a) *By Brian Mullin. Photo by Diana Bryant.*

Figure 6.33(b) *By Eugene MacDonald. Photo by Diana Bryant.*

Figure 6.34 *From Coxeter, 1961. © Wiley.*

Figures 6.35, 6.36, 6.37 *From Rivier, 1984. Reproduced with the permission of N. Rivier.*

Figure 6.39 *By N. G. De Bruijn. Software by G. W. Bisschop, Eindhoven Univ. of Tech., 1980.*

Figures 6.41 through 6.43 *Courtesy H. Lalvani.*

Figure 6.A.2 *From De Vries, V.,* Perspective, *Dover, 1968.*

Figure 6.A.3 and 6.A.4 *Communicated by Janos Baraos.*

Figure 7.1 *Courtesy of G. Segal.*

Figure 7.3(a) *From Kepler, J.,* Harmonices Mundi, Book II, *1619.*

Figure 7.3(b) *From Weyl, 1952.*

Figure 7.6 *From Beck, et al., 1969.*

Figure 7.7 *From Kepler, J.,* Mysterium Cosmographicum.

Figure 7.8 *From Ernst, 1976.*

Figures 7.18 and 7.20 *From Edmondson, 1987.*

Figure 7.22 *From Pugh, 1976.*

Figure 7.23 and 7.24 *From Chu, 1986. Reproduced with the permission of the editor.*

Figure 7.28 *By Patrick DuVal. DuVal, P.,* Homographies, Quaternions and Rotations. *London: Oxford Univ. Press, 1964.*

Figures 7.30, Figures 7.31(a), and Figure 7.32(a). *Photo by Nina Prantis.*

Figure 7.A.1 *Redrawn from Edwards, 1985.*

Figures 8.1, 8.2, and 8.3 *From Laycock, M.,* Bucky for Beginners: Synergetic Geometry. *Activities Resources, Box 4875, Hawyard, CA 94540.*

Figures 8.4 and 8.6 *From Loeb, 1986.*

Figure 8.7 *Courtesy of William Varney.*

Figure 8.8 *From Edmondson, 1987.*

Figure 8.9 *From Pugh, 1976.*

Figures 8.11 through 8.13 *From Edmondson, 1987.*

Figure 8.14 *From Senechal and Fleck, 1988.*

Figures 8.15 through 8.19 *Redrawn from figures in Holden, A.,* Shapes, Space, and Symmetry. *Copyright © 1971, 1973, Columbia University. Reproduced with the permission of Columbia University Press.*

Figure 8.20 *From Pugh, 1976.*

Figure 8.21 *By Dan Winter. Photo by Nina Prantis.*

Figure 8.22 Easy Landing *by Kenneth Snelson (located in Baltimore Harbor).*

Figure 8.23 Needle Tower *by Kenneth Snelson (Washington, D.C.: Hirschorn Museum).*

Figure 8.24(a) and (b) *From Pugh, 1976.*

Figure 8.25 *Construction by Bruce Brattstrom. Photo by Nina Prantis.*

Figure 8.26 *By Jeffrey Fleisher. Photo by Diana Bryant.*

Figure 8.27 *Drawing by Bruce Brattstrom based on photo in Pauling and Hayward. Photo by Nina Prantis.*

Figure 8.28(a) and (b) *Drawing by Bruce Brattstrom based on photos in Stevens, 1974.*

Figures 8.29 and 8.30 *From Edmondson, 1987.*

Figures 8.31 and 8.32 *Redrawn from Critchlow, 1987.* © *Thames and Hudson.*

Figures 8.33 through 8.37. *From Edmondson, 1987.*

Figure 8.39 *From Williams, 1972.*

Figure 9.1 *Redrawn by Bruce Brattstrom from Cundy and Rollett, 1961.*

Figure 9.3 *By Thomas Andrasz.*

Figure 9.5 *From Williams, 1972.*

Figure 9.6 *Redrawn from Loeb, 1976.*

Figure 9.7 *From Edmondson, 1987.*

Figure 9.8 *From Loeb, 1986.*

Figure 9.9 *By Michael Oren. Based on an original design of Arthur Loeb.*

Figure 9.10 *Redrawn from Steinhaus, 1969.*

Figures 9.11 and 9.12 *From Rotge, 1984.*

Figure 9.13 *Courtesy of Ron Resch.*

Figure 9.14 *From Pugh, 1976.*

Figure 9.15 *From Coxeter, 1988.*

Figures 9.17 and 9.18 *Redrawn from Williams, 1972.*

Figure 9.19 *From Ackland, 1972.*

Figures 9.20 and 9.21 *From Salvadori, M.,* Why Buildings Stand Up.

Figure 9.23 *By William Strauss.*

Figure 9.25 *By Francisco Rodriguez.*

Figures 10.1 and 10.2 *Redrawn from Loeb, 1966. Reprinted by permission of George Braziller, Inc.*

Figure 10.3 *From Edmondson, 1987.*

Figure 10.6 *Photo by Diana Bryant.*

Figure 10.7 *From Edmondson, 1987.*

Figure 10.8(a) *Drawing by Bruce Brattstrom.*

Figure 10.8(b) *From Thompson, 1966.* © *Cambridge Univ. Press.*

Figure 10.9 *From Williams, 1972.*

Figures 10.11 and 10.12 *From Loeb, 1970.*

Figure 10.13 *Redrawn from Pearce, 1978.*

Figure 10.14 *From Williams, 1972.*

Figures 10.15, 10.18(b), 10.19(b), 10.20(b), 10.21(a), and 10.22(a) *From Burt et al., 1974.*

Figure 10.16 *Photo by Diana Bryant.*

Figure 10.17 *Photo by Nina Prantis.*

Figure 10.18(a) *Photo by Diana Bryant.*

Figure 10.19(a) *Drawing by Bruce Brattstrom.*

Figure 10.20(a) *Photo by Diana Bryant.*

Figure 10.21(a) and 10.22(a) *Loeb, 1986.*

Figure 10.23 *From Williams, 1972.*

Figures 10.24 and 10.25 *From Edmondson, 1987.*

Figure 10.26 *From Baracs et al., 1979.*

Figure 10.27 *Redrawn from Williams, 1972.*

Figure 10.28 *Redrawn from Baracs et al., 1979.*

Figure 10.29 *Image by H. Lalvani. Software by P. Hanrahan and NYIT Computer Graphics Lab. © NYIT.*

Figure 10.30 *Drawn by H. Lalvani.*

Figure 10.31 *Image by H. Lalvani. Software by D. Sturman and NYIT Computer Graphics Lab. © NYIT.*

Figures 10.32, 10.33, 10.35, and 10.36 *Redrawn from Miyazaki, 1980.*

Figure 10.34 *Image by H. Lalvani. Software by P. Hanrahan and NYIT Computer Graphics Lab. © NYIT.*

Figure 11.3 *From M. Gardner, 1964.*

Figure 11.4 *From Kim, 1981.*

Figures 11.9, 11.11, and 11.14 *From Crowe, 1986.*

Figure 11.15 *From Dover pictorial archive.*

Figures 11.17 and 11.25 *From Martin, 1982.*

Figure 11.22 *Redrawn from Washburn and Crowe, 1989. © Univ. of Washington Press.*

Figure 12.1 *From Dover pictorial archive.*

Figure 12.2 *From Stevens, 1974.*

Figure 12.3 *From Baglivo and Graver, 1983. © Cambridge Univ. Press.*

Figure 12.4 *From Bentley and Humphreys, 1962.*

Figure 12.11 *From Stevens, 1974.*

Figures 12.12 and 12.14 *From Schattschneider, 1986.*

Figures 12.16 and 12.17 *From Dunham, 1986.*

Figure 12.18 *From Martin, 1982.*

Figure 12.19 *From Crowe, 1986.*

Figure 12.20 *From Findeli, 1986.*

Figure 12.21 *From Stevens, 1981.*

Figure 12.23 *From Stevens, 1981.*

Figure 12.24 *From Washburn and Crowe, 1989. © Univ. of Washington Press.*

Figures 12.25 and 12.26 *From Christie, 1989.*

Figures 12.27 and 12.28 *From Gombrich, 1979. © 1979 by Phaidon Press Ltd., Oxford. Used by permission of Cornell University Press.*

Figures 12.40 and 12.41 *From Lalvani, 1982.*

Figures 12.43 through 12.49 *From Shubnikov, 1988.*

Figure 12.50 *Redrawn by Bruce Brattstrom from Critchlow, 1984. Copyright © Thames and Hudson, Ltd.*

Proportion in Architecture

Number is the bond of the eternal
continuance of things.   PHILOLAUS

1.1 Introduction

The history of proportion in art and architecture has been a search for
the key to beauty. Is the beauty of a painting, a vase, or a building due
to some qualities intrinsic to its geometry or is it due entirely to the
craft of the artist and the eye of the beholder?

The architectural and artistic record indicates that a variety of systems of proportion have been used through the ages in an attempt to
create beautiful works. Subjective elements have also played a role;
here proportions of an object are modified to please the eye through a
slow process of evolution. In architecture this process may extend over
many generations in the gradual refinement of traditional forms. In
painting or sculpture the process may involve selecting the most ad-
mired proportions from nature. To a great extent each epoch of history
has expressed itself through the art and architecture of that age
[Panofsky, 1955]. As a result there has been vigorous debate as to
what constitutes "the best" approach to producing great works, with
each era discovering or rediscovering one part of the proverbial "elephant." This chapter will examine some of the approaches to proportion that have been used in the past and will show that they all can be
analyzed in a similar manner.

First we wish to state three canons that most practitioners would
agree underlie a good design. All good design should have

1. Repetition—some patterns should repeat continuously.

2. Harmony—parts should fit together.

3. Variety—it should be nonmonotonous (not completely predictable).

Many architects and artists would add to this a fourth requirement that the proportions of a design should relate to human scale.

Psychological studies of perception seem to indicate that the mind finds overly complex patterns burdensome and unpleasant although it enjoys patterns that embody order and symmetry—in other words patterns that repeat in an organized fashion [Alexander, 1959]. In practice, it also makes sense to use a small number of molds or modules over and over rather than fashioning numerous units of disparate size and shape. Once the modules from which to construct a design have been chosen, the various units must be capable of fitting together to make the finished form. The harmony of proportions should be achieved, according to the Renaissance architect Leon Battista Alberti, in such a manner that "nothing could be added, diminished or altered except for the worse [Gadol, 1969]." Finally, any system of proportions must be flexible enough to express the individual creativity of the artist or architect so that the unexpected may be incorporated into the design. There must always be an element of surprise to enliven the spirit of the beholder.

As for the preference for proportions of human scale, this reflects the desire of humans to feel personally connected to their art and their dwellings. People from primitive cultures are apparently more in touch with this wish, as can be seen in such direct anthropomorphic elements of architectural design as shown in Figure 1.1, which depicts the living compound of the Fali tribe of Africa and is shaped like the human torso [Guidoni, 1978]. We will show how people of various eras endeavored to satisfy these canons of design and will concentrate on how two systems succeeded to some measure in satisfying the canons of proportion. The first system was developed in antiquity and used by Roman architects, and the other was developed in the twentieth century by the French architect Le Corbusier.

1.2 Myth and Number

The nineteenth century mathematician Leopold Kronecker wrote: "The natural numbers came from God and all else was man made." In a sense Kronecker was echoing Plato's *Timaeus* [1977]: "And it was then that all these kinds of things thus established received the shapes of the ordering one, through the action of Ideas and numbers." As pointed out by Matila Ghyka [1978], Greek philosophers, and in particular Pythagoras, endowed natural numbers with an almost magical character. Pythagoras, a native of Samos on the western shores and islands of what is now Asiatic Turkey, took the advice of his teacher Thales, a rich merchant from Miletus who is known as the

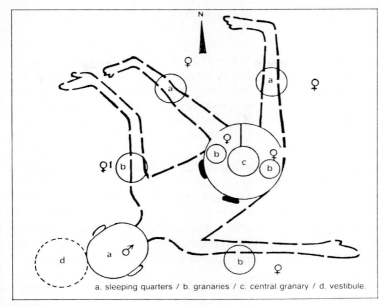

a. sleeping quarters / b. granaries / c. central granary / d. vestibule.

Figure 1.1

father of Greek mathematics, astronomy, and philosophy and who visited Egypt to learn its secrets [Turnbull, 1961], [Gorman, 1979].

Pythagoras singled out the triangular array of 10 points which he called the *tetraktys*. This pattern is the fourth in a series of triangular numbers.

The difference between each successive pair of triangular units is called a *gnomon*. In other words,

$$U = U + G$$

The basic units, U are (the empty set precedes the first dot):

The gnomons, G, are

.   . .   .

In mystical lore, according to John Michell [1988], the natural number 1 was called the *monad* (origin of all numbers). The dyad 2 was the first feminine number and represented the first stage of creation, the split into the mutually dependent opposites of positive-negative, hot-cold, moist-dry, etc. The number 3, the first masculine number, represented the second stage of creation, the productive union of negative and positive which follows the separation and refinement of these opposite elements. The sum of the first feminine and the first masculine number, 5, represented man, microcosmos, harmony, love, and health, while inanimate life was represented by the number 6. The tetraktys, 10, represented the cosmos and macrocosmos, while two interlocking tetraktyses, below, form a Star of David in which 12 evenly spaced dots, representing the signs of the zodiac, surround a thirteenth, representing the "source of all being."

.

.   .   .

.   .   .

.   .   .

.

Looking back from the present we can only speculate about the meaning of this cryptic symbolism. However, it is probably true that the prescientific mind found in the mystical mode of expression a concise way to convey the kernel of meaning in a mass of observations about the natural world. For example, the number 6 does seem to arise most frequently in inanimate forms such as snowflakes and other crystals. On the other hand the number 5 characterizes living forms such as the starfish and certain forms of radiolaria.

Number and geometry also lies at the basis of many sacred structures. Michell feels that certain sacred structures have the same underlying plan. In *Dimensions of Paradise*, Michell suggests that the layout of St. Joseph's settlement at Glastonbury (a sacred site in England rich in legend), Stonehenge, and the plan of the allegorical city in Plato's *Laws* all conform to the ground plan of the New Jerusalem described in *Revelation 21*. His construction is either an intriguing coincidence or, as Michell feels, evidence that ancient cultures may have possessed esoteric knowledge that has become lost to us. The reader must judge.

The New Jerusalem diagram, as Michell refers to it, is generated from a 3,4,5 right triangle. This, so-called, "Egyptian triangle" also had sacred significance to the Egyptians who used it in some of the key proportions of the Pyramid of Cheops (see Section 3.2). But what is so special about a 3,4,5 triangle? Well, the celestial sphere can be represented as a circle divided into 12 equal segments representing the regions of the Zodiac. Cut this circle open to a line with 12 equal segments. The line can then be folded up to a 3,4,5 right triangle with a perimeter of 12 units.

Next Michell uses the 3,4,5 triangle to create a large square with sides of 11 units surrounded by four small squares each with sides of 3 units as shown in Figure 1.2. Circles of diameter D_L = 11 and D_S = 3 are placed in the large and small circles, respectively. Michell has noticed that this ratio, when multiplied by a scale factor of 720, coincides with the ratio of the diameters of the earth and the moon, i.e.,

$$\frac{D_L}{D_S} = \frac{11}{3} = \frac{7920}{2160}$$

and to compound the "coincidence," 720 = (3 + 4 + 5)(3 × 4 × 5).

The circumference of a circle through the centers of the small squares (see Figure 1.2) equals the perimeter of the large square [as close as $22/7$ approximates pi (check this!)] and effectively squares the circle. This conforms with the ancient Greek unfulfilled wish to construct, using only compass and straightedge, a circle with the same perimeter as a given circle.

Finally, Michell creates his New Jerusalem diagram, shown in Figure 1.3, by arranging twelve "moon" circles around the periphery of

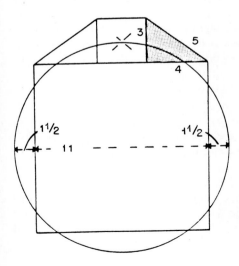

Figure 1.2 The underlying geometry of the New Jerusulem diagram.

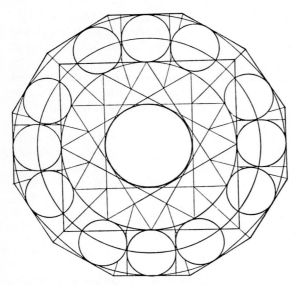

Figure 1.3   The New Jerusalem diagram of ancient cosmology.

the "earth" circle. He placed three to the north, three to the south, three to the east, and three to the west in line with the description of the twelve gates to the Holy City in *Revelation* (see Section 5.2.2). These twelve circles are positioned by the apexes of three double tetraktyses.

When Michell chooses a scale so that the dimensions of the large and small circles are 79.2 and 21.6 feet, respectively, key parts of this diagram closely coincide with the dimensions of Stonehenge and St. Joseph's Chapel. The circle through the center of the "moon" circles is 316.8 feet in circumference. But, according to Michell, this number repeats at a variety of scales as the 31,680-foot perimeter around the entire settlement of Glastonbury as originally constituted and as the 31,680-mile perimeter of New Jerusalem. Also, Pliny in his *Natural History*, gave 3,168,000 miles as the measure round the whole world.

In ancient tradition, the square, by its axial geometry symbolizing the directions of the compass, represented the earth and the dimensions of space while the circle, symbolizing the celestial sphere, represented the realm of the heavens and the dimension of time. Thus, ancient mathematics, architecture, astronomy, and, as we shall see in Section 1.4, music may have been all entwined to form a holistic view of the cosmos. If Michell's analysis has validity, it can be said that an attempt was made to bring heaven down to earth and replicate it at all scales and to synchronize space and time

To a great extent the history of the study of proportion is an attempt to recover the practical methods of producing the beautiful art and architecture of ancient cultures from the sketchy utterances that have survived the ages and the artifacts and structures that comprise the archaeological record.

1.3 Proportion and Number

Once the Greeks established a concept of *natural number*, i.e., the positive integers, they were faced with the task of generating the other numbers of the number system, i.e., the rational and irrational numbers. *Rational numbers* are numbers that can be expressed as the ratio of integers m/n where m and n are reduced to lowest terms and $n \neq 0$. Such numbers can always be represented as decimals whose digits repeat or terminate after some point. Numbers which cannot be expressed as the ratio of integers are called *irrational numbers*. These numbers have nonrepeating decimal equivalents. We who have grown up with a very convenient system for naming numbers such as 8.5, 2.735, .333..., etc., have a difficult time dissociating the concept of number from the symbol for number. However, in ancient Greece no symbols for numbers, as we know them, existed. The symbols that had been used previously by the Babylonians and Egyptians for the purpose of surveying or keeping records had long since been forgotten. Instead of representing numbers by symbols, Greek philosophers conceived of number as being the ratio of lengths. For example, if U is taken to be the basic unit or *monad*, the numbers ³⁄₂ and ²⁄₃ can be represented as shown in Figure 1.4. In other words, a group of three units stands in relation to a group of two units as 3:2 or 2:3 since three groups of two units equals two groups of three units. Any time a finite number of a group of units is exactly equal in length to the finite number of another group of units, we say that the two groups are commensurable. It was a common belief in the time of the Greeks that all pairs of lengths were commensurable. Great surprise and uneasiness resulted from the discovery that there existed pairs of lengths that were

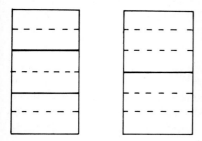

Figure 1.4 The proportional relation 3:2. Three pairs of two monads equals two triples of three monads.

not commensurable. In particular, the ratio of the diagonal of a square and the pentagon to their respective sides were the *incommensurable* ratios:

$$\sqrt{2}:1 \quad \text{and} \quad \phi:1$$

where $\phi = (1 + \sqrt{5})/2 = 1.618\ldots$ is the *golden mean*. This discovery represented a major intellectual stride forward since it had to have been made by pure reason rather than through measurement. The uneasiness was understandable since the problem of incommensurables threw into question the whole Greek system of representing numbers. How then could these incommensurable lengths be characterized? The brotherhood of Pythagoras dealt with this problem by banishing anyone who revealed their distressing secret, although Greek mathematicians developed great facility in constructing certain irrational numbers with compass and straightedge.

The following problem illustrates the profound difference between commensurable and incommensurable lengths. Try to solve it before reading on.

Problem 1.1 Subdivide rectangles with the following proportions into the fewest number of congruent squares: 3:2, 27:15, and $1\frac{1}{3}:\frac{5}{2}$. How many squares are needed to tile the rectangle in each case? Show that a rectangle with the proportions $\sqrt{2}:1$ cannot be tiled by a finite number of congruent squares. What can you say in general about the possibility of tiling a rectangle with proportions $a:b$?

It is obvious that for the first two rectangles 6 and 45 squares are needed with sides of 1 and 3 units, respectively. The third rectangle requires a minimum of 330 squares of $\frac{1}{6}$-unit sides, which can be seen by magnifying it by a factor of 6 to a rectangle of proportion 22:15 where 6 is the least common denominator of $1\frac{1}{3}$ and $\frac{5}{2}$. In Appendix 1.A, we will show that rectangles with commensurable sides can be tiled with a finite number of congruent rectangles while rectangles with incommensurable sides cannot.

Another problem of design that uses the concept of commensurable lengths is the problem of subdividing a given integer length L into numbers m and n of two modular lengths a and b units, respectively, where a and b are integers. This requires m and n to satisfy the equation

$$am + bn = L \quad \text{for } m \text{ and } n \text{ integers}$$

This is known as a *Diophantine equation* [Courant and Robbins, 1941]. Such equations have been studied since ancient times. The most exhaustive study of the application of Diophantine equations to design is

P. H. Dunstone's book, *Combinations of Numbers in Building* [1965].
More is said about this problem in Appendix 1.A.

It took until the latter part of the nineteenth century before mathematicians understood the nature of irrational numbers and could use them with confidence as part of the real number system. Nevertheless, the archaeological studies of Jay Hambridge [1979], which examined the proportions inherent in the structure of Greek vases and buildings such as the Parthenon, indicate that ϕ and $\sqrt{2}$ were very much used. The recent work of two historians of architecture, Professors Donald and Carol Watts [1986], has uncovered evidence that Roman architects may have based some of their art and architecture on a system (to be described later) derived from compass and straightedge constructions of a series of irrationals based on $\sqrt{2}$ and θ where $\theta = 1 + \sqrt{2} = 2.414\ldots$.

Greek mathematics also had a profound influence on artists and architects of the Middle Ages for whom the compass and straightedge were tools for organizing a canvas, often based on $\sqrt{2}$ and ϕ [Bouleau, 1963]. Although this carried over to the Renaissance to some degree (see Section 3.6), for the most part buildings and canvases of the Renaissance were organized by new principles of proportion based on commensurable ratios derived from the musical scale.

1.4   The Structure of Ancient Musical Scales

The aspect of Greek writings that had the greatest influence on Renaissance architecture was the emphasis of Plato in *Timaeus* on the importance of the ratio of small integers. These numbers are the basis for the seven notes of the *acoustic scale* and Plato's assumption that the musical scale also embodied the intervals between the seven known planets as viewed from an Earth-centered perspective (Mercury, Venus, Mars, Jupiter, Saturn, the Sun, and the Moon), which he later referred to (in the *Republic*) as the "harmony of the spheres." These connections deeply influenced the neoplatonists of the Renaissance who felt that, as a result of this connection, the soul must have some kind of ingrained mathematical structure.

Before we examine how the Renaissance architects were able to create a system of architectural proportions based on the musical scale, let us first look at the structure of ancient scales. The ancient scale of Pythagoras was based on the simple ratios of string lengths involving the integers 1, 2, 3, and 4 which made up the tetraktys; all ratios were expressible in terms of the first two primes, 2 and 3 (the first masculine and feminine numbers). Pythagoras understood that if a string is shortened to half its length by depressing it at its midpoint, the resulting bowed or plucked tone sounds identical to the tone of the whole string (or *fundamental tone*, as it is called) except that it is in the next higher register. This relationship, known to Pythagoras as a *diapason*, is what we now call an *octave*.

If a tone and its octave are simultaneously plucked, they give off a luminous sound caused by the anatomy of the ear [Benade, 1976]. (Of course, Pythagoras did not know the reason.) This is why the octave is called *consonant*. Pythagoras also knew that when a string is shortened to ⅔ and ¾ of its original length, other consonant tones are formed which also give off bright effects when they are simultaneously sounded with the fundamental. These special tones were known to Pythagoras as a *diapente* and a *diatessaron*, respectively. However, since they are the fifth and fourth notes of the scale, they are commonly known as a fifth and a fourth. Looking at this in a different way, if a length of string is subdivided into two parts by a bridge, the resulting tones will be an octave, fifth, and fourth when the corresponding ratio of the bowed length to the whole length is 1:2, 2:3, and 3:4 as shown in Figure 1.5.

The Greeks defined the string length corresponding to a whole tone as the ratio between the fourth and the fifth, or ⅞. The structure of the Pythagorean scale is described in *Timaeus*. It is formed by marking off a succession of whole tones while preserving the ratios corresponding to the fifth and the fourth, as shown in Figure 1.6. This leaves two intervals of ratio ²⁴³⁄₂₅₆ left in the octave, which correspond to halftones. Ratios of string length corresponding to powers of 2 introduce no new tones into the scale; they merely transform the fundamental tone to other octaves. The number 3 is needed to create new tones. For example, in Figure 1.7, G corresponds to the string length of ⅔ when the fundamental tone is C. When the string is shortened to

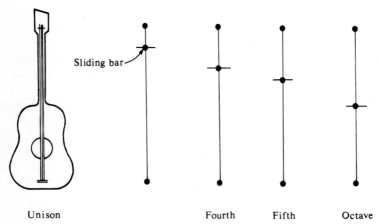

Unison   Fourth   Fifth   Octave

Figure 1.5 A length of string representing the fundamental tone or unison is divided by a bridge to form the musical octave, a fifth, and a fourth.

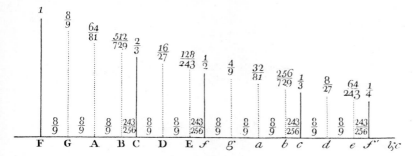

Figure 1.6   The Pythagorean scale derived from the primes 2 and 3.

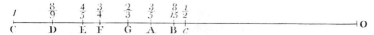

Figure 1.7   The Ptolemaic, or just, scale based on the primes 2, 3, and 5.

$(\frac{2}{3})^2$, the tone D one-fifth above the G (the tone obtained by counting G, A, B, C, D) occurs, which when lowered one octave, $\frac{4}{9} \times 2 = \frac{8}{9}$ (the string is doubled in length), yields the tone D, a whole tone above C. All the tones of the Pythagorean scale are gotten in this way by reducing successive fifths by the appropriate number of octaves.

It is in this context that origins are found for associating the archetypes of the "passive" feminine nature with the number 2 and the "creative" masculine nature with 3. The fact that it has taken thousands of years for these characterizations of male and female natures to begin to break down gives evidence to the power of archetypes as cultural forces.

Various intervals of the scale can be related to each other by splitting the octave by its arithmetical, geometrical, and harmonic means. In general, the *arithmetic mean* of an interval $[a,b]$ is the midpoint, c, of the segment and the points a, c, b form an arithmetic progression. The *geometric mean* is the point c such that $a/c = c/b$, i.e., $c = \sqrt{ab}$ and a, c, b form a geometric progression. The *harmonic mean*, which is less familiar, is the point c, such that the fraction by which c exceeds a equals the fraction by which b exceeds c, i.e., $(c - a)/a = (b - c)/b$. As a result,

$$\frac{1}{c} = \frac{1}{2}\left(\frac{1}{a} + \frac{1}{b}\right)$$

or

$$c = \frac{2ab}{a + b} \tag{1.1}$$

and the series *a, c, b* is referred to as a *harmonic* series. For example, the interval [6,12] represents the octave 2:1. The arithmetic and harmonic means of 6 and 12 are 9 and 8, respectively. That 9 divides the interval into two ratios, 3:2 and 4:3, the musical fifth and fourth, while 8 divides the interval reciprocally into the ratios 4:3 and 3:2 is shown as follows:

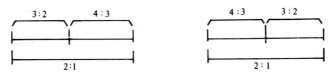

Thus we see that the combination of arithmetic and harmonic means duplicates proportions within an interval, which can be a way of satisfying the first canon of architectural proportion, namely, repetition.

1.5   The Musical Scale in Architecture

Now we turn to the manner in which Renaissance architects applied the Pythagorean scale. The Renaissance architect most influential in applying the musical scale to design was Alberti [Wittkower, 1971], [Scholfield, 1958]. He restricted the lengths, widths, and heights of his rooms to the ratios related to the ancient Greek scale that are shown in Table 1.1.

TABLE 1.1

Ratio	Musical interval
1:1	Unison
4:3	Fourth (diatesseron)
3:2	Fifth (diapente)
16:9	
2:1	Octave (diapason)
9:4	
8:3	Eleventh (fourth above octave)
3:1	Twelfth (fifth above octave)
4:1	Fifteenth (next octave)

All were consonant (or pleasant sounding) except for 9:4 and 16:9, which were compound ratios composed of successive fifths and fourths. To understand how these ratios are all related by a common system, we must first consider the series upon which all systems of proportion are built, the *geometric series*.

In *Timaeus*, Plato conceived of the geometric series as being the binding force of the universe:

> When God put together the body of the universe, he made it of fire and earth. But it is not possible to combine two things properly without a third to act as a bond to hold them together. And the best bond is one that effects the closest unity between itself and the terms it is combining, and this is done by a continued geometrical proportion,...so God placed water and air between fire and earth; and made them so far as possible proportional to each other, so that air is to water as water is to earth—so by these means and from these four constituents the body of the universe was created to be at unity owing to proportion.

The geometric series referred to in the above passage is

fire air water earth

where

$$\frac{\text{fire}}{\text{air}} = \frac{\text{air}}{\text{water}} = \frac{\text{water}}{\text{earth}}$$

Mathematically, $a\ b\ c\ d$ forms a double geometric series if

$$\cdots = \frac{a}{b} = \frac{b}{c} = \frac{c}{d} = \cdots \tag{1.2}$$

where the dots indicate that the series may be continued in both directions. Thus, $a = 1$ and $b = 2$ generates the forward series

$$1\ 2\ 4\ 8\cdots$$

while $a = 1$ and $b = 3$ generates

$$1\ 3\ 9\ 27\cdots$$

These two geometric series arise from the prime numbers 2 and 3 (the first feminine and masculine numbers), which lie at the basis of the Pythagorean scale, and they were arranged into a lambda configuration (λ) by ancient commentators to Plato's work:

1
2   3
4   9
8   27

We shall now see how this double geometric series relates to Alberti's musical proportions. The first of these series is based on the octave (2:1). Another geometric series is formed by the arith-

metic means of each successive pair restricted to integer values
only:

$$1 \quad 2 \quad 4 \quad 8 \quad 16 \quad 32 \; \ldots$$

$$3 \quad 6 \quad 12 \quad 24 \; \ldots$$

Notice that while each number of the second series is the arithmetic
mean of the two numbers that brace it in the upper series, each number of the upper series is the harmonic mean of the pair of numbers
that brace it from below. Also, each series cuts the other in the ratio
3:2 and 4:3 (the musical fifth and fourth). This may be continued
again and again to form endless geometric series in the ratio 2:1 from
left to right, 3:2 along the left-leaning diagonal, and 4:3 along the
right-leaning diagonal involving integers only:

$$
\begin{array}{cccccc}
1 & 2 & 4 & 8 & 16 & 32 \; \ldots \\
 & 3 & 6 & 12 & 24 \; \ldots \\
 & & 9 & 18 & 36 & 72 \; \ldots \\
 & & & 27 \; \ldots
\end{array}
\tag{1.3}
$$

Thus Plato's lambda is formed by the boundary of these geometric series.

P. H. Scholfield [1958] points out that this double series acts like a
chessboard on which horizontal moves represent octaves and moves
along the diagonal represent fifths and fourths. Alberti's ratios (see
Table 1.1) are all represented by any group of numbers from the series
forming the pattern:

$$
\begin{array}{ccc}
 & & 8 \quad 16 \\
\text{such as} & & 6 \quad 12 \quad 24 \\
 & & 9 \quad 18
\end{array}
$$

with the addition of the major whole tone 9:8. Alberti selected any
three numbers from this subscale to represent the breadth, height,
and length of a room. He generally took the height of a room to be
either the geometric, arithmetic, or harmonic means of the length and
breadth. It is easy to see that the subscale gives a convenient guide to
selecting appropriate combinations of this kind. Thus Alberti's system
followed the Pythagorean musical scale.

Followers of Alberti such as Andreas Palladio based their architecture on a revision of the Pythagorean scale that was the work of the

Alexandrian astronomer Ptolemy. This scale, shown in Figure 1.7, achieved a higher order of consonance by considering ratios of the first five integers, which included the prime 5 in addition to 2 and 3. Thus Palladio's architecture included the ratio 3:5 corresponding to the musical sixth (instead of the Pythagorean ratio 16:27), 4:5 (instead of the Pythagorean ratio 64:81), and 5:6 corresponding to the major and minor thirds (a minor tone is one-half interval below the major tone) as Figure 1.7 shows.

The double Series (1.3) can also be related to human dimensions in which a scale of modules is derived from submultiples of the height of a 6-foot person, or 72 inches. Each of these submultiples can then be added together in an arithmetical progression to form the whole. Thus the factors of 72 are arranged in Table 1.2.

TABLE 1.2

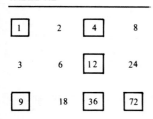

For example, if the module m is taken to be $\frac{1}{12}$ of the whole, six of these make up the whole:

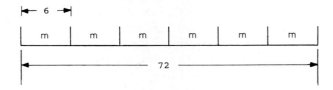

Scholfield has pointed out the surprising fact that six of the twelve subintervals in Table 1.2 (in boxes) result in English measures, namely, the inch, the hand (4 inches), the foot (12 inches), the span (9 inches), the yard (36 inches), and the fathom (6 feet, or 72 inches).

It was actually the Roman architect Vitruvius who spoke of the desirability of basing systems of proportion on the human body. For example, he specified that the entire body, when erect with arms outspread, fits into a square and when spread-eagled, into a circle described around the navel. His 10 books on architecture [1960] com-

prise the only surviving record of the architecture of antiquity, and
these books greatly influenced the architecture of the Renaissance. In
fact, Alberti's, *Ten Books on Architecture* were modeled after
Vitruvius' books. In these books Alberti related the design of the classical Greek columns, ionic, doric, and corinthian, to dimensions of the
human body [Gadol, 1969]. Vitruvius' system was based on subdividing the human form into 120 modules and considering its factors,
listed in Table 1.3, which include series derived from the prime 5. The

TABLE 1.3

1	2	4	8
3	6	12	24
5	10	20	40
15	30	60	120

measurements of various parts of the body were then expressed as an
appropriate fraction of the whole body. Thus not only could repetitions
of proportions be incorporated in a design with the aid of this system
but so also could modules of the same size be repeated to form the
whole, often in symmetric patterns.

Palladio took this system one step further by applying it to architectural interiors. Not only did he apply the Renaissance system of
proportion to the dimensions of a room but he designed the sequence of
rooms in geometric progressions. Although Palladio claimed that
"beauty will result from the form and correspondence of the whole
with respect to the several parts...that the structure may appear an
entire and complete body" [Wittkower, 1971], the limitation of these
geometric progressions prevented him from achieving this worthy objective. The problem was that, in general, geometric progressions do
not possess additive properties, i.e., the sum of two elements in each
geometric progression of Series (1.3) is never equal to another element
of the progression. Thus the second canon of proportion fails and the
system is limited in its application to proportioning only parts of the
whole plan. Along with criticisms concerning the validity of the claim
that what pleases the ear must also please the eye, the lack of additive
properties led to the demise of the system.

1.6   Systems of Proportion Based on $\sqrt{2}$, θ, and φ

The collapse of the Renaissance theory of proportion left architectural
theory in a state of confusion. Without an adequate system, architects
resorted solely to subjective judgments in their designs, often with
dreadful results. However, in the nineteenth century architects, stim-

ulated by an examination of proportions observed in nature during the process of self-similar growth of organisms (see Section 2.10), began to reexamine systems of proportionality in architecture.

In this section we shall show why three proportions, $\sqrt{2}:1$, $\theta:1$, and $\phi:1$, can be singled out as having special properties for use as the basis of architectural systems of proportion. Also, for reasons that we now state, it is unlikely that other proportions can satisfy our three canons of proportion as well.

1.6.1 Additive properties

First of all, it is easy to verify that the golden mean has the property

$$1 + \phi = \phi^2 \tag{1.4}$$

Multiplying Equation (1.4) by powers of ϕ yields the series of expressions

$$\cdots \frac{1}{\phi^2} + \frac{1}{\phi} = 1, \frac{1}{\phi} + 1 = \phi, 1 + \phi = \phi^2, \phi + \phi^2 = \phi^3, \ldots \tag{1.5}$$

where the powers of ϕ form a double geometric series which we shall refer to as the ϕ series:

$$\ldots, \frac{1}{\phi^2}, \frac{1}{\phi}, 1, \phi, \phi^2, \phi^3, \ldots \tag{1.6}$$

Because of Equation (1.5), the ϕ series also has the property that each term is the sum of the two preceding terms. Generally, such a series is called a *Fibonacci series*. That is,

$$a_0 \, a_1 \, a_2 \cdots a_{n-2} \, a_{n-1} \, a_n \cdots \tag{1.7a}$$

is a Fibonacci series if

$$a_n = a_{n-1} + a_{n-2} \tag{1.7b}$$

That is, the F series is

$$1 \ 1 \ 2 \ 3 \ 5 \ 8 \ 13 \ 21 \ 34 \ 55 \ 89 \cdots \tag{1.8}$$

generated by 1, 1, and Equation (1.7b).

All Fibonacci series have the property that ratios of successive terms approach ϕ in the limit, alternating above and below this number (see Section 3.2), i.e.,

$$\lim_{n \to \infty} \frac{a_{n+1}}{a_n} = \phi$$

For example, from Series (1.8),

$$\tfrac{2}{1} = 2.0, \tfrac{3}{2} = 1.5, \tfrac{5}{3} = 1.667, \tfrac{8}{5} = 1.6, \tfrac{13}{8} = 1.625 \qquad \text{etc.} \qquad (1.9)$$

Using the Fibonacci properties of the ϕ series, this series can be constructed with compass and straightedge as we shall show in Section 3.4.

The golden mean and Fibonacci series are the basis of a useful system of architectural proportions developed by the French architect Le Corbusier, known as the *Modulor*. This system will be discussed in the next section. The golden mean and Fibonacci series also have other interesting mathematical properties, some of which will be discussed in the next two chapters. They are connected with certain natural processes such as plant growth, which will be discussed in Section 3.7.

As we did for the golden mean, we can show that θ satisfies the equation

$$1 + 2\theta = \theta^2 \qquad (1.10)$$

and that the powers of θ form a double geometric series

$$\cdots \frac{1}{\theta^2}\, \frac{1}{\theta}\, 1\, \theta\, \theta^2 \cdots \qquad (1.11)$$

with the property that each term is the sum of twice the previous term and the term before that. Such a series is called a *Pell's series*. In general, Pell's series have the property

$$a_n = a_{n-2} + 2a_{n-1}$$

That is,

$$1\ 2\ 5\ 12\ 29\ 70 \cdots \qquad (1.12)$$

It can be shown that the ratio of successive terms in any Pell's series approaches θ as a limit:

$$\lim_{n \to \infty} \frac{a_{n+1}}{a_n} = \theta$$

For example, from Series (1.12)

$$\tfrac{2}{1} = 2.0, \tfrac{5}{2} = 2.5, \tfrac{12}{5} = 2.4, \tfrac{29}{12} = 2.416, \ldots \qquad (1.13)$$

In Section 1.8, the ratio $\theta{:}1$ will be shown to lie at the basis of a system of proportions used by the Romans during the first and second centuries.

1.6.2 Subdividing rectangles

In Section 2.11, we shall describe the gnomic breakdown of rectangles into proportional units by a method known historically as the *principle of repetition of ratios*, which accomplishes in the realm of geometry what the musical scale did in the realm of sound, namely, to provide a means to reproduce proportions within a design. In this section we consider the more general question of how to subdivide a rectangle into subrectangles exhibiting the fewest number of different proportions.

The rectangle in Figure 1.8 is subdivided most generally by a vertical and horizontal line into nine different subrectangles: the four evident in the figure, four additional ones gotten by combining adjacent rectangles, and the outer rectangle enclosing all of the others. However, Figure 1.8(b) and (c) shows that this can be reduced to only two or three, respectively, if the rectangle has proportions $\sqrt{2}{:}1$ or $\phi{:}1$. This will be reconsidered in Section 2.11 in connection with the principle of repetition of ratios.

A similar analysis can be carried out for rectangles subdivided by two horizontal and two vertical lines. The 36 different rectangles for the general case can be reduced to 4 and 5 different rectangles if θ, $\sqrt{2}$, and ϕ are used for the proportions. If four vertical and four horizontal lines are used, the 225 different rectangles can be reduced to only 11.

Thus we see that proportions based on $\sqrt{2}$, θ, and ϕ facilitate the repetition of ratios that fit together to form a whole in aesthetically pleasing ways which satisfy our three canons of proportion.

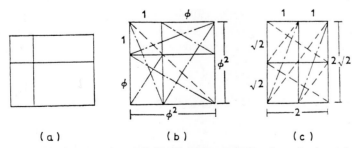

(a)   (b)   (c)

Figure 1.8 A rectangle subdivided by one vertical and one horizontal line into nine subrectangles. The rectangles have (a) all different proportions; (b) two different proportions based on $\sqrt{2}$; (c) three different proportions based on ϕ.

1.6.3 Continued fraction expansions

Perhaps the most convincing evidence of the mathematical pedigrees of ϕ, θ, and $\sqrt{2}$ is given by expanding them in what is known as a *continued fraction* (see Appendix 1.A) [Khinchin, 1964], [Olds, 1963]. Since ϕ satisfies

$$\phi^2 = \phi + 1$$

we can solve for ϕ:

$$\phi = 1 + \frac{1}{\phi}$$

Replacing ϕ repeatedly in this expression yields

$$\phi = 1 + \cfrac{1}{1 + \cfrac{1}{1 + \cfrac{1}{\cdots}}}$$

Likewise,

$$\theta^2 = 2\theta + 1$$

and so,

$$\theta = 2 + \frac{1}{\theta} = 2 + \cfrac{1}{2 + \cfrac{1}{2 + \cfrac{1}{\cdots}}}$$

Also, since $\sqrt{2} = \theta - 1$, it too can be expanded as the continued fraction.

By terminating these fractions at different stages, the ratios called *convergents* given by Equations (1.9) and (1.13) for ϕ and θ are obtained. The series of partial fractions for $\sqrt{2}$ is

$$\tfrac{1}{1} = 1.0,\ \tfrac{2}{1} = 2.0,\ \tfrac{3}{2} = 1.5,\ \tfrac{7}{5} = 1.4,\ \tfrac{17}{12} = 1.4166,\dots \quad (1.14)$$

From the theory of continued fractions, these ratios are the best approximations to ϕ, θ, and $\sqrt{2}$ possible with denominators no larger than the given ones.

Now that we have established ϕ, θ, and $\sqrt{2}$ as the cornerstone of a satisfactory system of proportion, we will study in more detail the system based on θ and $\sqrt{2}$ used by Roman architects of the first and second century and the Modulor system of Le Corbusier based on ϕ.

1.7 The Golden Mean and Its Application to the Modulor of Le Corbusier

Le Corbusier created the first modern system of proportion, which he called the Modulor [1968a]; [1968b], [Martin, 1982]. This system satisfies the three canons of proportion in addition to being built to the measure of the human body. Unlike the Renaissance system, which used a static series of commensurable ratios to proportion the length, width, and height of rooms, Le Corbusier's system developed a linear scale of lengths based on the irrational number ϕ, the golden mean, through the double geometric and Fibonacci ϕ series:

$$\cdots \frac{a}{\phi^2}\ \frac{a}{\phi}\ a\ a\phi\ a\phi^2\ a\phi^3 \cdots \tag{1.15}$$

for some convenient unit a.

In general, the ratios involved in this system were incommensurable, although Le Corbusier often used an integer Fibonacci series approximation to this series, enabling him to operate in the realm of commensurable ratios. However, the fact that Series (1.15) is a Fibonacci series satisfying Equation (1.5) enables the Modulor system to be manipulated analytically in terms of ϕ and its powers rather than through its decimal equivalent. In this section we will study the Modulor.

1.7.1 The red and blue series

Le Corbusier created a double scale of lengths which he called the *red and blue series*. The blue series was simply a ϕ series. This series is constructed by cutting an arbitrary length in the *golden section*, i.e., two segments with lengths in the ratio ϕ:1. A method for doing this will be described in Section 3.4. Since Series (1.15) is a Fibonacci series, all lengths of the double series can be constructed with compass and straightedge. The sequence of elements of the blue series is shown in Series (1.16), with $2d$ replacing a in the ϕ series for arbitrary d (not drawn to scale).

$$\frac{2d}{\phi^2}\quad \frac{2d}{\phi}\quad 2d\quad 2d\phi\quad 2d\phi^2\ 2d\phi^3$$

Blue series: $\cdots$ x   x   x   x   x   x $\cdots$

Red series: $\cdots$ x   x   x   x   x $\cdots$

$$\qquad\qquad d\qquad d\phi\quad d\phi^2\ \ d\phi^3\ \ d\phi^4 \tag{1.16}$$

The red series is constructed according to the pattern of Series (1.3); each length is the arithmetic mean of successive lengths of the blue series that brace it. Therefore, the resulting sequence of elements of the red series is interspersed between lengths of the blue series as shown in Series (1.16). According to Section 1.4, each length of the blue series is the harmonic mean of the two successive lengths that brace it from the red series. The following computation shows that the harmonic mean divides the difference between each pair of lengths of the red series in the golden section $1:\phi$. Consider the interval $[\phi^2, \phi^3]$ from the red series. Using Equations (1.1) and (1.5), the harmonic mean of this interval is

$$c = \frac{2\phi^2\phi^3}{\phi^2 + \phi^3} = 2\phi$$

which is the element from the blue series that intersperses the interval. By using the additive properties of the ϕ series, it is easy to show that 2ϕ cuts the interval in the golden section. (Show this!)

Another relationship between the red and blue series can be seen by considering any length from the blue series, say $2\phi^n$. It equals the difference between the lengths ϕ^{n+2} and ϕ^{n-1} from the red series as we shall show in Section 3.3, i.e.,

$$2\phi^n = \phi^{n+2} - \phi^{n-1} \tag{1.17}$$

The series are drawn to scale in Figure 1.9 which shows how the two series work together with lengths of one interspersed with lengths of the other. This mitigates the effect of the too-rapid geometric growth of either series taken by itself.

Figure 1.10 shows a set of rectangular tiles whose lengths and widths are measurements from either the red or blue series or both. Represented among these tiles are squares, double squares, and golden mean rectangles. This figure also shows that since the lengths and widths are members of a Fibonacci series, if two rectangles having the same width and two successive lengths from either the red or the blue series are joined, a rectangle with the next length in the red or blue series emerges.

To get some experience with the many relationships between these tiles, the reader is encouraged to construct a set of rectangles and try to find interesting ways to combine them. Figure 1.11 shows several

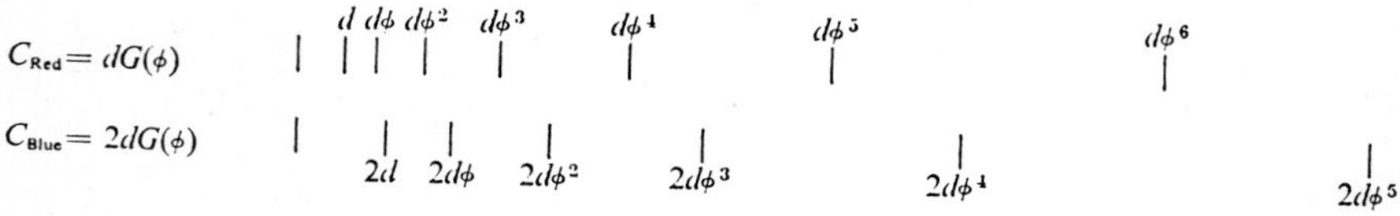

$C_{Red} = dG(\phi)$

$d \quad d\phi \quad d\phi^2 \quad d\phi^3 \qquad d\phi^4 \qquad\qquad d\phi^5 \qquad\qquad\qquad d\phi^6$

$C_{Blue} = 2dG(\phi)$

$2d \quad 2d\phi \quad 2d\phi^2 \qquad 2d\phi^3 \qquad\qquad 2d\phi^4 \qquad\qquad\qquad 2d\phi^5$

Figure 1.9   The Modulor red and blue scale of lengths measured from a common origin.

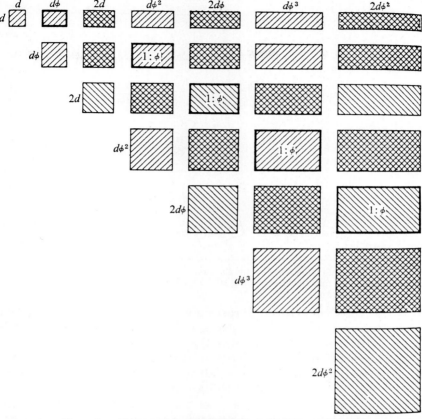

Figure 1.10 The red and blue series of Le Corbusier's Modulor and the tiles of various proportions which the system gives rise to. There are three groups: rectangles whose side lengths are drawn only from the blue series (shaded in one direction), rectangles whose side lengths are drawn only from the red series (shaded in the opposite direction), and those rectangles produced from pairs of dimensions, one red and one blue (both shadings superimposed).

tilings of rectangles by red and blue tiles, found in Le Corbusier's book *Modulor* [1968a].

Problem 1.2 Use the Fibonacci properties of the ϕ series shown in Equation (1.5) to show that the sum of the lengths across the top edges of the rectangles of Figure 1.11 agree with the sum of the lengths across the bottom edges. Also check the sum of the right and left edges for agreement.

The Modulor system is extremely versatile. Once a rectangular area has been tiled by the Modulor, the tiles can be rearranged in many different ways to form new tilings of the rectangle. It can also be used to tile rectangles of arbitrary dimensions to within any preset tolerance (see Section 3.3).

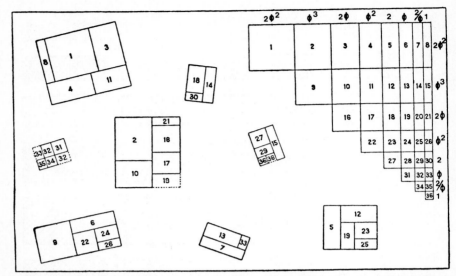

Figure 1.11   A Modulor exercise by Le Corbusier. Eight rectangles are subdivided by Modulor rectangles and coded according to the table in the upper right-hand corner.

Figure 1.12 shows some interesting breakdowns of a square $2\phi^3$ on a side into rectangles of the Modulor. In the last column a 5-inch square is tiled to ¼-inch tolerance. Each column rearranges the same tiles in three different ways. Thus the Modulor satisfies the three canons of proportion. It provides a small number of modules (the rectangles in Figure 1.10) capable of tiling a given rectangular space; the modules all have proportions based on the golden mean, ensuring repetition; and the system has sufficient versatility to enable the designer to find aesthetically interesting subdivisions.

Construction 1.1   Construct your own set of modules and find your own breakdowns of a $2\phi^3$ square. Also test the versatility of the Modulor system by tiling a 5-inch square with red and blue rectangles to within a ¼-inch tolerance.

Despite these satisfactory properties, the Modulor was useful to Le Corbusier and other architects primarily as a theoretical tool, and only rarely has it been used for designing complete buildings. As Le Corbusier suggests in *Modulor 2* [1968*b*], this is to some extent because the scale is too coarse and leaves large gaps at significant points in the design. However, we now show one way in which these gaps can be filled.

1.7.2   Filling in the gaps

Since any gap between two successive lengths of either the red or blue scale equals the preceding length of the scale, an exact scaled-down

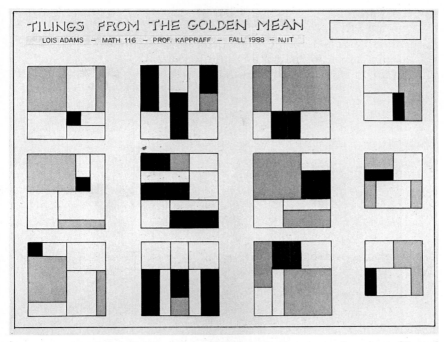

TILINGS FROM THE GOLDEN MEAN
LOIS ADAMS – MATH 116 – PROF. KAPPRAFF – FALL 1988 – NJIT

Figure 1.12 A Modulor tiling by Allison Baxter. In the first three columns a $2\phi^3$ by $2\phi^3$ square is subdivided into three different sets of of tilings. Each set uses the same tiles but is arranged in three different ways. The last column presents the tiling of a 5- by 5-inch square to within ¼-inch tolerance by the same tiles arranged in three different ways.

replica of the red and blue series up to this length fits exactly into the gap. For example, the gap of the length ϕ between ϕ^2 and ϕ^3 of the red series can be filled in by the red and blue series up to ϕ. In this way the Modulor can be extended to a series that is self-similar at every scale, much as we shall see in the next chapter for the fractal patterns of Section 2.12 and the biological patterns of growth of Section 2.10. This can be done without leaving the Modulor system. We shall refer to such a self-similar system as being *closed*.

1.7.3 Human scale

Renaissance artists were well aware that the golden mean modulates the parts of the human body. For example, the Botticelli Venus shown in Figure 1.13 was subdivided by Theodore Cook into a sequence of powers of the golden mean [1979]. For example, ratio

$$\frac{\text{Navel to top of head}}{\text{Navel to feet}} = \frac{\phi^5}{\phi^4 + \phi^5} = \frac{\phi^5}{\phi^6} = \frac{1}{\phi}$$

Figure 1.13   Cook's analysis of a Botticelli Venus.

In other words 1:ɸ. This appears to be close to the average value for this ratio in the adult population at large [Huntley, 1970]. It is also the proportion that seems to have been chosen, both consciously and unconsciously, by artists in all ages to scale human figures in their paintings.

The trademark of the Modulor is shown in Figure 1.14. A 6-foot British policeman with arms upraised provides the determining points of the red and blue series. If the policeman's upraised arm is given the value $2d/\phi$ on the blue scale while the top of his head is d, the remainder of the scale is completely determined and can be constructed by compass and straightedge. (Try it!)

Le Corbusier made these lengths concrete by choosing d so that it is the height of the 6-foot policeman (or 183 centimeters in the metric system). His upraised hand was then set at 226 centimeters. The other lengths of the scale are then approximated rather well by constructing two integer Fibonacci series based on these values, as shown in Figure 1.14:

$$\text{Red: } \ldots 27, 43, 70, 113, 183, \ldots$$

$$\text{Blue: } \ldots 54, 86, 140, 226, \ldots \tag{1.18}$$

Since Le Corbusier worked on both sides of the Atlantic, he found it to be of great practical importance and quite miraculous that when the red and blue scales based on the 6-foot policeman were converted to English units, the corresponding lengths were, to a close tolerance, either an integral number of inches or on the half inch [March and Steadman, 1974]:

$$\text{Blue: } \ldots 8 \text{ in, } 13 \text{ in, } 21 \text{ in, } 34 \text{ in, } 55 \text{ in, } 89 \text{ in, } 144 \text{ in}$$

$$\text{Red: } \ldots 6\tfrac{1}{2} \text{ in, } 10\tfrac{1}{2} \text{ in, } 17 \text{ in, } 27\tfrac{1}{2} \text{ in, } 44\tfrac{1}{2} \text{ in, } 72 \text{ in} \tag{1.19}$$

Equally good design results can be obtained by using the abstract scale of Series (1.16) or its Fibonacci approximations in Series (1.18) and (1.19).

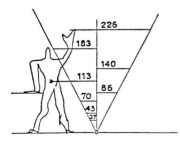

Figure 1.14 The "trademark" of the Modulor. A man-with-arm-upraised provides, at the determining points of his occupation of space—foot, solar plexis, head, tips of fingers of the upraised arm—three intervals which give rise to a Fibonacci series.

1.8   An Ancient System of Roman Proportion

Near the mouth of the Tiber River stand the excavated brick remains of the ancient Roman port of Ostia. Among the ruins of this rich archaeological site are those of a neighborhood of apartments, shops, and garden houses. According to the Watts, who studied the system of proportions that underlies the Garden Houses of Ostia, "even in their ruined state they convey a palpable sense of order and design."

According to the Watts, the key to the design of the Garden Houses is a single geometric pattern based on the square and a particular way of dividing it came to be called the *sacred cut*. By ensuring proportional relations among the parts of the complex and the parts to the whole, the sacred cut lends unity and harmony to the design. The sacred cut works as follows [Watts and Watts, 1986]:

> A sacred cut of a reference square is constructed by drawing arcs that are centered on the corners and pass through the center of the square [as shown in Figure 1.15(*a*)]. By connecting the points where the arcs cut the side, one obtains a nine-part grid, whose central square is called the sacred-cut square. The length of each arc *AB* is equal, to within .6 percent, to the length *CD* of (the diagonal of) half the reference square [see Figure 1.15(*b*)]. Hence the sacred cut provides an approximate method of squaring the circle. The perimeter of a square composed of four lines *CD* is nearly equal to that of a circle composed of four sacred cuts [see Figure 1.15(*c*)].

It is evident from Figure 1.15(*a*) that the ratio of the side of the square to the radius of the sacred cut is $\sqrt{2}:1$ while the ratio of the diagonal of the large square to the radius of the sacred cut is 2:1. The problem of squaring the circle, which was one of the central problems of Greek mathematics, probably marks the influence of the Greeks on this system of Roman proportions. As a matter of fact, it was the Danish

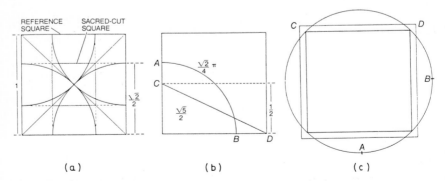

Figure 1.15   Sacred cuts of a reference square. (*Reprinted from "A Roman Complex" by Donald J. and Carol M. Watts. Illustrated by Tom Prentiss.* © *Scientific American.*)

scholar Tons Brunes who coined the term *sacred cut*. Brunes claims that the sacred cut was transmitted from Egypt to Greece in the sixth century B.C. by Pythagoras and then through the Romans to medieval Europe. As the Watts point out in their article,

> To ancient geometers, the circle symbolized the unknowable part of the world (since its circumference was proportional to the irrational number π) while the square represented the comprehensible world. Squaring a circle was a means of expressing the unknowable through the knowable, the sacred through the familiar. Hence the term sacred cut.

According to Watts,

> The geometric order of Ostia's Garden House complex is established by three successive sacred cuts. In Figure 1.16, a square roughly congruent with the perimeter of the complex encloses a circle that touches the corner of the courtyard (*a*). Sacred cuts of the east and west sides of this reference square determine the position of the outer walls of the courtyard buildings (*b*). The second reference square, concentric with the first, is defined by the width of the courtyard and the position of the fountains; the sacred cuts of its east and west sides guide the placement of the party walls along spines of the courtyard buildings (*c*). The third reference square is the sacred-cut square of the second and its cuts define the innermost walls of the courtyard buildings (*d*). The buildings are precisely five times as long as the final sacred-cut square, and their width is equal to its diagonal (*e*). A superposition of all sacred cuts shows how they unfold from a common center, thereby emphasizing the major east-west axis of the complex (*f*).

The sacred cut appears to have been used to proportion the design at all scales from the overall dimensions of the courtyard to the individual buildings to the rooms within each building and even to the tapestries on the wall.

1.8.1   A double series based on the sacred cut

The sacred cut can be related to a double scale quite similar to the red and blue series. Here each scale is the double geometric and Pell's series with common ratio θ discussed in Section 1.6.1, where as before $\theta = 1 + \sqrt{2}$. The ratio of adjacent elements from Series 2 to Series 1 is $\sqrt{2}:1$. Thus,

$$\sqrt{2}\,\frac{1}{\theta} \qquad \sqrt{2} \qquad \sqrt{2}\theta \qquad \sqrt{2}\theta^2 \qquad \sqrt{2}\theta^3$$

Series 2: $\cdots$	x	x	x	x	x	$\cdots$
Series 1: $\cdots$		x	x	x	x	$\cdots$
		1	θ	θ^2	θ^3	(1.20)

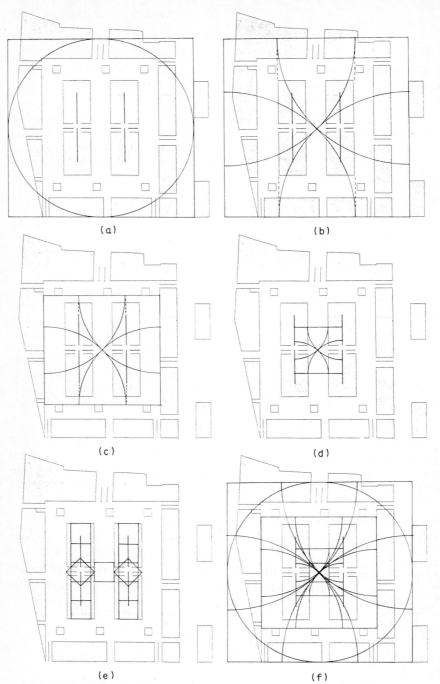

(a)　　　　　　　　　　　(b)

(c)　　　　　　　　　　　(d)

(e)　　　　　　　　　　　(f)

Figure 1.16　Geometric order of Ostia's Garden House complex is established by three sacred cuts. *(Reprinted from "A Roman Apartment Complex" by Donald J. and Carol M. Watts. Illustration by Tom Prentiss.* © Scientific American.)*

It can be verified that the lengths of Series 1 are the arithmetic means of the lengths from Series 2. Thus, according to the results of Section 1.4, the lengths of Series 2 supply the harmonic means between adjacent pairs of lengths from Series 1. This breaks the distance between pairs of lengths from Series 1 in a ratio of 1:θ, or 0.414 (as compared to 1:ϕ or 0.618 for the red and blue series).

A double series of commensurable lengths which are good approximations to Series 1 and 2 can be derived from the convergents of the continued fraction expansion of $\sqrt{2}$, given by Equation (1.14). The resulting double series:

Series 2:

$$
\begin{array}{ccccc}
1 & 3 & 7 & 17 & 41 \\
x & x & x & x & x & \cdots
\end{array}
$$

Series 1:

$$
\begin{array}{ccccc}
x & x & x & x & x & \cdots \\
1 & 2 & 5 & 12 & 29
\end{array}
\qquad (1.21)
$$

are each Pell's series with ratios closely approximating $\sqrt{2}$. Series 1 and 2 of Series (1.19) and (1.20) also possess the following additive properties:

1. The sum of two successive elements of Series 1 is an element of Series 2, e.g., 1 + 2 = 3.
2. The sum of an element of Series 1 and the corresponding element of Series 2 results in the next element of Series 1, e.g., 2 + 3 = 5.
3. The difference between two successive elements of Series 2 is twice an element of Series 1, e.g., 7 − 3 = 2 · 2.

Using these additive properties and beginning with the two lengths 1, $\sqrt{2}$, all other lengths of Series 1 and 2 can be constructed with compass and straightedge. These properties can also be used to subdivide squares or rectangles into lengths from Series 1 and 2 (see Figure 1.8).

Of course this subdivision can be repeated at different scales. In fact courtyard buildings at Ostia are regulated by the sacred cuts of a square whose sides are 41 Roman feet and whose diagonal is equal to the interior width of the building, or 58 feet (twice 29 feet from Additive Property 3). Also, gaps between lengths of Series 1 or Series 2 can be subdivided into scaled-down, self-similar replicas of these series just as was done for the Modulor. However, there is an important difference, discussed below.

1.8.2 Filling in the gaps

Just as the lengths of gaps in the red and blue series were found within the same series, Series (1.21) shows that the length of a gap in Series 1 is found in Series 2 of Series (1.21). As a result, gaps from Series 1 can be tiled by a self-similar replica of the entire double series up to this length. However, gaps from Series 2 are not found in the double series; if elements of Series 2 are doubled, a third series is obtained which contains the gap lengths of Series 2:

$$\text{Series 3: } \cdots 2 \quad 2\theta \quad 2\theta^2 \quad 2\theta^3 \cdots$$

$$\text{Series 2: } \cdots \sqrt{2} \quad \sqrt{2}\theta \quad \sqrt{2}\theta^2 \cdots$$

$$\text{Series 1: } \cdots 1 \quad \theta \quad \theta^2 \quad \theta^3 \cdots$$

Series 2 and 3 now fill gaps from Series 2 with a double scale that is self-similar to the original pair. This process can be continued at any scale; however, it will require an infinite progression of scales obtained by doubling the preceding scale to get the next. We refer to such a system as being *open*.

In conclusion, the Roman system based on the sacred cut appears to have been extremely successful as a system of proportionality. The system shows that the Romans and the Greeks were quite comfortable dealing with incommensurable proportions and that it was Renaissance architects who lost this knack. As a result of the Renaissance architects' insistence on limiting themselves to commensurable proportions only, their systems lacked the additive properties needed for this whole design to be the sum of its parts.

APPENDIX 1.A

Under what conditions can a rectangle of proportions $a:b$ be tiled by a finite number of congruent squares? If this rectangle is to be tiled by a finite number of squares, a and b must be divisible into a finite number of segments of equal length, i.e.,

$$a = mp \tag{1.A.1}$$

and

$$b = np \tag{1.A.2}$$

for m and n nonzero integers, where p is maximized in order to tile with the fewest number of squares. The number of congruent squares N is then

$$N = mn$$

Dividing Equation (1.A.1) by (1.A.2) yields

$$\frac{a}{b} = \frac{m}{n} \tag{1.A.3}$$

from which it follows that lengths a and b must be commensurable. If they are not, finite numbers m and n do not exist and the rectangle cannot be tiled by a finite number of squares.

In the event a and b are commensurable, it follows from Equation (1.A.3) that

$$a = km \quad \text{and} \quad b = kn \tag{1.A.4}$$

Thus, from Equations (1.A.1) and (1.A.2), p is maximized when m and n are the smallest positive integers satisfying Equation (1.A.3). In other words, k is the largest number that divides both a and b to yield integer quotients. If a and b are both integers (and they can always be taken to be integers by scaling the rectangle), k is what mathematicians call the *greatest common divisor* (GCD) symbolized by $k = \{a,b\}$. When integers a and b have no common divisor but 1, $k = 1$ and a and b are said to be *relatively prime*. So we see that m and n are merely the integers in the representation of a/b in lowest terms. It is also evident from Equation (1.A.4) that k is the side length of the congruent squares.

It can be shown that if positive integers a and b are relatively prime and d is a positive integer, the Diophantine equation,

$$am + bn = d \tag{1.A.5}$$

for m and n integers, always has solutions when d is a multiple of the GCD $\{a,b\}$ [Courant and Robbins, 1941]. In the event that m and n are constrained to be positive numbers as they would be if they represented the numbers of two modular lengths subdividing an overall length L (see Section 1.3) and a and b are relatively prime, i.e., $\{a,b\} = 1$, it can be proven that there exists a *critical number* (CN) that equals $(a - 1)(b - 1)$ such that Equation (1.A.5) has at least one solution for $d \geq \text{CN}$ and that there are exactly $\text{CN}/2 - 1$ solutions for values of d less than CN.

To complete this cycle of ideas, the GCD of any two integers a and b can be determined by expanding a/b in a special class of compound fractions known as *continued fractions*. Rather than give a lengthy explanation of how to carry out this expansion, we will generate it for one typical example and leave it to the reader to generate examples of his or her own or study more extensive treatises on this subject [Khinchin, 1964], [Olds, 1963]:

$$\frac{840}{611} = 1 + \frac{229}{611} = 1 + \frac{1}{611/229}$$

$$\frac{611}{229} = 2 + \frac{153}{229} = 2 + \frac{1}{229/153}$$

$$\frac{229}{153} = 1 + \frac{76}{153} = 1 + \frac{1}{153/76}$$

$$\frac{153}{76} = 2 + \frac{1}{76}$$

Since 76/1 leaves no remainder, this sequences of quotients ends and the GCD can be shown to be equal to the denominator of this quotient, or 1, which shows that 611 and 229 are relatively prime.

Putting these results together,

$$\frac{840}{611} = 1 + \cfrac{1}{2 + \cfrac{1}{1 + \cfrac{1}{2 + \cfrac{1}{76}}}}$$

This continued fraction method of finding the GCD is equivalent to a procedure known as *Euclid's algorithm* [Courant and Robbins, 1941].

Problem 1.A.1 Find the continued fraction developments of 2/5, 43/30, and 27/15. What is the GCD of the numerator and denominator in each case?

2

Similarity

To see a World in a Grain of Sand
And a Heaven in a Wild Flower
Hold Infinity in the palm of your hand
And Eternity in an hour.   WILLIAM BLAKE
                    "Auguries of Innocence"

2.1 Introduction

The natural world presents itself to us with a great multiplicity of forms. The shapes of plants, animals, forests, mountains, clouds know no bounds. Yet something in the human mind has sought to tame this great diversity and reduce its orders of complexity to a few general principles. All religions and mythologies begin by creating a world of order from the surrounding chaos. The words of Blake express a yearning to see through the diversity of nature to the underlying connectedness of all things.

Mathematics and science have introduced ways of naming, then classifying, and finally understanding our observations of the natural world in order to gain mastery over it for better or worse. Much of this book is about how geometry presents us with ways of understanding the diversity of forms. In this chapter we shall see how the geometrical notion of similarity gives a way of describing the process of growth in nature.

We begin with a discussion of the the mathematics of similarity and then show how this relates to self-similar forms. We also present a brief introduction to the fractals of Benoit Mandelbrot, which are ultimate generalizations of the notion of self-similarity and present us with a way of literally "holding infinity in the palm of your hand." We conclude this chapter with a brief discussion of some of the ideas of D'Arcy Thompson from his classic study, *On Growth and Form* [1966], in which he describes some of the factors that influence the growth of biological structures and cause organisms to alter their forms to fit their sizes.

2.2 Similarity

Perhaps the most elementary transformation of a geometrical figure is a *similarity* in which the shape of a figure is preserved but its size is altered. Two figures are similar if corresponding lengths have the same ratio, that is, if one is either a magnification or a reduction of the other. We shall refer to the common ratio between lengths as the *magnification* or *growth factor*.

Figure 2.1 shows lines drawn between corresponding points of two similar figures intersecting at a common point, *P*, called the *center of similitude*. This point is familiar; it is the point between the object and image in a pinhole camera. In Figure 2B.1 the object and image are placed side by side and corresponding points are stretched away from *O* by a stretching factor *k*. Such transformations are called *dilatations*, and *k* is the growth factor of the two similar figures, since

$$\frac{MP}{NQ} = \frac{OM}{ON} = k$$

Dilatations will be discussed further in Appendix 2.B where we will show that they are related to another important geometrical transformation called *inversion in a circle*.

If the corresponding lengths are all equal, i.e., the growth factor is unity, the two figures not only have the same shape but also have the same size although they may have different positions and orientations in space. Therefore, they can be matched point for point by moving them rigidly in space as we shall describe in more detail in Chapter 11. In elementary geometry such figures are called *congruent*. In this book we will consider two kinds of congruence. When two figures can be matched point for point by a rigid body motion, they will be called *directly congruent*; when they can be matched by some combination of a rigid body rotation followed by reflection in a mirror, they will be called *indirectly congruent*, or *enantiomorphic*.

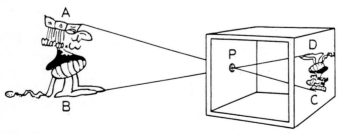

Figure 2.1 Object and image of a pinhole camera are similar figures.

2.3   Families of Similar Figures

Figure 2.2 shows three similar squares, triangles, and sombreros. It is clear by the definition of similarity that any two circles or squares are similar, whereas for two triangles to be similar the lengths of their corresponding sides must be proportional and their angles must be equal. For two sombreros or for any two forms, in general, to be similar, a much larger number of proportional lengths may have to be specified. The following important theorem governs the areas of families of similar figures.

Theorem 2.1   The areas of a family of similar two-dimensional figures are proportional to the square of any characteristic length within the figures; the constant of proportionality depends on the shape of the figure and the characteristic length, i.e.,

$$A = c\ell^2 \qquad (2.1)$$

Thus for any pair of shapes from such a family,

$$A_1 = c\ell_1{}^2 \qquad \text{and} \qquad A_2 = c\ell_2{}^2$$

or

$$\frac{A_2}{A_1} = \left(\frac{\ell_2}{\ell_1}\right)^2 \qquad (2.2a)$$

In this equation, $k = (\ell_2/\ell_1)$, the growth factor. For example, if the characteristic length of a square is taken to be the length of its side, $c = 1$, which is consistent with the common definition of the area of a

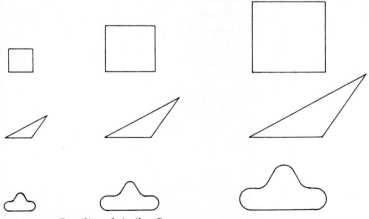

Figure 2.2   Families of similar figures.

square, $A = s^2$. However, if the characteristic length is taken to be the diagonal, $A = \frac{1}{2} d^2$ and $c = \frac{1}{2}$.

Also the areas of families of circles and equilateral triangles are given in terms of the diameters d of the circle and sides s of the triangles by

$$A = \frac{\pi}{4} d^2 \quad \text{and} \quad A = \frac{\sqrt{3}}{4} s^2$$

From Equation (2.2a), we notice that if we double a length, i.e., $\ell_2/\ell_1 = 2$, the area is multiplied by 4, i.e., $A_2/A_1 = 4$, as Figure 2.3 illustrates for squares and triangles.

In Section 8.10 we shall use a version of this theorem generalized to volumes. The volumes of a family of similar three-dimensional figures are proportional to the cube of any characteristic length, i.e.,

$$V = c\ell^3$$

from which we conclude that the volumes of any pair of figures from the family satisfy

$$\frac{V_2}{V_1} = \left(\frac{\ell_2}{\ell_1}\right)^3 \tag{2.2b}$$

2.4   Self-Similarity of the Right Triangle

The dissection of a right triangle results in a family of similar right triangles. To see this, construct two congruent right triangles ABC, of any shape, as shown in Figure 2.4(a). Cut one of them along the altitude BD of length b, drawn to its hypotenuse AC to obtain the right triangles ABD and BCD, respectively. The altitude cuts the hypotenuse of triangle ABC into line segments AD and DC of lengths a and c, respectively. That these two triangles along with the original are a family of similar figures can be seen by superimposing their common right angles, as shown in Figure 2.4(b). The common ratio between corresponding sides is

$$\text{Growth factor} = \frac{a}{b} = \frac{b}{c} \tag{2.3}$$

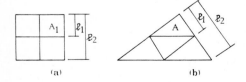

(a)   (b)

Figure 2.3 When a characteristic length is doubled, the area multiplies by four, illustrated for (a) a square and (b) a triangle.

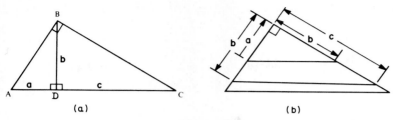

Figure 2.4 Dissection of a right triangle into a family of three similar right triangles.

Thus the right triangle embodies *self-similarity*.

Also, by Equation (2.3), the altitude of the right triangle of length b divides the hypotenuse into two segments of lengths a and c, where b is the *mean proportional* (same as the geometric mean) between a and c. We refer to this as the *theorem of the mean proportional*. Equation (2.3) plays an important role in describing self-similar forms in nature, as we shall see in Section 2.10. Johannes Kepler fully recognized the importance of the self-similarity of the right triangle when he wrote:

> Geometry has two great treasures; one is the Theorem of Pythagoras, the other, the division of a line into extreme and mean ratio. The first we may compare to a measure of gold, the second we may name a precious jewel.

Many proofs of the *pythagorean theorem* have been given, including one by President Garfield, another by Leonardo da Vinci, and an ancient proof given in Section 5.13.3, based on rotational symmetry. One of the most elegant proofs is based on the similarity of triangles ABC, ABD, and BCD obtained by dissecting triangle ABC [see Figure 2.5(*a*)]. In order to get a better picture of these similar triangles, we reflect them in mirrors lying on each of their hypotenuses as shown in Figure 2.5(*b*). If we denote the respective areas of these triangles by A_1, A_2, and A_3,

$$A_3 = A_1 + A_2 \qquad (2.4)$$

From Equation (2.1),

$$A_3 = c\,(\ell_3)^2, \ A_2 = c\,(\ell_2)^2, \ A_1 = c\,(\ell_1)^2 \qquad (2.5)$$

where ℓ_1, ℓ_2, and ℓ_3 are the sides of triangle ABC and the hypotenuses of the three similar right triangles.

Replacing Equation (2.5) in (2.4), it follows that

$$(\ell_3)^2 = (\ell_1)^2 + (\ell_2)^2.$$

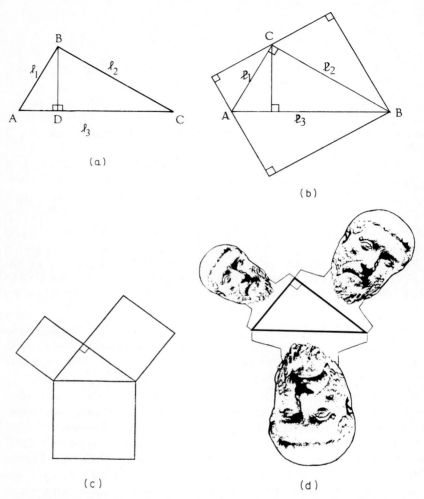

Figure 2.5 Similar families placed on the sides of (a) a right triangle ABC, (b) right triangles reflected from within ABC, (c) squares, and (d) busts of Pythagoras.

From this proof, we see that if three squares are erected on the three sides of a right triangle [see Figure 2.5(c)], the sum of the areas of the squares equals the area of the square constructed on the hypotenuse. But Theorem 2.1 also shows this to be true of any similar figures constructed on the three sides of the triangle, such as the busts of Pythagoras that H. Jacobs whimsically illustrates in Figure 2.5(d) [1987].

2.5 Line Choppers

A family of similar triangles can be used to divide a given length into fractional parts using only compass and straightedge. Such a *line chopper* can be constructed with an arbitrary number of division points, as Figure 2.6 shows for a line chopper with six equally spaced division points $A_0, A_1, A_2, \ldots, A_5$ and parallel line segments $A_1B_1, \ldots, A_5B_5$. Here lines A_0A_5 and A_1B_1 are drawn arbitrarily and we use the fact that through any point a line may be drawn parallel to the given line A_1B_1.

Now if we want to divide a line segment of length L into three equal parts, we merely place the line segment with one end on A_0 and the other end along A_3B_3 as shown in Figure 2.6. By similar triangles, L is subdivided into thirds. To create a line segment of length $4/3L$, merely use a compass to mark off one additional length of magnitude $L/3$.

In this way, segments of length $(m/n)L$ for m and n positive integers can be constructed from a line segment of length L using only compass and straightedge.

2.6 A Circle Chopper

A pair of intersecting lines can be cut by a circle in six distinct ways, two of which are shown in Figure 2.7. In Figure 2.7(a), the intersection point O lies interior to what we call a *circle chopper* and AOB, COD are chords of the circle. In the other figures, O either lies exterior to the circle chopper and OAB, OCD are either secant lines or tangent lines to the circle.

That the two intersecting lines are cut by the circle chopper into two pairs of proportional line segments is given by the following remarkable theorem.

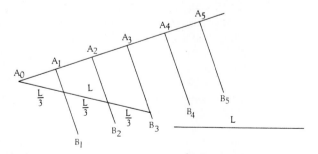

Figure 2.6 A line chopper subdivides a length L into a rational proportion $(m/n)L$ illustrated for $\frac{1}{3}L$ and $\frac{2}{3}L$.

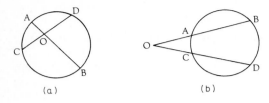

(a)   (b)

Figure 2.7 A circle chopper divides a length so that $OA \cdot OB = OC \cdot OD$ where (a) O is an interior point to the circle and (b) O is an exterior point.

Theorem 2.2   The circle chopper subdivides any pair of intersecting lines so that

$$\frac{OA}{OC} = \frac{OD}{OB} \tag{2.6}$$

A limiting case of this theorem states that the two tangent lines drawn from a circle to their point of intersection are equal.

The proof of this theorem for the case in which O is interior or exterior to the circle, as it is in Figure 2.7(a) and (b), follows from the fact that triangle AOD is similar to triangle BOC. These triangles are similar because the intersecting angles are equal, i.e.,

$$\sphericalangle DOA = \sphericalangle COB$$

$$\sphericalangle BAD = \sphericalangle BCD$$

and $$\sphericalangle ADC = \sphericalangle ABC$$

because of Theorem 2.3 (also referred to in Appendix 2.A as Theorem 2.A.1).

Theorem 2.3   Inscribed angles to a circle that intercept equal arcs on the circumference of the circle are equal.

For a proof of this theorem, see Appendix 2.A. Another proof of Theorem 2.2 following a radically different logic is developed in Appendix 2.B along with a cycle of ideas leading to a formulation of *hyperbolic geometry*. As a corollary to Theorem 2.2, when O is interior to the circle, it follows from Equation (2.6) that the products of the segments of the two intersecting chords are equal, i.e., $OA \cdot OB = OD \cdot OC$. This corollary can lead to alternate ways of solving geometrical problems. Consider Martin Gardner's [1978] two problems, following, which can be solved either by this corollary or by other means.

Problem 2.1   In the middle of a park there is a large circular play area. The city council would like to put a diamond-shaped wading pool inside the circular area, as shown in Figure 2.8(a). How long is each side of the pool?

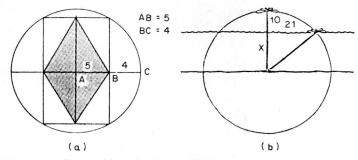

AB = 5
BC = 4

(a)   (b)

Figure 2.8 Two problems by Martin Gardner.

Problem 2.2   There is a famous puzzle about a water lily that the poet Henry Longfellow introduced into his novel, *Kavenaugh*. When the stem of the water lily is vertical, the blossom is 10 centimeters above the surface of the lake. If you pull the lily to one side, keeping the stem straight, the blossom touches the water at a spot 21 centimeters from where the stem formerly cut the surface. How deep is the water? Figure 2.8(*b*) helps to visualize this problem. Your task is to solve for *x*.

2.7 Construction of the Square Root of a Given Length

In Section 2.5 we were able to construct, with compass and straight-edge, any length m/n that is a rational fraction of a given unit. A length equal to $\sqrt{L}$ can also be constructed with the aid of Figure 2.9 as follows:

1. Construct a circle with diameter AB where DB is taken to be one unit and AD is a line segment of length L.

2. Draw a line through D perpendicular to AB.

3. The length of line segment CD, where the circle cuts the perpendicular, has magnitude $\sqrt{L}$.

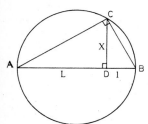

Figure 2.9 Construction of the square root of a given length L.

This construction follows from the observation that *ACB* is a right triangle by the corollary to Theorem 2.A.2. As a result, the theorem of the mean proportional of a right triangle [Equation (2.3)] states that

$$\frac{AD}{CD} = \frac{CD}{BD}$$

But if *AD* = *L, BD* = 1 while we let *CD* = *x,*

$$\frac{L}{x} = \frac{x}{1}$$

from which it follows that $x = \sqrt{L}$.

2.8   Archimedes Spiral

In Section 2.4 we showed that the right triangle can be subdivided into self-similar right triangles. But the right triangle is also connected to the more general theme of self-similar growth through the geometry of the spiral. The spiral is an archetypical symbol found in the art and metaphysics of people in every age. For example, spiral patterns appear on the walls of the cave dwellers, in the sacred symbols of the Buddhists and Hopi Indians, and in the mazes found on the doors of early Gothic cathedrals. Jill Purce [1974], Anne Tyng [1969], and Jay Kappraff [1990] have explored the cultural and metaphysical meaning of the spiral.

There are two fundamentally different kinds of spirals, the *Archimedes spiral* and the *logarithmic spiral*. The Archimedes spiral is rarely found in natural forms although it does correspond to the foraging pattern of certain shellfish. It is the pattern formed on the ground by a horse tethered to a tree as it walks round and round the tree letting out its rope as it walks [see Figure 2.10(*a*)] or by a coiled snake. We represent this schematically as shown in Figure 2.10(*b*), where only the labeled points actually lie on the spiral.

We see from Figure 2.10(*b*) that each time the horse walks around the tree it increases its distance from the tree by *k* units. Thus, since 2π radians equals the angle of one revolution, $\theta / 2\pi$ gives the total number of revolutions that the horse has made, and Table 2.1 shows the relation between the number of revolutions and the distance *r* from the tree.

The Archimedean spiral leads to an *arithmetic series* in *r*. Thus we see from Table 2.1 that $\theta/2\pi$ and *r* both increase in arithmetical progression and we obtain the following relationship between them:

$$r = \frac{k}{2\pi}\,\theta \qquad \text{or} \qquad r = a\theta.$$

where $a = k/2\pi$

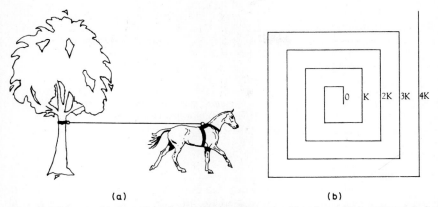

(a)   (b)

Figure 2.10 (a) A horse tethered to a tree walks an Archimedes spiral as it unwinds the rope but keeps it taut; (b) a schematic diagram of the Archimedes spiral.

TABLE 2.1

$\theta/2\pi$	r
0	0
1	k
2	$2k$
3	$3k$
.	.
.	.
.	.

2.9 Logarithmic Spiral

Now let us consider the more important *logarithmic spiral*. Interestingly, this spiral is built up from a right triangle. Consider any right triangle to which an altitude has been drawn to the hypotenuse from the opposite vertex, such as the one shown in Figure 2.4(a).

Restating the theorem of the mean proportional, given by Equation (2.3),

$$\frac{a}{b} = \frac{b}{c}$$

Now consider a sequence of right triangles arranged to form a spider web plotted on polar coordinates, as shown in Figure 2.11. The vertices of these triangles lie on a logarithmic spiral. By repeatedly applying the theorem of the mean proportional to these right triangles,

$$\cdots = \frac{c'}{b'} = \frac{b'}{a} = \frac{a}{b} = \frac{b}{c} = \frac{c}{d} = \frac{d}{e} = \cdots$$

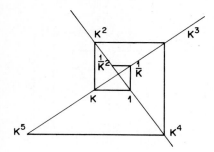

Figure 2.11 A schematic diagram of a logarithmic spiral.

Now, if we let the distance from the center of the spider web to two vertex points of the spiral displaced by 90 degrees be $a = 1$ and $b = k$ units, the above ratios all equal $1/k$, making the successive points $c = k^2$, $d = k^3$, etc., and $b' = 1/k$, $c' = 1/k^2$, etc. Thus we generate the *double geometric series* of numbers,

$$\cdots \frac{1}{k^2}\frac{1}{k}\ 1\ k\ k^2\ k^3 \cdots \tag{2.7}$$

shown in Table 2.2. Since $\pi/2$, or 90 degrees, represents a quarter of a revolution in radians,

$$\frac{\theta}{\pi/2}$$

records the number of quarter revolutions from point to point in the sequence.

From Table 2.2 we see that the distance r from the center of the spider web forms a double geometric series as the number of quarter revolutions,

$$\frac{\theta}{\pi/2}$$

forms an arithmetic series.

From the table we obtain the following relationship between θ and r:

$$r = k^{\theta/(\pi/2)} \qquad \text{or} \qquad r = a^{\theta} \tag{2.8}$$

where $a = k^{2/\pi}$. Taking logarithms of both sides of Equation (2.8),

$$\log r = (\log a)\theta \tag{2.9}$$

Therefore, on semilog graph paper, r versus θ is a straight line connecting $(\theta, r) = (0, 1)$ to $(\pi/2, k)$.

TABLE 2.2

Vertex	$\dfrac{\theta}{\pi/2}$	r
.	.	.
.	.	.
.	.	.
c'	-2	$1/k^2$
b'	-1	$1/k$
a	0	1
b	1	k
c	2	k^2
d	3	k^3
.	.	.
.	.	.
.	.	.

Problem 2.3 Spirals grow at different rates. With the help of a semilog plot, draw four spirals on polar coordinate graph paper in which $k = 2$, ϕ, $\sqrt{\phi}$, and 1, where ϕ stands for the golden mean. Notice how the growth rates of the spirals depend on k.

Using the *growth principle* for the logarithmic spiral that the *radial distance squares as the central angle doubles* and the mean property of the right triangle given by Equation (2.3), other points of the logarithmic spiral can be constructed with compass and straightedge. (Try this!)

Gardner uses Problem 2.4 involving logarithmic spirals to demonstrate the value of insightful mathematical thinking [1978]:

Problem 2.4

Tom Pizza has trained his four turtles so that Abner always crawls toward Bertha, Bertha toward Charles, Charles toward Delilah, and Delilah toward Abner. One day he put the four turtles in *ABCD* order at the four corners of a square room. He and his parents watched to see what would happen.

"Very interesting son," said Mr. Pizza. "Each turtle is crawling directly toward the turtle on its right. They all go the same speed, so at every instant they are at the corners of a square." (See Figure 2.12.)

"Yes Dad" said Tom, "and the square keeps turning as it gets smaller and smaller. Look! They're meeting right at the center!"

Assume that each turtle crawls at a constant rate of 1 centimeter per second and that the square room is 3 meters on the side. How long will it take the turtles to meet at the center? Of course, we must idealize the problem by thinking of the turtles as points.

Mr. Pizza tried to solve the problem by calculus. Suddenly Mrs. Pizza shouted: "You don't need calculus, Pepperone! It's simple. The time is 5 minutes."

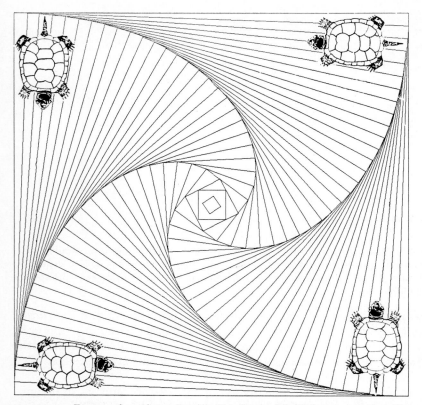

Figure 2.12   Four turtles, Abner, Bertha, Charles, and Delilah, traverse the sides of a square but are constrained to follow each other at all times. Their paths must be logarithmic spirals whose common center is the center of the square.

What was Mrs. Pizza's insight? If you cannot provide the requisite insight to solve this problem, you can always diagram the paths of the turtles in small increments of time, drawing four sides of the square at the end of each interval. The result is the pattern shown in Figure 2.12.

2.10   Growth and Similarity in Nature

The logarithmic spiral is commonly found in nature, for example, in the form of the nautilus shell or the striations of the shells of other sea animals, as shown in Figure 2.13. This follows from an important property of spirals. Any arc of the spiral between two radii separated by an angle θ is similar. In other words, one such arc can be magnified or reduced to form the others, as shown in Figure 2.14. It was pointed out by D'Arcy Thompson [1966] that the nautilus shell and the horns of a steer grow by accretion according to the genetic code of the ani-

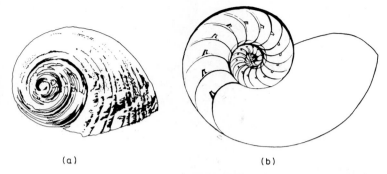

(a)   (b)

Figure 2.13   Natural forms illustrating logarithmic spiral growth. (a) Shell forms; (b) nautilus.

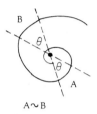

A∿B

Figure 2.14   The central angle of a logarithmic spiral intercepts similar arcs on the spiral.

mal. Thus, if the outer part of the horn grows at a constant rate but faster than the inner part, a logarithmic spiral results as shown in Figure 2.15(a) for a sequence of wooden chips that approximate the annual growth of the horn [Stevens, 1974]. Furthermore, the above property ensures that each section of shell or horn will be self-similar, preserving the identity of that aspect of the organism. If the wooden chips are cut so that the cross sections of the cuts are not perpendicular to the horizontal plane, as they are in Figure 2.15(b), the spiral will wind into three-dimensional space and is called a *helix*. Horns and teeth actually grow in helices whose projections onto the horizontal are logarithmic spirals.

Problem 2.5   The helix shown in Figure 2.16(a) can be thought to represent a spiral ramp rising on the surface of a cylindrical building with radius R and height H and constant pitch α, where the pitch is defined as the angle between the direction of the spiral and the horizontal as seen in the edge view. If the height of the cylinder is $H = 100$ feet and the pitch is $\alpha = 30$ degrees, how far must a person walk up the ramp compared to the distance straight up the side of the wall? Show that the distance up the ramp does not depend on the radius R of the cylinder. The following experiments with spirals supply a hint for the solution of Problem 2.5.

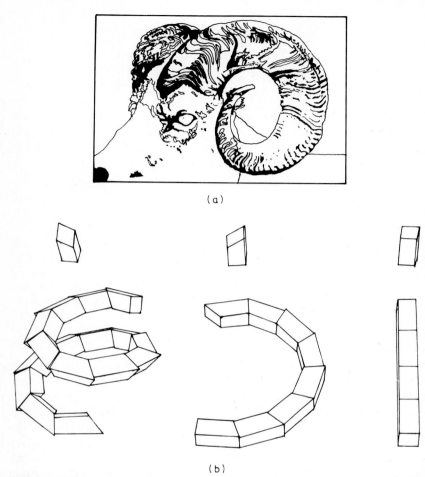

(a)

(b)

Figure 2.15   A logarithmic spiral is formed when a horn grows faster on the outside than the inside illustrated with rectangular wooden blocks cut by a perpendicular plane. If the plane cuts the block at an angle, the growth pattern is helical.

Experiment 2.1.   Get hold of the cardboard cylinder from a roll of paper towels. Mark the spiral ridge of this roll with a red pencil. Cut open the roll along a vertical line AB to form a *period rectangle* of height H, width $2\pi R$, and pitch α, as shown in Figure 2.16(b) for the spiral ramp. Measure R, H, and α for this spiral. Since the points on both vertical sides of this period rectangle are considered to be identical, i.e., $A = A'$, $B = B'$, $C = C'$, the line of constant pitch α continues to

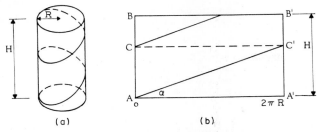

(a)   (b)

Figure 2.16 *(a)* A helical curve on the surface of a cylinder; *(b)* the cylinder is opened to a period rectangle by cutting along a generator. The trace of the spiral is shown.

rise from point C on the left side after reaching the identical point C' on the right side of the period rectangle.

Also note that the spiral revolves about the cylinder in a counter-clockwise manner as in the threads of a standard screw. Such a spiral is called a *right-handed spiral* because the fingers close in a counter-clockwise direction when the right hand is closed into a fist. Right-handed spirals are distinguished from spirals that slope in the oppo-site direction, left-handed spirals.

Can a right-handed spiral be moved in space and matched up point for point with a left-handed spiral? Look at a right-handed spiral in a mirror and notice that it is different from a left-handed spiral.

Construct a double helix as illustrated on the period rectangle of Figure 2.17(a). The configuration of the DNA molecule (the double helix) was discovered by Crick and Watson [see Figure 2.17(b)]. An-other property demonstrating the self-similarity of logarithmic spi-rals can be shown using calculus; namely, the angle between the radius and the tangent at any point is the constant angle ψ as shown in Figure 2.18. For this reason this spiral is sometimes called an *equiangular spiral*. This property is used by certain in-sects that fly toward a light along a logarithmic spiral. They may

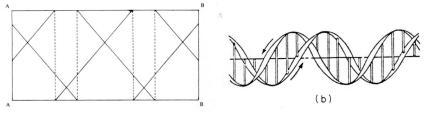

(b)

Figure 2.17 *(a)* Double helix drawn on a period rectangle. *(b)* the DNA double helix.

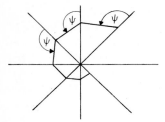

Figure 2.18 A fly moves toward a light source by intercepting light rays at equal angles ψ. The path is a logarithmic spiral.

be thought to possess sensing mechanisms which cause them to intersect light rays at a constant angle. It can be shown, using calculus, that

$$\tan \psi = 1/\ln a \qquad (2.10)$$

where $a = k^{2/\pi}$ as before. These angles were also studied by Theodore Cook [1979] who correlated them with the spiral growth of various natural forms.

2.11 Growth and Similarity in Geometry

We have seen in Section 2.10 that spiral forms generally comprise dead tissue such as shells or horns in which new growth adds to old growth in just such a way as to maintain similarity. Let us investigate this process of growth geometrically.

Begin with some geometric form or pattern, which we call a *unit*, and add to it another form or pattern, called a *gnomon* (see Section 1.2), which is required to enlarge the unit while preserving its form. For example consider the following sequence of units:

and gnomons

The units are square patterns of dots while the gnomons are the L-shaped patterns of dots which must be added to one unit to get the next largest unit in the sequence.

If we consider any rectangle whose sides are in proportion $a{:}b$ and draw a line from one vertex that intersects the diagonal at right angles, the rectangle can be divided into two rectangles. The smaller of the rectangles, whose sides are in proportion $b{:}c$, is similar to the parent, as shown in Figure 2.19(a). This subdivision was known historically as the *principle of the repetition of ratios* and was used by architects during the Renaissance [Scholfield, 1958].

Referring to the similar right triangles *AOB, BOC,* and *COE* and using the theorem of the mean proportional given by Equation (2.3), it follows that the hypotenuses are in proportion,

$$a/b = b/c \tag{2.11}$$

and thus rectangle *ABCD* is similar to *BCEF*.

We may state this in another way. Represent the class of similar rectangles with sides in ratio $a{:}b$ by the symbol U, in which case

$$U = U + G$$

where G is the leftover portion, or gnomon, that remains when a similar rectangle U is removed from the parent [see Figure 2.19(b)]. This process can be repeated over and over again to yield a decomposition of U into an indefinite number of gnomons G and one similar unit U:

$$U = G + U$$
$$U = G + G + U$$
$$\cdots$$
$$U = G + G + \cdots + G + U$$

as shown in Figure 2.19(c). Successive units in this decomposition satisfy the geometric Series (2.7). For example, if the unit U is the rectangle with proportions $\sqrt{2}{:}1$, shown in Figure 2.20,

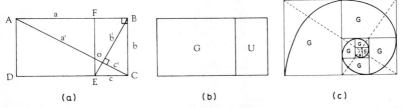

(a)   (b)   (c)

Figure 2.19 Illustration of the principle of repetition of ratios. (a) Diagonal AC of rectangle *ABCD* is intersected at O by a line segment *EB* at right angles to AC; (b) rectangle *ABCD* is divided into a proportional unit U and a leftover part, or gnomon, G; (c) the process is repeated. Corresponding points of G form a logarithmic spiral with center at O.

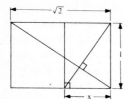

Figure 2.20 The unit U and gnomon G of a square of proportion $\sqrt{2}$:1 are equal.

$$\frac{/\sqrt{2}}{1} = \frac{1}{x} \quad \text{or} \quad x = \frac{\sqrt{2}}{2}$$

Thus $G = U$, and if we have a rectangle of proportion $\sqrt{2}$:1, folding it in half and in half again must yield only rectangles of the same proportion.

Sacred architecture is an area of study in which architects try to recover the geometrical ideas that have gone into the creation of certain revered structures of antiquity. The $\sqrt{3}$:1 rectangle occurs in one such sacred form known as the Vesica Piscis. As Figure 2.21 shows, the Vesica Piscis is the fish-shaped region in common to two intersecting circles of equal radii whose centers lie on each others circumference. The common radius AB and the intersection points C and D form two inverted equilateral triangles. As a result, the surrounding rectangle has proportions $\sqrt{3}$:1.

Problem 2.6   If the parent rectangular unit has ratio $\sqrt{3}$:1, use the principle of repetition of ratios to find the gnomon (G).

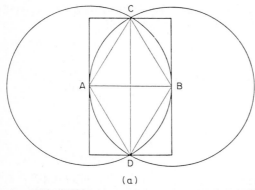

(a)

(b)

Figure 2.21   (a) The Vesica Piscis; (b) marble relief of Christ in a vesica.

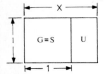

Figure 2.22

Now consider the inverse problem: Given that the gnomon G is a square (i.e., $G = S$), as shown in Figure 2.22, what is the unit U? First note from Equation (2.3) that

$$\frac{x}{1} = \frac{1}{x - 1} \quad \text{or} \quad 1 + x = x^2$$

Solving for x, $x = \phi$ as we saw by Equation (1.4) where ϕ is the golden mean.

Thus the rectangle whose gnomon is a square has the proportion $\phi{:}1$, and a breakdown of this rectangle by the principle of repetition of ratios results in a logarithmic spiral of "whirling squares." Also, the proportions of successive units in this breakdown satisfy the double geometric and Fibonacci ϕ Series (1.6) and form the basis of the Modulor series of Le Corbusier, discussed in the last chapter.

As we did for the golden mean rectangle, $\phi{:}1$, we can show that the unit (U) whose gnomon (G) is a double square, i.e., two squares situated side by side $(G = DS)$ has ratio $\theta{:}1$ where $\theta = 1 + \sqrt{2} = 2.414....$(Do this!) (See Section 1.6.1.)

2.12   Infinite Self-Similar Curves

In recent years, Benoit Mandelbrot, a Polish-born mathematician, has made a study of a strange-looking class of self-similar curves known as *fractals* [1982], [Kappraff, 1986]. He discovered that these curves and certain variants of them are a basic tool for analyzing an enormous variety of natural phenomena such as the shape of mountain ranges, coastlines, rivers, trees, star clusters, and cloud formations. In this section we will examine some of these self-similar curves a little more closely.

2.12.1   Length and scale of a curve

Viewing a curve at a given scale and the definition of its length are two intimately connected notions. There are many different ways to represent a curve at a given scale. One method is illustrated in Figure 2.23, where the curve on the left, spanning the unit interval [0,1], is

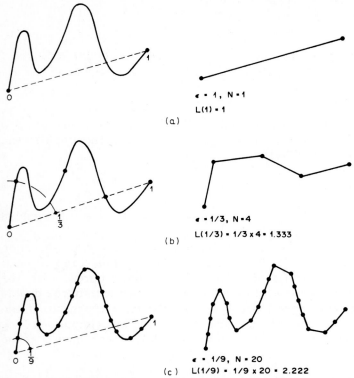

Figure 2.23 Determination of the length L of a curve spanning $|0, 1|$ by approximating the curve with N line segments. (*a*) Representation of curve at scale of $\epsilon = 1$; (*b*) representation of curve at scale of $\epsilon = \frac{1}{3}$; (*c*) representation of curve at scale of $\epsilon = \frac{1}{9}$.

shown on the right at scales of 1, ⅓, and ⅑ in Figures 2.23(*a*), (*b*), and (*c*), respectively. The scaled curves are derived from the actual curve by subdividing the curve with dividers set to intervals of length equal to one-third and one-ninth of the unit, starting at the beginning of the curve as illustrated by the arcs. Each new point is gotten by setting the compass point on the previous point and marking the intersection of the arc of the compass and the curve. The marked points are then connected with line segments. The length of the curve, $L(\epsilon)$ at scale ϵ is then defined by

$$L(\epsilon) = \epsilon N(\epsilon) \tag{2.12}$$

where $N(\epsilon)$ is the number of segments of length L that span the curve. The *total length* L of the curve is then defined as the limiting value that $L(\epsilon)$ approaches as ϵ approaches zero or, mathematically,

$$L = \lim_{\epsilon \to 0} L(\epsilon)$$

A British meteorologist, Lewis Richardson, applied this definition to determine the coastal length of many different countries, and he discovered that, for each of them, the number of segments at scale ϵ satisfied the empirical law

$$N(\epsilon) = K \, \epsilon^{-D} \tag{2.13}$$

where K and D are constants depending on the country. Inserting Equation (2.13) in (2.12),

$$L(\epsilon) = K \, \epsilon^{1-D} \tag{2.14}$$

which yields straight lines when L is plotted against ϵ on log-log graph paper.

Richardson's data indicate that the configuration of coastlines is derived from a general law of nature, and Mandelbrot's analysis of Richardson's data led to the following expression of that law:

> Each segment of a coastline is statistically similar to the whole, i.e., the coastline is statistically self-similar.

2.12.2 Geometrically self-similar curves

Curves are called *geometrically self-similar* if they appear the same at every scale. In other words, if we look at the curve from afar, it appears the same as it does in a closeup view, in terms of its details. In his book *The Fractal Geometry of Nature* [1982], Mandelbrot presents a procedure for constructing curves that are geometrically self-similar. To understand how self-similar curves relate to Richardson's law, it is sufficient to set $K = 1$ and rewrite Equation (2.14) as

$$L(\epsilon) = \epsilon \left(\frac{1}{\epsilon} \right)^{D} \tag{2.15}$$

First, consider a trivial example of a self-similar curve, the straight-line segment of unit length shown in Figure 2.24. This segment is self-similar at any scale. For example, at the scale ⅓, three similar editions of the segment replicate the original. Thus, from Equation (2.15),

$$L(\tfrac{1}{3}) = \tfrac{1}{3} \times 3$$

or

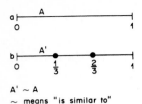

A' ∼ A
∼ means "is similar to"

Figure 2.24   The unit interval: A trivial example
of a self-similar curve with dimension $D = 1$.

$$L\left(\frac{1}{3}\right) = \frac{1}{3} \times \frac{1}{(1/3)^1}$$

and consequently $D = 1$.

Now consider a less trivial example of a curve, self-similar at a sequence of scales $(1/3)^n$, $n = 0, 1, 2, 3, \ldots$ known as the *Koch snowflake*. Since the curve is infinite in length, continuous, and nowhere smooth, it cannot be drawn. However, it can be generated by an infinite process, each stage of which represents the curve as seen at one of the scales in the above sequence. Figure 2.25(*a*), (*b*), and (*c*) shows views of the Koch snowflake at scales of 1, 1/3, and 1/9, respectively, both as linear segments on the left and incorporated into triangular snowflakes on the right. The snowflake is generated iteratively by replacing each segment of one stage with four identical segments one-third the original in length in the next stage. Thus, whereas for stage 1,

$$L(1) = 1$$

for stage 2,

$$L\left(\frac{1}{3}\right) = \frac{1}{3} \times 4 \tag{2.16}$$

or

$$L\left(\frac{1}{3}\right) = \frac{1}{3} \times \frac{1}{(1/3)^D} \tag{2.17}$$

Solving for D from Equations (2.16) and (2.17),

$$D = \frac{\log 4}{\log 3} = 1.2618\ldots$$

For each successive stage in the development of the snowflake, the length is determined from Equation (2.15) for the same value of D.

Each segment of a given stage is seen to be similar to a segment 3 times as large as in the previous stage. Thus, in the limit, each seg-

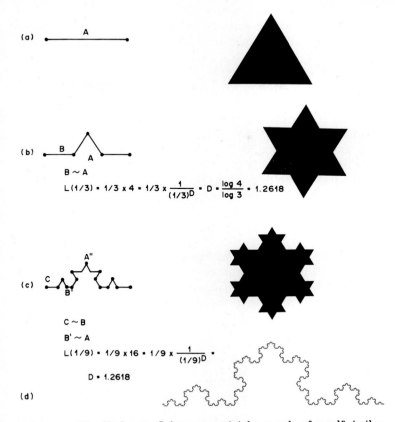

Figure 2.25 The Koch snowflake: a nontrivial example of a self-similar curve with dimension $D = 1.2618$. (a) Koch snowflake at scale of $\epsilon = 1$; (b) Koch snowflake at scale of $\epsilon = \frac{1}{3}$; (c) Koch snowflake at scale of $\epsilon = \frac{1}{9}$; (d) Koch snowflake at an advanced stage in its generation.

ment of length $(\frac{1}{3})^n$ of the Koch snowflake must be geometrically similar to the whole, satisfying both Richardson's data and Mandelbrot's interpretation of it. This property of self-similarity at a sequence of scales is more evident in Figure 2.25(d), which shows a Koch snowflake at an advanced stage in its development.

Mandelbrot shows that, as for the Koch snowflake, any geometrically self-similar curve satisfies

$$D = \frac{\log N}{\log (1/r)} \tag{2.18}$$

where N is the number of congruent segments of length r, the *contraction ratio*, that replaces the unit interval in the initial stage of the iteration. Thus, for the Koch snowflake, $N = 4$ and $r = \frac{1}{3}$.

Mandelbrot refers to D as the *dimension* of the curve, and he shows that for curves of infinite length on a plane surface spanning a finite distance

$$1 < D \leq 2$$

where 1 is the dimension of a line and 2 is the dimension of a surface. The magnitude of D is a measure of the *roughness* of the curve.

The relationship between N and r, expressed by Equation (2.18), is quite general and is illustrated for other geometrically self-similar structures in Figures 2.26 and 2.27. Figure 2.26 is an analogous curve to the Koch snowflake with dimension $D = \frac{3}{2}$, while Figure 2.27 is the third stage of a space-filling Peano curve of dimension 2 that fills up the interior of the Koch snowflake. In its final stage, it would be a non-self-intersecting curve that touches every point within its outer boundaries.

Mandelbrot coined the term *fractal curves* to refer to curves with dimension $1 < D \leq 2$, the term *fractal surfaces* to refer to surfaces with dimension $2 < D \leq 3$, and the term *fractal point sets* for point sets with $0 < D \leq 1$. Although, according to this definition, fractals need not be

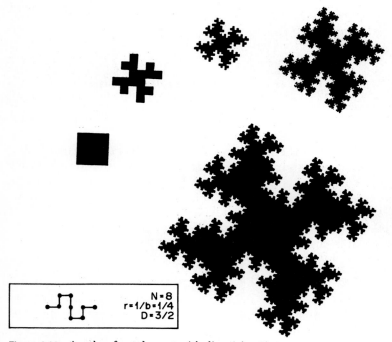

$$N = 8$$
$$r = 1/b = 1/4$$
$$D = 3/2$$

Figure 2.26   Another fractal curve with dimension $\frac{3}{2}$.

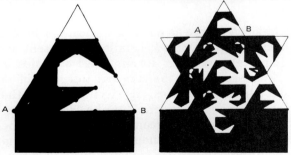

Figure 2.27   The third stage in the generation of a space-filling Peano curve filling the interior of a Koch snowflake. ("*Mandelbrot's Space Filling.*" © *1978 by Benoit B. Mandelbrot. Reprinted by permission of Scientific American, Inc.*)

self-similar, it is the class of self-similar fractals that are generated by Mandelbrot's recursive procedure.

2.12.3 Fractals and scale

Let's revisit the Koch snowflake shown in Figure 2.25, but this time imagine the curve to represent a spectrum of amplitudes of sound over an interval of time. Generally, a phonograph recording of sound, such as the sound of a violin, changes if the record is played fast or slow. In fact, a record of whale sounds is inaudible until the record is played at a sufficiently high speed. However, Koch snowflake music would clearly sound the same played at one-third the speed and then amplified three times. More precisely, if the amplitude at time t is represented by the function $B(t)$, the scaling property of snowflake music is a statement about the identity of the functions $B(t)$ and $B(rt)/t$, where r is the contraction ratio. Such sounds are called *scaling noises* and have been studied by a colleague of Mandelbrot's, Richard Voss, a physicist at IBM Watson Research Center [Gardner, 1978*d*].

It is this so-called *scaling invariance* which is the most important property of fractals. To the degree to which a fractal represents a naturally occurring form or process, virtually all of the relevant information about the fractal model of this form or process is already present in the initial stages of its generation. This includes its self-similar unit, its mode of transformation from stage to stage, and its dimension. Thus, a realistic image of the form or simulation of the process can be obtained with relatively little information about it. This has extremely important implications for image processing as Michael Barnsley shows in *Fractals Everywhere* [1988].

2.12.4 Statistical self-similarity

Although the Koch snowflake can serve as a mathematical model of a coastline, as we mentioned in Section 2.12.1, it fails to represent actual coastlines in two important respects. Its sequence of scales is bound to powers of ⅓. Thus, examining the curve at intervals of ¼ would yield none of its self-similar properties. Also as irregular as the snowflake is, its structure is completely ordered, unlike that of coastlines. Both of these shortcomings can be overcome by randomizing the fractals. It was Mandelbrot's discovery that many natural phenomena such as coastlines and mountain ranges are *statistically self-similar*. For example, no matter at what distance the mountain range shown in Figure 2.28 is viewed, a similar pattern is reproduced, in a statistical sense.

Mandelbrot also discovered that such things as fluctuations in the

Figure 2.28   A computer-generated landscape with dimension $D = 2.5$.

levels of rivers; variations in the brightness of sunspots; changes in the rhythm, variations, and pitch of music; and fluctuations in the stock market all appear to be statistically self-similar. This has led Voss to make the daring conjecture: "The changing landscape of the world seems to be statistically self-similar [Gardner, 1978d]."

2.13   On Growth and Form

In Lilliput, "His Majesty's Ministers, finding that Gulliver's stature exceeded theirs in the proportion of twelve to one, concluded from the similarity of their bodies that his must contain at least 1728 (or 12^3) of theirs (by volume), and must needs be rationed accordingly [Thompson, 1966]." But as Galileo showed in great detail, creatures with dimensions one-twelfth those of a human's body would not be able to survive unless their entire form changed appropriately. In order to confront the forces of their environments, organisms spanning the scale from very little to very big, e.g., from ants to elephants, must evolve different forms. The connection between growth and form is a subject of great fascination. Thompson's classic, *On Growth and Form* [1966], and J. T. Bonner's *Morphogenesis* [1963] are devoted to investigations of these and related issues.

Similarity is also a concept of crucial importance to architects who must design buildings to large and small scales. Before the relation of size to form was understood, the architect had to be satisfied with copying examples of successful architecture without altering its dimensions or else risk the collapse of the structures. History is replete with structures that failed after they were scaled up. In the background we always have Galileo's warning that as a structure becomes larger it gets weaker. He cites as examples [1954]:

> Who does not know that a horse falling from a height of three or four cubits will break his bones, while a dog falling from the same height will suffer no injury? Equally harmless would be the fall of a grasshopper from a tower or the fall of an ant from the distance of the moon. And just as smaller animals are proportionately stronger and more robust than larger, so also smaller plants are able to stand up better than larger

In this chapter we have described the mathematics behind similarity and discussed how certain biological structures are able to maintain similarity during growth, which let us consider the question of why organisms generally must alter their forms to fit their sizes.

In Section 2.2, we showed that

$$V = c_1 \ell^3 \qquad \text{while} \qquad S = c_2 \ell^2$$

Thus,

$$\frac{V}{S} = \frac{c_1 \ell^3}{c_2 \ell^2} = c\ell$$

so that the ratio of volume to area is proportional to the characteristic length of a given form. We give two examples of how this relation influences the form of living organisms.

First consider a cylinder, shown in Figure 2.29, which may be thought of as a crude model of a limb. Its volume and cross-sectional area is given by $V = c_1 d^3$ and $S = c_2 d^2$ where d is the diameter of the cylinder. Thus,

$$\frac{V}{S} = cd \tag{2.19}$$

But V is proportional to the weight of the cylinder so that V/S is proportional to the force per unit area, or stress, upon the base of the limb. Equation (2.19) thus states that doubling the size of an animal

Figure 2.29 A cylinder of volume V, cross-sectional area S, and diameter d exerts a stress of $V/S = cd$.

has the effect of doubling the stresses experienced by its limbs. To compensate, the elephant has developed very thick limbs.

As a second example of Equation (2.19), let S be the surface area of a warm-blooded animal, while V is its bulk (volume). The rate of heat loss from a warm-blooded animal is proportional to its surface area, while the rate of heat gain is proportional to its bulk (larger animals tend to burn a greater amount of energy per unit time), and the rate of heat loss must equal the rate of heat gain so as to keep its temperature constant, according to Equation (2.19). Since the rate of heat loss for small animals (low values of ℓ) relative to rate of heat gain is greater than for larger animals, small animals must consume many more calories in the course of a day, relative to their weight, than large animals. According to Thompson,

> Man consumes a fiftieth part of his own weight of food daily, a mouse will eat half its own weight in a day; its rate of living is faster, it breeds faster, and old age comes to it much sooner than to man. A warm-blooded animal much smaller than a mouse become an impossibility; it could neither obtain nor digest the food required to maintain its constant temperature.

Appendix 2.A

Theorem 2.A.1   Inscribed angles that intersect equal arcs on a circle are equal.

proof   The following proof was communicated to me privately by Amos Franceschelli, a mathematics teacher retired from the Rudolf Steiner School in New York. It differs from the standard proof in that it is not analytical but, rather, it makes use of the symmetry of the circle and calls upon the reader to use quiet contemplation along with the logic of Euclidean geometry. We sketch the proof and leave the details (or quiet contemplation) to the reader.

intuitively accepted or preproved properties (IAP)

1. A circle has *perfect symmetry* by which we mean it can be rotated into itself about its center through any chosen angle (see Figure 2.A.1). A circle rotated

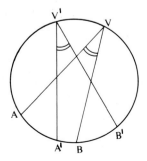

Figure 2.A.1   A circle has perfect symmetry: any inscribed angle AVB can be rotated through an arbitrary arc VV' to a congruent angle $A'V'B'$.

into itself about its center will carry an inscribed angle AVB into a new position $A'V'B'$. Arcs VV', AA', BB' will all be equal.

2. The perpendicular bisector of any chord to a circle goes through the center of the circle and divides the circle into two symmetric halves related to each other by reflection. As a result of this reflection symmetry, two parallel lines intercept equal arcs on a circle. Conversely, if the endpoints of two equal arcs AB and CD are connected by chords AC, BD, the lines AC, BD will be parallel, i.e., $AC \parallel BD$ (see Figure 2.A.2).

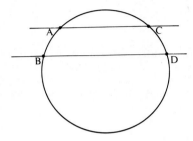

Figure 2.A.2 Two pairs of parallel lines intersect equal angles.

3. If two angles have their sides respectively parallel and in the same sense, the angles are equal (Figure 2.A.3).

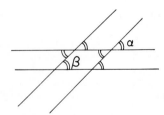

Figure 2.A.3 $\angle \alpha = \angle \beta$

proof proper Given a circle with inscribed angles AVB and $AV'B$ intercepting the same arc AB, rotate the circle about itself, together with angle $AV'B$ only, say counterclockwise, until AV' takes on the position $A'V'' \parallel AV$ and angle $AV'B$ moves into the position of angle $A'V''B'$ (by first IAP) (see Figure 2.A.4). Then arc $A'B' = $ arc AB and $\angle A'V''B' = \angle AV'B$. Now

$$\text{arc } AA' = \text{arc } VV'' \qquad \text{(by second IAP)} \qquad (2.A.1)$$

Also, arcs $A'B' - A'B = AB - A'B$ (since arc $A'B' = $ arc AB). Thus, arc $BB' = $ arc AA' and it follows from Equation (2.A.1) that arc $BB' = arc \ VV''$. Hence $BV \parallel B'V''$ (by the second IAP) and $AV \parallel A'V''$ (by the rotation we made). Therefore $\angle AVB = \angle A'V''B' = \angle AV'B$ (by the third IAP and by our rotation). Q.E.D.

Theorem 2.A.2 Inscribed angles equal one-half of the central angle that intercepts the same arc.

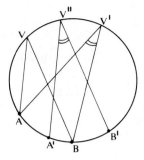

Figure 2.A.4 Angle $AV'B$ is rotated through arc $V'V''$ to $A'V''B'$ where $AV' = A'V''$.

proof

1. Given an arc AB of a circle, draw the central angle 2θ, extend one of the radii to a diameter of the circle at C, and consider the inscribed angle $\sphericalangle ACB$ as shown in Figure 2.A.5.

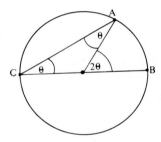

Figure 2.A.5 Inscribed angle θ equals one-half the central angle intercepting the same arc.

2. If the central angle is taken to be 2θ, the inscribed angle must be θ, making use of the fact that the exterior angle of a triangle equals the sum of the alternate interior angles. This proves the theorem.

corollary Any angle inscribed in a semicircle is a right angle.

Appendix 2.B

2.B.1 Centers of similitude and inversion

If two similar figures are placed side by side with the same orientation, the joins of any two corresponding points P and Q or M and N define a center of similitude O as shown in Figure 2.B.1. Similarly, corresponding points are stretched away from O by a factor k where

$$\frac{OM}{ON} = \frac{OP}{OQ} = k$$

In general, transformations in which points are stretched away from a center are called *dilatations* [Coxeter, 1955].

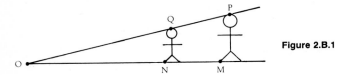

Figure 2.B.1

Now consider what happens if two circles with centers at M and N are transformed by a dilatation as shown in Figure 2.B.2 where O is the center of similitude and P and Q are two points on the circle that correspond to each other under the dilatation

$$\frac{ON}{OM} = \frac{OP}{OQ} = \frac{OP' \cdot OP}{OP' \cdot OQ} = \frac{OP' \cdot OP}{OA \cdot OB} \qquad (2.B.1)$$

where $OP'OQ = OA\,OB$ follows from Theorem 2.2.

From Equation (2.B.1) it follows that,

$$OP \cdot OP' = \frac{OA \cdot OB \cdot ON}{OM} = k^2 \qquad (2.B.2)$$

Now, any pair of points lying on the same half of a line through O and satisfying Equation (2.B.2) are said to be related by inversion in a circle of radius k and center at O. Thus, Theorem 2.B.1.

Theorem 2.B.1 Any two circles, the corresponding points of which are related by dilatation, are also related by inversion.

2.B.2 Another proof of Theorem 2.2

Now let's look at a proof of the "circle chopper" Theorem 2.2 in Section 2.6. Consider the circle of radius r and center at M that cuts a ray drawn from O at points A and B as shown in Figure 2.B.3. Drop a perpendicular MS to OB. It is evident from the figure that

$$OA \cdot OB = (OS - s)(OS + s)$$
$$= (OS^2 - s^2) = OS^2 - (r^2 - h^2)$$
$$= (OS^2 + h^2) - r^2 = OM^2 - r^2$$
$$= t^2 \qquad (2.B.3)$$

where t depends only on point O and the circle but not points A and B. Thus any other line through O would yield the same value of t^2.

Much more can be gotten from the proof of this theorem than we bargained for. Since from Equation (2.B.3), $t^2 + r^2 = OM^2$, we can see from Figure 2.B.4 and the fact that the tangent to a circle is perpendicular to its radius, by use of the pythagorean theorem, that t must be the length of the tangent to the circle from O. But lots of circles

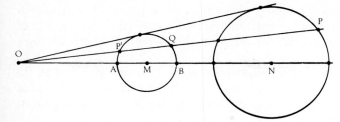

Figure 2.B.2   Two circles related by dilatation.

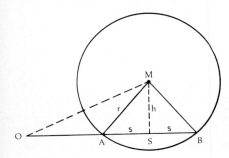

Figure 2.B.3 Alternative proof of Theorem 2.2 that $OA \cdot OB =$ constant.

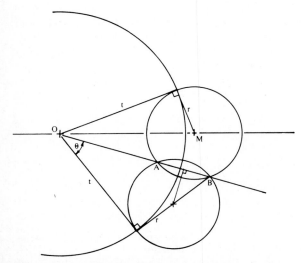

Figure 2.B.4   Construction of a circle orthogonal to a family of circles that share a common chord.

have AB for a chord. Therefore, there is a whole pencil of circles with tangent lines of the same length t from O. Thus, a circle through O of radius t is orthogonal to all circles from this pencil of circles through AB.

Also, if point O is moved along line AB to O', a new circle with its center at O' cuts the pencil of circles through AB orthogonally. In this way, we can construct the set of mutually orthogonal circles shown in Figure 2.B.5.

As a final piece in this web of ideas, since $OA \cdot OB = t^2$, A and B are related by inversion in each of the orthogonal circles to the pencil of circles through AB, as stated in Theorem 2.B.2.

Theorem 2.B.2   Any circle through two points that are inverse with respect to a given circle intersects that circle orthogonally.

2.B.3   The Poincaré plane and stereographic projections

The parallel axiom of euclidean geometry says that there is only one line parallel to a given line. This axiom was the subject of much discussion throughout the history of mathematics. For centuries mathematicians tried to deduce it from the other axioms. Finally, in 1823 Bolyai and Lobachevsky gave an example of another geometry called *hyperbolic geometry* that satisfied all the other axioms of euclidean geometry except the parallel axiom. This established that the parallel

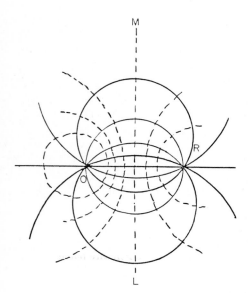

Figure 2.B.5   Two families of orthogonal circles.

axiom is independent of all the other axioms. In Bolyai's geometry, through any point there can be an infinity of lines parallel to a given line [Coxeter, 1961].

Poincaré constructed a "model" of this hyperbolic geometry in which the euclidean plane is replaced by the interior of a circle called the *Poincaré plane* and lines are represented by arcs of circles that cut the circle orthogonally. Figure 2.B.6(*a*) shows how an infinite number of lines (arcs) can be parallel to a given line (arc) in hyperbolic geometry. A pencil of arcs in the Poincaré circle correspond to the set of lines intersecting at a common point in euclidean geometry [see Figure 2.B.6(*b*)]. By Theorem 2.B.1 each of these arcs shares the chord through the intersection point and its inverse in the Poincaré circle. Likewise, using Theorem 2.B.2, given two points *P,Q* in the Poincaré plane, the unique line (arc of a circle) between them is an arc of the unique circle through *P,Q* and the inverse of either *P* or *Q* with respect to the Poincaré circle.

Poincaré's hyperbolic universe is as different from the euclidean universe of our geometric experience as we can imagine. In Figure 12.17, Douglas Dunham has generated, by computer, a print in the style of Escher's famous woodcut, *Circle Limit I*. All the fish in Dunham's print are "congruent." In what sense is this true? The Poincaré circle is considered to be the "infinitely distant" edge of the "universe" and the fishes' "apparent" sizes decrease as they approach this circle. In other words, if you lived in this universe and wanted to walk from a point within it toward the edge, your footsteps would, in the euclidean view, seem to diminish to length zero as you approached the Poincaré circle so that you would never be able to reach it. More generally, in hyperbolic geometry, all similar figures are congruent in the sense that if they are transformed one to the other, their lengths in the metric (formula for measuring length) of hyperbolic geometry are equal. There are many other geometric curiosities exhibited by the Poincaré model of hyperbolic geometry such as the "idealized" equilateral triangle shown in Figure 2.B.7. It is the largest triangle in the hyperbolic plane. It has finite area, infinite perimeter, and angles of

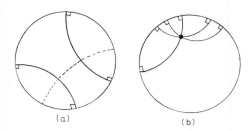

(a)   (b)

Figure 2.B.6 (*a*) Two parallel lines; (*b*) a pencil of lines intersecting at a point in the Poincaré plane.

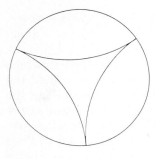

Figure 2.B.7 The largest triangle in the Poincaré plane.

zero degrees (although its vertices have a specious existence since they lie outside the "universe").

The Poincaré model of the hyperbolic plane can be made more transparent by looking at it as the stereographic projection of the points on a sphere to points on a plane T tangent to the sphere at the south pole S from a projection point at the north pole N of the sphere (see Figure 2.B.8).

In this projection, the image of a typical point P is the intersection Q of NP with T. Thus points on the equator l map to the circle k in the plane. The south pole is at the center of k. Points in the northern hemisphere map to points outside of k while points in the southern hemisphere map to points inside k. The most notable property of the stereographic projection is that it maps circles on the sphere to circles on the plane and preserves angles between arcs that intersect on the sphere [Coxeter, 1961].

As P moves toward the north pole, its image Q moves further away from the south pole. Also, concentric small circles on the sphere of decreasing radius around N map to circles in the plane of increasing radius. N is a singularity point of the projection, but sometimes it is said that N maps to the "circle at infinity." In this way, the "infinitely dis-

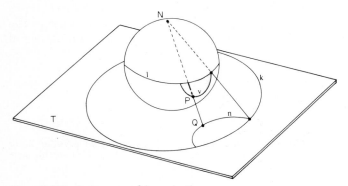

Figure 2.B.8 A stereographic projection.

tant" points of the plane are made palpable by associating them with the north pole of the sphere.

Now take the projection k of the equator l in Figure 2.B.8 to be the boundary of the Poincaré circle or plane. Intersect the sphere with a plane perpendicular to plane T. The curve of intersection between this plane and the sphere is a small circle v orthogonal to l. Since stereographic projections preserve angle, the arc of the small circle maps to an orthogonal arc of the Poincaré circle, i.e., a line from hyperbolic geometry.

The Golden Mean

*Behind the wall, the gods play; they play
with numbers, of which the universe is made
up.*   LE CORBUSIER

3.1 Introduction

Fine artists, composers, architects, scientists, and engineers have often created their best works by keeping an open dialogue with the natural world. The natural world consists of a wonderful duality between order and chaos. Careful study of a cloud formation or a running stream shows that what at first appear to be random fluctuations in the observed patterns are actually subtle forms of order. Mathematics is the best tool that humans have created to study the order in things.

Despite the infinite diversity of nature, mathematics and science have always attempted to reduce this complexity to a few general principles. In this chapter we investigate some of the many ways in which one enigmatic number, the golden mean φ, appears and reappears throughout works of art and science [Huntley, 1970], [Doczi, 1981], [Ghyka, 1952], [Tyng, 1975], [Kappraff, 1990]. Much of Chapter 1 is devoted to describing the Modulor, an architectural system of proportion based on the golden mean, while Section 2.11 shows how φ is related to patterns of spiral growth. In Chapters 5 and 6, the golden mean is shown to form the basis of a special kind of tiling that is now being used to explain the phenomenon of quasicrystals. In Chapter 8 we shall see that the golden mean lies at the mathematical basis of the platonic solids. In this chapter, we shall see that this number, which proportions the Pyramid of Cheops and the Parthenon, also orchestrates the growth of plants and serves as a key organizing element in the music of Béla Bartók.

3.2 Fibonacci Series

Special consideration was given by the Greeks to the two *harmonizing numbers*, 10 and 6. Their ratio is

$$\frac{10}{6} \quad \text{or} \quad \frac{5}{3}$$

Why is this ratio so special? According to Ghyka [1978], the cross section of the Pyramid of Cheops shown in Figure 3.1 has a hypotenuse and semibase of 89 ells and 55 ells, respectively (the ell was an ancient Egyptian measure).

The F series of Section 1.6.1, rewritten below,

$$1\ 1\ 2\ 3\ 5\ 8\ 13\ 21\ 34\ 55\ 89 \cdots \tag{3.1}$$

is a Fibonacci series, and thus the ratio of successive terms approximates the golden mean (see Section 1.6.1). Notice that $5/3 = 1.667$ and $89/55 = 1.619$ are two such approximations. Actually, the right triangle shown in Figure 3.1(*a*), which approximates the measurement of the Pyramid of Cheops, has sides: 1, $\sqrt{\phi}$, ϕ. Up to similarity, this is the only right triangle with sides in a geometric series just as the 3,4,5 right triangle is the only right triangle with sides in an arithmetic series.

Problem 3.1 How far out in the F series must one go for the ratio of successive terms to get within five decimal place accuracy to ϕ? Answer this question for the Fibonacci series that begins with 1, 3 (the Lucas series, 1, 3, 4, 7, 11, ...).

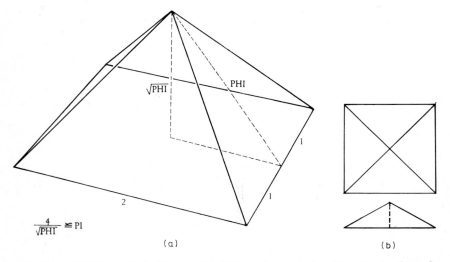

Figure 3.1 (*a*) The perfect phi pyramid; (*b*) top view and elevation 89 ells along the hypotenuse with a semibase of 55 ells.

Anne Tyng [1975] has looked at a Fibonacci *chain* of linked forms as a model for neuron chains. Such a chain is shown in Figure 3.2.

At each level a *link* acts either as an element in an ongoing chain or as one of the initiators of a new chain. These series within series may then be included as hierarchies within hierarchies of patterns. The spacing of elements in each row has a high degree of randomness; at the same time order is achieved through the overall pattern of proportional linkages.

Similar hierarchical arrangements result in countless other patterns such as the one shown in Figure 3.3. According to Tyng:

> Here the growth of the trunk and branches of a tree is shown, where the sleeve of the cambium adds a new layer of wood annually, thickening each part of trunk and limbs in proportion to the amount of new growth of twigs and branches above it.

Tyng has hypothesized that this tree diagram may be analogous to the clustering of nerve bundles in the brain. In the input from twigs to trunk, the Fibonacci size ratios increase from 1 to 13 in increased *wire* size, reducing friction for increased current, so that in the high Fibonacci numbers the size ratios would correspond to the φ:1 amplitude ratio for observed values of neuron stability.

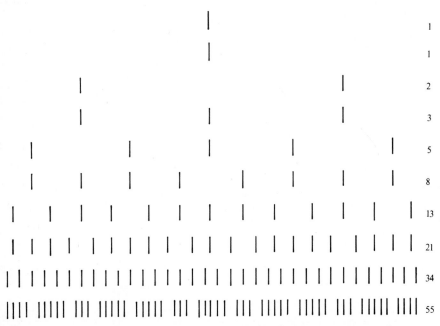

Figure 3.2   An illustration of pattern, order, and hierarchy in Fibonacci growth.

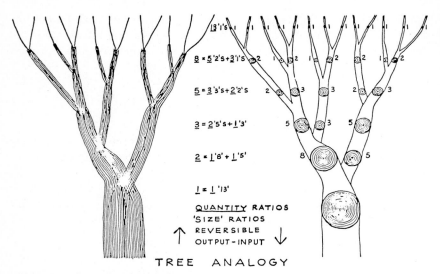

Figure 3.3   A tree illustrating Fibonacci growth.

Exercise 3.1   Create a hierarchical pattern from the Fibonacci series {1, 2, 3, 5, 8, ...}. Your fundamental pattern can be dots, lines, or anything else of your choosing. Order your modules to give a geometrical rendering of the Fibonacci series. It might be useful to use graph paper to help organize your work at first. One result of this exercise is shown in Figure 3.4.

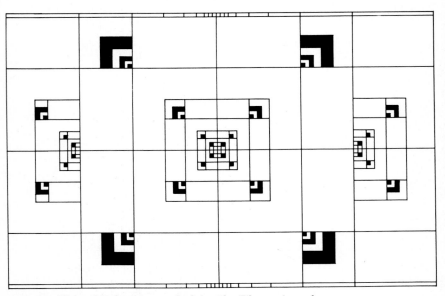

Figure 3.4   Hierarchical pattern embodying the Fibonacci numbers.

The Fibonacci series had its origin in *Liber Abaci*, written in 1202 by the mathematician Leonardo of Pisa, alias Fibonacci. In this book, Fibonacci posed the following problem:

> Rabbits always give birth to a pair of rabbits of opposite sexes. A pair of newborn rabbits must wait for a month to pass before they are mature enough to reproduce. Starting with a pair of rabbits determine the progression of rabbit pairs as time goes on.

Figure 3.5(*a*) shows the population of rabbit pairs on a tree graph (not the usual genealogical tree graph). Of course, this graph can be continued indefinitely (continue it for two more months). Notice that the graph of the F series, which is an approximate geometric series, has more of the organic quality of an actual tree than the symmetric tree graph corresponding to the geometric progression shown in Figure 3.5(*b*):

$$1\ 2\ 4\ 8\ 16\cdots$$

Actually, the F series is an approximate geometric series. In fact, any number in the series is approximately the geometric mean (see Section 1.4) of the numbers directly preceding and succeeding it, in the sense that

$$F_n^2 = F_{n-1}F_{n+1} + (-1)^{n+1} \tag{3.2}$$

where F_n is the nth number in the F Series (3.1). This equation can be rewritten

$$\frac{F_n}{F_{n-1}} - \frac{F_{n+1}}{F_n} = \frac{(-1)^{n+1}}{F_n F_{n-1}} \tag{3.3}$$

This equation is also the consequence of the fact that the ratios of successive terms of the F series are convergents of a continued fraction [Khinchin, 1979].

As a consequence of Equation (3.3), the ratios of successive terms approach the limiting value ϕ by approximating it successively from above and below. In Section 6.9 we will see how this limiting process manifests itself in patterns of plant growth.

Because the ratio of successive terms in a Fibonacci series approaches ϕ in the limit, the golden rectangle, whose sides are the ratio ϕ:1, is the most "stable" of all rectangles in the sense that starting with any rectangle, a Fibonacci sequence of rectangles must approach a golden rectangle. A sequence of rectangles whose sides have ratios of successive terms from the F series beginning with a square (ratio 1:1) is shown in Figure 3.6.

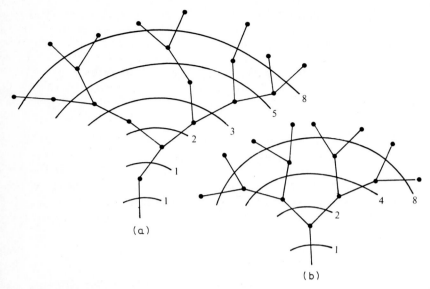

Figure 3.5 A tree pattern (*a*) from the Fibonacci series and (*b*) from a geometric series.

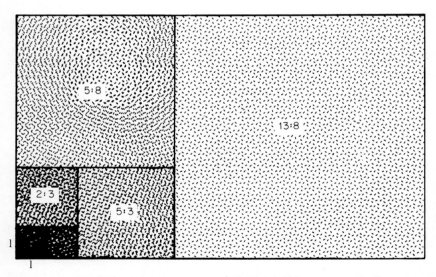

Figure 3.6 A rectangle of any proportions may be expanded to asymptotically approximate a golden rectangle illustrated for a 1:1 square.

3.3   Some Tiling Properties of ϕ

Since the Modulor scale, introduced in Section 1.7.1 is constructed either with lengths from the F series or powers of ϕ, it is important to know whether these lengths can fit together to form lengths of arbitrary dimensions. Otherwise, this series will be restricted to constructing a special class of linear dimensions.

One mathematical result along these lines is given by Theorem 3.1.

Theorem 3.1   Any positive integer can be written uniquely as the sum of nonconsecutive Fibonacci numbers from the F series.

Furthermore, the first number in the decomposition is obtained by extracting the largest number of the F series less than the given number. The second number is the largest number from the F series less than the remainder and so on. For example,

$$32 = 21 + 8 + 3$$

As a matter of fact, this decomposition gives the winning strategy for a game known as *Fibonacci Nim* [Gardner, 1978a]. A stack of pennies is placed on the table. One player removes an arbitrary number of them. The other player can then remove up to and including twice as many pennies as the preceding number. The person to remove the last penny wins. The winning strategy is to always withdraw a number of pennies equal to the smallest number in the decomposition of Theorem 3.1.

Another theorem of this kind is more directly applicable to our requirements; it is Theorem 3.2.

Theorem 3.2   Any positive real number can be represented uniquely as a sum of nonconsecutive numbers from the ϕ series:

$$\ldots \frac{1}{\phi^2}, \frac{1}{\phi}, 1, \phi, \phi^2, \phi^3, \ldots$$

As a result of Theorem 3.2, any length can be constructed to within an arbitrary preset tolerance by a sum of lengths from this series. Thus, lengths from the red and blue series can be arranged to fit any realistic measurements, e.g., the 5-inch square of Construction 1.1.

In other words, the ϕ series works much like the number system base 2. In this system, every number up to and including 2^N can be written uniquely as a sum of all or some of the numbers from the series:

$$\ldots \frac{1}{2^2}, \frac{1}{2}, 1, 2, 2^2, 2^3, \ldots, 2^{N-1}$$

where

$$\sum_{k=-\infty}^{N-1} 2^k = 2^N$$

Likewise, it can be shown that

$$\sum_{k=-\infty}^{N-1} \phi^k = \phi^{N+1} \tag{3.4}$$

Problem 3.2   Using the fact that the sums of infinite and finite geometrical progressions with common ratio r are $1/(1 - r)$ and $(1 - r^{n+1})/(1 - r)$, respectively, prove Equation (3.4).

Another important tiling property of ϕ is due to the additive properties of the ϕ series. Any positive power of ϕ can be decomposed into a combination of ϕ and ϕ^2, e.g.,

$$\phi^3 = 1\phi + 1\phi^2$$

$$\phi^4 = 1\phi + 2\phi^2$$

$$\phi^5 = 2\phi + 3\phi^2$$

$$\phi^6 = 3\phi + 5\phi^2$$

$$\cdot \quad \cdot \quad \cdot$$

$$\cdot \quad \cdot \quad \cdot$$

$$\phi^n = F_{n-2}\phi + F_{n-1}\phi^2$$

where the pattern of coefficients follows the numbers of the F series and F_n denotes the nth number in the series. We leave to the reader the task of using this series to verify Equation (1.17).

3.4   The Golden Rectangle and the Golden Section

In Section 3.2, a *golden rectangle* was built up from a rectangle of arbitrary proportions. On the other hand, in Section 2.11 we showed that a golden rectangle could be broken down arbitrarily into a sequence of many whirling squares and one similar golden rectangle. For reasons mentioned in Sections 1.6 and 1.7, the golden rectangle has aesthetic qualities that have singled it out as an ideal geometric element with which to apportion space on an artist's canvas or proportion the doorways, windows, and facades of buildings, from the Parthenon to brownstones in Brooklyn. The following procedure can be used to construct a golden rectangle with compass and straightedge:

1. Start with a square.

2. Add the semilength of a side to the length from a vertex to the midpoint of the opposite side.

The resulting length, along with the side of the original square, constitutes a golden rectangle, as shown in Figure 3.7.

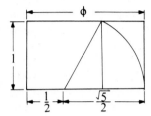

Figure 3.7 Construction of a golden rectangle using compass and straightedge.

It is useful to be able to section a line into two subintervals with golden mean ratio, φ:1. A simple construction of the *golden section* is found in the artist Paul Klee's *Notebooks* [1961]. To subdivide a line segment in the golden section, Klee suggests the following procedure (see Figure 3.8):

1. Start with line segment AB.

2. Draw $AC = \frac{1}{2} AB$ perpendicular to AB.

3. Circular arc CA intersects CB at F.

4. Circular arc BF intersects AB at G, breaking AB into the golden section.

Once a pair of lengths 1 and φ are determined, the φ series can be constructed with compass and straightedge by making use of its Fibonacci properties given by Equation (1.5) as shown in Figure 3.9.

Strange as it may seem, the 3,4,5 right triangle can be found inside of a square and related to the golden mean as follows: bisect the sides of the square and connect three of the square's vertices to the mid-

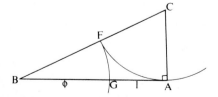

Figure 3.8 Dividing a line into its golden section with compass and straightedge.

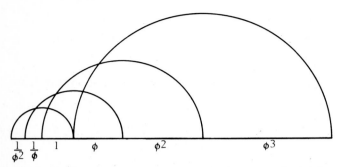

$\frac{1}{\varphi^2}$ $\frac{1}{\varphi}$ 1 φ φ^2 φ^3

Figure 3.9   Construction of a φ series with compass and straight-edge beginning with lengths 1 and φ.

points of the sides by diagonals as shown in Figure 3.10. Each of these diagonals has length $\sqrt{5}/2$, or $\varphi - \frac{1}{2}$, and the resulting triangle is a 3,4,5 right triangle [Lawlor, 1982].

Construction 3.1   Euclid showed in Book XIII of *The Elements* that the golden mean is closely related to the structure of a set of symmetric polyhedra known as the platonic solids. These solids are the subject of Chapters 7 and 8, and their relation to the golden mean is discussed in Section 8. 7. In the meantime, you can construct one of the platonic solids called the *icosahedron* by cutting slits through three golden rectangles and arranging them to form three mutually or-thogonal, self-intersecting rectangles as shown in Figure 3.11(*a*) and (*b*). Three-

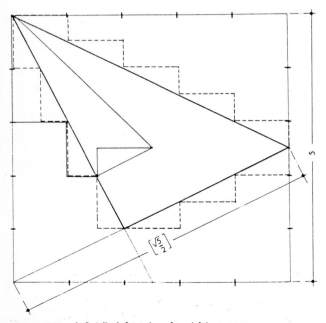

Figure 3.10   A 3,4,5 right triangle within a square.

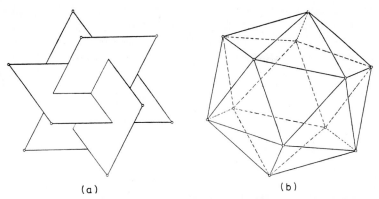

Figure 3.11 An icosahedron, a polyhedron with 20 equilateral triangle faces, defined by the 12 vertices of three mutually orthogonal golden rectangles.

by-five index cards are very close approximations to golden rectangles and can be used for this construction. The icosahedron is formed by connecting the corners of the rectangles by 30 equal lengths of string corresponding to the 30 edges of the icosahedron. H. F. Verheyen has also related the structure of the Pyramid of Cheops to the icosahedron (see Section 9.9).

3.5   The Golden Mean Triangle

In Figure 3.12, an isosceles triangle *ABC* is shown with base angles of 72 degrees. Using a compass, *AD* is marked off so that *AD = AB* and triangles *ABD ~ ABC*. Thus *AD* cuts triangle *ABC* into a unit and a leftover isosceles triangle *ADC*, or gnomon. If we let *AB* = 1 and *AC* = *x*, by similar triangles

$$\frac{x}{1} = \frac{1}{x - 1}$$

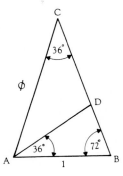

Figure 3.12   The base angle of a golden isosceles triangle of type 1, *ABC*. Angle *CAB* is bisected to form a similar golden triangle *ABD* and a gnomon *ADC*, also a golden triangle of type 2.

from which it follows, by algebraically solving for x, that $x = \phi$. Since triangles ABC and ADC have sides in the ratio ϕ:1, they are both called *golden triangles*. The process can be repeated to form a sequence of whirling triangles, as Figure 3.13 shows.

Construction 3.2 Make designs using modules which are golden triangles and their gnomons, the sizes of which are related to each other by powers of the golden ratio.

3.6 The Pentagon and Decagon

The pentagon is totally governed by the golden mean since it may be subdivided into three golden triangles as shown in Figure 3.14. This figure also illustrates that for a pentagon

$$\text{Diagonal:side} = \phi:1$$

As a result, given lengths 1 and ϕ, a pentagon can be constructed with compass and straightedge. (Do this!)

There are many equivalent ways of constructing a regular pentagon, but if you really want to construct one painlessly, you can do it with a loop of a strip of paper [see Figure 3.15(*a*) and (*b*)]. Draw the loop tightly and crease along the edges neatly, and there you have it! Did you ever realize that every time you tie a knot, you are construct-

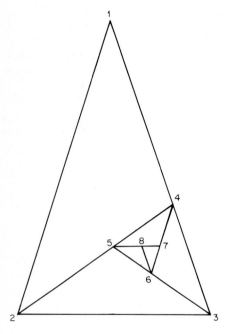

Figure 3.13 Whirling golden triangles.

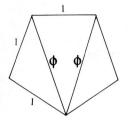

Figure 3.14 A pentagon subdivides into one type 1 and two type 2 golden triangles.

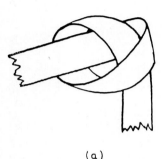

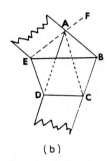

(a)   (b)

Figure 3.15 A pentagon can be constructed from a knotted strip of paper.

ing a regular pentagon in the sense of Figure 3.15? See if you can come up with a geometric proof that this figure is a regular pentagon [Davis and Chinn, 1969].

According to John Michell, "an important exercise in sacred geometry is to combine the hexagon [symbolic of inanimate life; see Section 1.2] and pentagon [symbolic of animate life] in one synthetic figure" [1988]. How this is achieved with tolerable accuracy using the Vesica Piscis (see Section 2.11) is shown in Figure 3.16. This figure was published by the artist Albrecht Dürer in his *Course in the Art of Mea-*

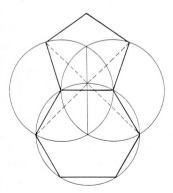

Figure 3.16 Approximate construction of a pentagon from the Vesica Piscis.

surement with Compasses and Rulers and was reproduced in C. Bouleau's *The Painter's Secret Geometry* [1963]. Michell relates how the Dutch artist Franz Deckwitz explained Dürer's enigmatic print rich in geometric imagery, *Melancholia* (see Figure 3.17), by using this combination of hexagon and pentagon [1988]. D. Crowe also traces the history and geometry of the print [1990].

Figure 3.17   *Melancholia I*, 1514. Engraving 243 × 187 mm. Centennial gift of Landon J. Clay. (*Courtesy, Museum of Fine Arts, Boston*)

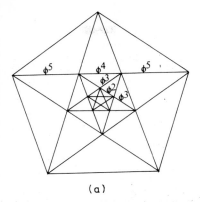
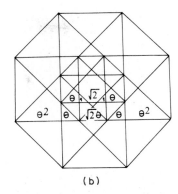

(a)   (b)

Figure 3.18 (a) Star pentagons at decreasing scales. The ratio of lengths from scale to scale is ϕ^2:1. The edges of the star cut each other in the golden section. (b) Star octagons at decreasing scales. The ratio of lengths from scale to scale is θ:1. The edges of the star cut each other in the ratio $\sqrt{2}$:1.

The star pentagram, a pentagon along with all its diagonals, was an ideal symbol for the Brotherhood of Pythagoras since the diagonals of a pentagon cut each other in the golden section as shown in Figure 3.18(a). Notice how the envelope of the star pentagram forms another pentagon of edge length $1/\phi^2$. Contrast this with the diagonals of a star octagon, shown in Figure 3.18(b), which cut themselves in either $\sqrt{2}$ or θ. Medieval and Renaissance artists and architects, influenced by the compass and straightedge constructions of Greek mathematics, based some of their art on star pentagon and octagon constructions. Figure 3.19 shows a reproduction of Raphael's *The Crucified Christ* with Ghyka's analysis to show its structure based on the pentagon and the decagon [1952]. A sketch (not shown) from the *Notebooks* of Leonardo da Vinci shows a church, never built, that has the structure of a star octagon [Scholfield, 1958].

The decagon can also be inscribed in a circle. For such a figure,

$$\text{Radius:side} = \phi\text{:}1$$

The decagon shows up in the natural world as the shape of the DNA molecule (see Figure 3.20). The vertices of a decagon star are also evident in Dan Winter's star crystal illustrated in Figure 8.22.

3.7 The Golden Mean and Patterns of Plant Growth

3.7.1 The geometry of plant growth

As a young man, Le Corbusier studied the elaborate spiral patterns of stalks, or paristiches as they are called, on the surface of pine cones, sunflowers, pineapples, and other plants (see Figure 3.21). This led

Figure 3.19 (a) Raphael's *The Crucified Christ (courtesy of the National Gallery, London)*; (b) a structural diagram of it due to M. Ghyka

him to make certain observations about plant growth that have been known to botanists for over a century.

Plants, such as sunflowers, grow by laying down leaves or stalks on an approximately planar surface. The stalks are placed successively around the periphery of the surface. Other plants such as pineapples

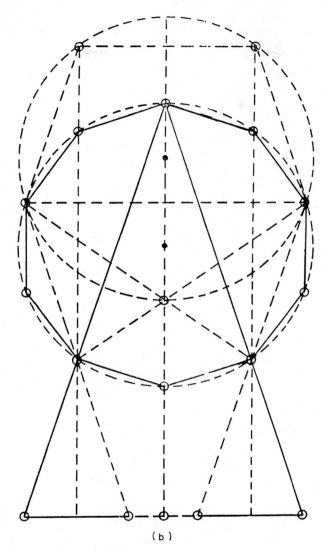

(b)

Figure 3.19   (*Continued*)

or pinecones lay down their stalks on the surface of a distorted cylinder. Each stalk is displaced from the preceding stalk by a constant angle as measured from the base of the plant, coupled with a radial motion either inward toward or outward from the center for the case of the sunflower [see Figure 3.21(*b*)] or up a spiral ramp as on the surface of the pineapple [see Figure 3.21(*a*)]. The angular displacement λ is called the *divergence angle* and is related to the golden mean. The

Figure 3.20 Detailed computer-generated model of DNA seen from above.

(a)

Figure 3.21 Spiral growth in plants. (*a*) Pineapple and pinecone; (*b*) sunflower.

(b)

Figure 3.21   (*Continued*)

radial or vertical motion is measured by the *pitch h*. The dynamics of plant growth can be described by λ and *h*; we will explore this further in Section 6.9 [Coxeter, 1953].

Each stalk lies on two nearly orthogonally intersecting logarithmic spirals, one clockwise and the other counterclockwise. The numbers of counterclockwise and clockwise spirals on the surface of the plants are generally successive numbers from the F series, but for some species of plants they are successive numbers from other Fibonacci series such as the Lucas series (see Problem 3.1). These successive numbers are called the *phyllotaxis* numbers of the plant. For example, there are 55 clockwise and 89 counterclockwise spirals lying on the surface of the sunflower; thus sunflowers are said to have 55,89 phyllotaxis. On the other hand, pineapples are examples of 5,8 phyllotaxis (although, since 13 counterclockwise spirals are also evident on the surface of a pineapple, it is sometimes referred to as 5,8,13 phyllotaxis). We will analyze the surface structure of the pineapple in greater detail in Section 6.9.

3.7.2   Nature responds to a physical constraint

After more than 100 years of study, just what causes plants to grow in accord with the dictates of Fibonacci series and the golden mean remains a mystery. However, recent studies suggest some promising hypotheses as to why such patterns occur [Jean, 1984], [Marzec and Kappraff, 1983], [Erickson, 1983].

A model of plant growth developed by Alan Turing states that the elaborate patterns observed on the surface of plants are the consequence of a simple growth principle, namely, that new growth occurs in places "where there is the most room," and some kind of as-yet-undiscovered growth hormone orchestrates this process. However, Roger Jean suggests that a phenomenological explanation based on diffusion is not necessary to explain phyllotaxis. Rather, the particular geometry observed in plants may be the result of minimizing an *entropy function* such as he introduces in his paper [1990].

Actual measurements and theoretical considerations indicate that both Turing's diffusion model and Jean's entropy model are best satisfied when successive stalks are laid down at regular intervals of $2\pi/\phi^2$ radians, or 137.5 degrees about a growth center, as Figure 3.22 illustrates for a celery plant. The centers of gravity of several stalks conform to this principle. One clockwise and one counterclockwise logarithmic spiral wind through the stalks giving an example of 1,1 phyllotaxis.

The points representing the centers of gravity are projected onto the circumference of a circle in Figure 3.23, and points corresponding to the sequence of successive iterations of the divergence angle, $2\pi n/\phi^2$, are shown for values of n from 1 to 10 placed in 10 equal sectors of the circle. Notice how the corresponding stalks are placed so that only one stalk occurs in each sector. This is a consequence of the following spac-

Figure 3.22   A plant, such as the celery plant, lays down new stalks where there is the most room.

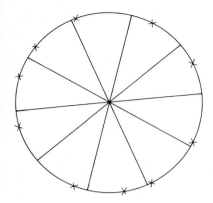

Figure 3.23   Points $2\pi n/\phi^2$ placed on a circle for $n = 1, 2, \ldots, 10$.

ing theorem that is used by computer scientists for efficient parsing schemes [Knuth, 1980].

Theorem 3.3   Let x be any irrational number. When the points $[x]_f$, $[2x]_f$, $[3x]_f, \ldots, [nx]_f$ are placed on the line segment $[0,1]$, the $n + 1$ resulting line segments have at most three different lengths. Moreover, $[(n + 1)x]_f$ will fall into one of the largest existing segments ($[\ ]_f$ means "fractional part of").

Here clock arithmetic based on the unit interval, or mod 1 as mathematicians refer to it, is used, as shown in Figure 3.24, in place of the interval mod 2π around the plant stem. It turns out that segments of various lengths are created and destroyed in a first-in-first-out manner. Of course, some irrational numbers are better than others at spacing intervals evenly. For example, an irrational that is near 0 or 1 will start out with many small intervals and one large one. Marzec and Kappraff [1983] have shown that the two numbers $1/\phi$ and $1/\phi^2$ lead to the "most uniformly distributed" sequence among all numbers between 0 and 1. These numbers section the largest interval into the golden mean ratio, $\phi{:}1$, much as the blue series breaks the intervals of the red series in the golden ratio.

Thus nature provides a system for proportioning the growth of plants that satisfies the three canons of architecture (see Section 1.1). All modules (stalks) are isotropic (identical) and they are related to the whole structure of the plant through self-similar spirals proportioned by the golden mean. As the plant responds to the unpredictable elements of wind, rain, etc., enough variation is built into the patterns to make the outward appearance aesthetically ap-

Figure 3.24   Points $[n\phi]_f$ placed in the unit interval for $n = 1, 2, \ldots, 10$.

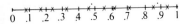

pealing (nonmonotonous). This may also explain why Le Corbusier
was inspired by plant growth to recreate some of its aspects as part
of the Modulor system.

3.7.3 Wythoff's game

Theorem 3.3 is the key to describing the mathematics of plant growth.
I made my own personal discovery of this theorem as the result of
playing a Nim-type game known as *Wythoff's game* [Coxeter, 1953]
with my students in a course called *The Mathematics of Design* that I
teach at the New Jersey Institute of Technology [Kappraff, 1986a].
This game begins with two stacks of pennies. A proper move is to remove any number of pennies from one stack or an equal number from
both stacks. The winner is the person removing the last penny. The
winning strategy is based on Theorem 3.4 due to S. Beatty.

> **Theorem 3.4** If $1/x + 1/y = 1$, where x and y are positive irrational numbers,
> the sequences $[x], [2x], [3x],\ldots$ and $[y], [2y], [3y],\ldots$ together include every pos-
> itive integer taken once ([ ] means "integer part of," for example, $[\phi] = 1$).

For a proof, see [Coexter, 1953]. Since $1/\phi + 1/\phi^2 = 1$ from Equation
(1.5), Beatty's theorem shows that $[n\phi], [n\phi^2]$ exhausts all of the natural numbers with no repetitions as n takes on the values $n = 1, 2,\ldots$.
Table 3.1 shows results for $n = 1, 2, \ldots, 6$. Can you notice a pattern in
these numbers that will enable you to continue the table without computation? The pairs are also winning combinations for Wythoff's
game. At any move a player can reduce the number of counters in
each stack to one of the pairs of numbers in Table 3.1. The player who
does this at each turn is assured victory.

After playing Wythoff's game a number of times with my students,
I noticed that if I considered the fractional parts of n rather than the

TABLE 3.1

n	$[n\phi]$	$[n\phi^2]$
1	1	2
2	3	5
3	4	7
4	6	10
5	8	13
6	9	15
.	.	.
.	.	.
.	.	.

integer parts, these satisfied Theorem 3.3. This led me to my work on plant growth.

3.8   The Music of Bartók: A System Both Open and Closed

It is understandable that architects should look to music in search of a system to proportion their buildings, as Alberti and Palladio did (see Section 1.5). After all, musical composition superimposes its emotional and aesthetic elements on a structure of supreme order. The music of Bartók, as analyzed by the Hungarian musicologist Ernö Lendvai, embodies perhaps the fullest interplay between emotional content and structure [1966], [Bachmann and Bachmann, 1979]. Bartók based his music on the deepest layer of folk music. He believed that every folk music of the world can finally be traced to a few primeval sources. Through these sources, according to Lendvai, "[Bartók] discovered and drew into his art the laws governing the depths of the human soul which have not been touched by civilization." He was also greatly influenced by Impressionism and the atonal trends of his day, and combined the Western structures of harmony with folk music into an organic whole.

Artists must create a system in which to frame their work. It is interesting to me that to achieve these primitive or "natural" effects, Bartók based the entire structure of his music on the golden mean and Fibonacci series—from the largest elements of the whole piece, whether symphony or sonata, to the movement, principal, and secondary themes and down to the smallest phrase. In this regard his music resembles the organic wholeness of the Modulor, exemplifying a closed system (see Section 1.7.2). He contrasts this *closed golden mean system* with a dual system based on the *overtones* ascending from a fundamental tone—an *open system* analogous to the system of proportions at the basis of the Garden Houses of Ostia (see Section 1.8.2). It is beyond the scope of this book to examine Bartók's music in detail. We will, however, give three examples of his use of Fibonacci series:

1. From Lendvai:

In the first movement of *Music for Strings*, from the pianissimo (soft) the movement reaches the boiling point by a gradual rise to forte-fortissimo (very loud), then gradually recedes to piano-pianissimo (very soft) as shown in Figure 3.25. The 89 bars of the movement are divided into parts of 55 and 34 bars by the pyramid-like peak of the movement.... The form is proportioned within these units by cancellations of the sordino (or mute) in the 34th bar and its repeated use in the 89th bar.... Positive and neg-

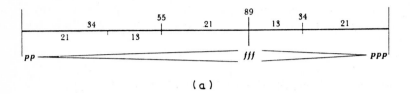

(a)

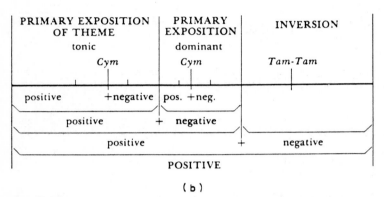

(b)

Figure 3.25 (a) Important transitions in the first movement of *Music for Strings, Percussion and Celesta* by Bartók; (b) the theme is divided into positive and negative sections.

ative sections embrace each other like the rising and sinking of a single wave. [Here, positive sections are long followed by short sections while negative sections are short followed by long sections as shown in Figure 3.25(b).] It is no accident that the exposition ends with the 21st bar and that the 21 bars concluding the movement are divided into parts of 13 + 8 [all elements of the F series].

2. In order to understand the other two applications of the Fibonacci series and the golden mean to the structure of Bartók's music, some understanding of musical notation would be helpful. Nevertheless, even without such a background the ideas can still be appreciated, and they give a striking example of the utility of the golden mean. The ideas involve Bartók's use of the pentatonic scale.

Pentatony is perhaps the most ancient human sound system. It rests on a pattern reflected by the melody steps of the major second (2), minor third (3), and the fourth (5). The numbers in parentheses are the number of semitone intervals separating a note from the fundamental tone in the 12-tone chromatic scale. (The well-tempered scale, the scale upon which the piano is based, divides the frequency length corresponding to an octave into an increasing geometric sequence with a common ratio, $2^{1/12}$. The tones of the well-tempered scale are harsh when compared with tones corresponding to the ratio of small integers of the just, or Ptolemaic scale (see Figure 1.7). There is unfortunately no way to define the chromatic scale so that the change from one key to another does not introduce new tones slightly different from the corresponding tones of the former key while also preserving the ratios. While players of stringed instruments can change the position of their fingers slightly to compensate for these changing pitches, the well-tempered scale was a necessary compromise in order to accommodate the invention of the piano, which can have only a set number of keys.) Don't confuse these numbers with the notion of the musical "second," "third," etc., which denotes the number of notes that separate a given note from the base note in the seven-tone scale in any key, i.e., do, re, mi, ...(see Section 1.4). The black notes on the piano make up a pentatonic scale. Successions of two and three halftones are the intervals between the black notes, and almost any succession of notes played on the piano using only the black notes leads to a pleasant sounding tune.

The *pentatonic scale* lies at the basis of the oldest folk melodies and the simplest nursery songs, which follow a la, sol, mi (2,3,5) form. The interval from la to sol is two halftones, thus "sol" breaks the interval from la to mi in the ratio 3:5—a Fibonacci approximation to the golden section. This golden-section cell division pervades all of Bartók's music. Bartók's use of this Fibonacci progression of tones can be followed in the last movement of the *Divertimento*. According to Lendvai, the principal theme appears in the variations (see Figure 3.26). The intervals of the pentatonic scale demark the rising and falling of the musical line about a center located at the golden section just as the musical dynamics (loudness and softness) were centered by the

Figure 3.26 Golden section cell division in the last movement of *Divertimento*.

Fibonacci series in example 1. In line 1, the notes rise one halftone above the central note and fall one beneath it, i.e., (1,1). In line 2 they rise two halftones above and fall one below the center (2,1). Line 3 is (2,3), while lines 4 and 5 continue this progression to (3,5) and (5,8).

3. Bartók also used Fibonacci numbers in an another way. Roughly speaking, the musical tissue of his music may be imagined to be built up of cells 2, 3, 5, 8, and 13 in size, i.e., the minor second (2), minor third (3), fourth (5), minor sixth (8), and the augmented octave (13). Such a progression, starting with C as the fundamental tone, C, D, E flat, F, A flat, C sharp, is represented in musical notation in Figure 3.27.

Bartók contrasted this Fibonacci scale with a scale based on the sequence of overtones of the fundamental note. To explain what is meant by the overtones of a tone, we must consider another aspect of the tone, namely, its wave properties. For example, a plucked string sets up condensations and rarefactions in the air that travel with the speed of sound. If a fundamental tone vibrates with a frequency of 100 cycles per second, its octave vibrates at 200 cycles per second (2:1), its fifth at 150 cycles per second (3:2), etc. In other words, the frequency of musical interval is the inverse of the ratio of string lengths corresponding to that interval. It is well known that when a tone is sounded loudly, the ear manufactures all multiples of the tone, with the lower multiples more audible than the higher ones [Benade, 1976], (i.e., tones in the frequency ratios 2:1, 3:1, 4:1, 5:1, etc.). The first is the octave. The second is the fifth if it is lowered by one octave (i.e., $3/1 \times 1/2 = 3/2$). The third is a double octave. The fourth overtone is a major third when lowered by two octaves (i.e., $5/1 \times 1/2 \times 1/2 = 5/4$). Continuing in this manner, we find that the overtone scale is given by the increasing sequence of ratios along with their corresponding tonal names as follows:

1	$\dfrac{9}{8}$	$\dfrac{5}{4} = \dfrac{10}{8}$	$\dfrac{11}{8}$	$\dfrac{3}{2} = \dfrac{12}{8}$	$\dfrac{13}{8}$	$\dfrac{7}{4} = \dfrac{14}{8}$	$\dfrac{15}{8}$	$\dfrac{2}{1} = \dfrac{16}{8}$
C	D	E	F$^{\#}$	G	A	B$^{\flat}$	B	C

The ratios are named from the tones on the well-tempered scale that they closely approximate. With the exception of A, there are no ap-

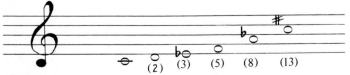

Figure 3.27 The Fibonacci scale of Bartók. The successive tones of the scale increase in a Fibonacci series of halftones.

proximations for those notes that use ratios of smaller integers. It is also notable that the frequencies of the overtone series form an arithmetic series with common difference of ⅛. Thus, removing the arithmetic progression of the tones that form the overtone series from the geometric progression of tones that comprise the 12-tone chromatic scale leaves an F series of halftones, the only exception being the major second, D, that appears in both series. These two worlds of harmony complement each other to such a degree that the Bartokian scale can be separated into Fibonacci and overtone scales, much as were the red and blue series of Le Corbusier and the pair of scales of the Garden Houses (see Sections 1.7 and 1.8). Separately, each is merely a part of a whole and neither can exist without the other, as shown in the following table:

	(0) = (12)	(2)	(3)	(5)	(8)	(13) = (1)		
Fibonacci scale:	C	D	E flat	F	A flat	C sharp		
		(2)	(4)	(6)	(7)	(9)	(10)	(11)
Acoustic scale:	C	D	E	F sharp	G	A	B flat	B

First of all, this system decouples all the notes of the 12-tone scale into two scales (although the D appears on both scales). Furthermore, the two systems reflect each other in the octave, or as musicians say, the fifth, 3:2, reflects the fourth, 4:3, since the fifth breaks the octave into a fifth and a fourth:

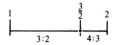

Similarly, the major third, 5:4, breaks the octave into the major third and minor sixth, 8:5 (see Section 1.4):

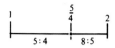

For the most part, Bartók builds his compositions on this system although he deviates from it occasioñally to create special effects. For example, in the finale of the *Sonata for Two Pianos and Percussion*, the acoustic scale C, E, F sharp, G, A, B flat, C contrasts with the golden mean section of the piece, C, E flat, F, A flat that dominates the first movement (see Figure 3.28).

Systematically, the two scales are related by organically complementing and reflecting each other. Each is the other's negative reflec-

Figure 3.28 Fibonnaci and acoustic scales in "Sonota for Two Pianos and Percussion."

tion in the 12-tone system. These two scales also complement each other in terms of the emotional content of the music. The overtone system can only admit consonant intervals (by nature of the overtone harmonies). In other words, chords made up of notes from this scale are all pleasing to the ear. On the other hand, chords from the Fibonacci system are "tense" and "dissonant." Thus each system is capable of disclosing one aspect of life.

As Lendvai explains, Bartók was able to use his double scale to set up a duality between both the structural and emotional elements of his music. The essence of this duality lies in the closed nature of the Fibonacci scale, in contrast to the open nature of the overtone scale. While the dissonant golden mean harmonies move around the circle of fifths (a circle of progressively increasing fifths upon which Western music is built) and modulates from key to key, according to the particular laws of harmony developed by Bartók, the overtone scale rises linearly from a common fundamental note. In this way tensions developed in the first movement of a piece by golden mean harmonies are resolved in the last movement by the familiar chords of Western music based on the overtone scale. A striking example of this organic relation between the dual systems is shown by the opening and concluding bars of the *Cantata Profana* (see Figure 3.29) in which the two scales mirror each other tone for tone—a Fibonacci scale and a pure overtone scale.

Figure 3.29 The opening and closing of *Cantata Profana* shows how the Fibonacci and acoustic scales mirror each other tone for tone.

TABLE 3.2

Fibonacci scale	Overtone scale
Golden-section system	Acoustic system
Closed world	Open world
Circular pattern of melody	Straight pattern of melody
Uneven meter	Even meter
Asymmetries	Periodicity
Demonaic world	Serene, festive world
Organic	Logic
Inspiration	Thought
Augmentation-diminution	Stabilized forms
Finite (circular motion)	Infinite (straight line)

Lendvai says much more about the duality of Bartók's two scales. Several of his dualities are illustrated in Table 3.2.

The closed nature of the golden-section harmonies can be likened to the emblem of Dante's Inferno—the circle or ring—while the overtone scale is akin to the symbol of his Paradisio—the straight line, the arrow, the ray. Lendvai dramatizes this notion with the following illustration:

> The golden-section can easily be (constructed) with the aid of a simple "knot" [as shown in Figure 3.15]; every proportion of this knot will display the golden-section. It is this property of the pentagram that Goethe alludes to in Faust, Part I:
>
> | Mephistopheles: | Let me admit; a tiny obstacle
Forbids my walking out of here:
It is the druid's foot upon your
threshold. |
> |---|---|
> | Faust: | The pentagram distresses you?
But tell me, then, you son of hell.
If this impedes you, how did you come in?
How can your kind of spirit be deceived? |
> | Mephistopheles: | Observe! The lines are poorly drawn:
That one, the angle pointing outward,
Is, you see, a little open. |

4

Graphs

*The crucial quality of shape, no matter of
what kind, lies in its organization, and when
we think of it in this way we call it form.*
CHRISTOPHER ALEXANDER
Notes on the Synthesis of Form

4.1 Introduction

An artist or architect usually captures the earliest stage of an idea through a sketch depicting its raw outline. As work progresses, the rendering of the idea reveals more and more structure. Objects appear in their proper perspective, and length and angle become more definite. This range of visual thinking also pervades mathematics through the subject of geometry. Like the artist or architect's finished product, euclidean geometry—the geometry most of us studied in high school—considers line segments to be of definite lengths and to meet each other at precise angles. However, not all geometries have these metric properties of length and angle.

In this chapter, we discuss a freewheeling geometry of dots and lines called *graph theory* [Baglivo and Graver, 1983], [Trudeau, 1976], [Ore, 1963]. As for the artist's or architect's rough sketch, graph theory preserves geometrical relationships only in their most general outlines. In graph theory, polygons are defined as cycles of lines connecting two or more dots as shown in Figure 4.1. However, a line does not have to be straight in dot and line geometry, nor are there such things as perpendicular or parallel lines, and it does not make sense to talk about bisecting lines or measuring lengths and angles.

The power of graph theory is that it can be used to model many patterns in nature from the branching of rivers to the cracking of brittle surfaces to subdivisions of cellular forms (see Figure 4.2) as well as many abstract concepts. The free-form geometry of dots and lines can be used to study these structures, and we shall see that this geometry

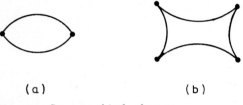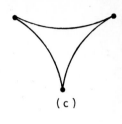

(a)   (b)   (c)

Figure 4.1   Some graphical polygons.

Figure 4.2   (*a*) Pattern in soap bubbles; (*b*) patterns observed on the shore of a river when the mud has been dried up by the sun; (*c*) branching patterns of rivers.

has a rich underlying foundation. We are going to start by investigating what happens in a freewheeling situation. Before reading on, try this exercise.

> **Exercise 4.1**   Place dots on a piece of paper and then connect them with lines. Lines begin and end at dots and may loop around to begin and end at the same dot; however, two lines will not be permitted to intersect except at a dot. Can you find any pattern to the results? At first thought it would seem impossible for any order to come out of such an unstructured exercise. But is it?

In order to make it easier to analyze things, let

$$D = \text{number of dots}$$
$$L = \text{number of lines}$$
$$A = \text{number of enclosed areas}$$

Observer 1 carrying out this exercise made two conjectures:

$$L = D - 1 \qquad \text{if all the dots are connected with a minimum of lines} \tag{4.1a}$$

$$A + D - L = 1 \tag{4.1b}$$

Let's look at the diagrams in Figure 4.3 from which the observer made his conjectures. From the results it appears as though he is correct. But wait! observer 2 came up with the diagrams in Figure 4.4. These diagrams appear to each have more than one segment and so Equation 4.1*b* does not apply. However, this equation can be modified so that it is true in every case that

$$A + D - L = \text{number of pieces in the diagram} \qquad (4.2)$$

It may appear at first that this exercise could have been made even more freewheeling if we permitted lines to cross at points other than the dots. However, Figure 4.5 shows that the same number of dots and lines can give rise to any number of enclosed areas if lines are permitted to cross. It may help to think of the lines as strings connecting a set of tacks—the problem is to untangle the strings so they don't cross in order to discover what A is. As we will see later, it's not always possible to untangle the strings, so there are some diagrams in which A is not well defined.

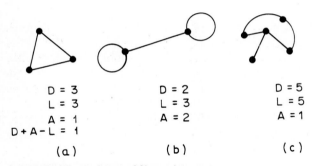

$$
\begin{array}{ccc}
D = 3 & D = 2 & D = 5 \\
L = 3 & L = 3 & L = 5 \\
A = 1 & A = 2 & A = 1 \\
D + A - L = 1 & & \\
(a) & (b) & (c)
\end{array}
$$

Figure 4.3 Some dots and lines pictures.

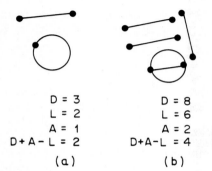

$$
\begin{array}{cc}
D = 3 & D = 8 \\
L = 2 & L = 6 \\
A = 1 & A = 2 \\
D + A - L = 2 & D + A - L = 4 \\
(a) & (b)
\end{array}
$$

Figure 4.4 Some more dots and lines pictures.

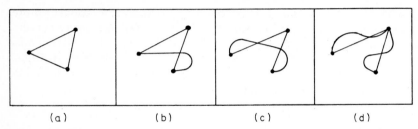

(a)   (b)   (c)   (d)

Figure 4.5   If edges are permitted to cross, an arbitrary number of enclosed areas are possible.

Equation (4.2) has a very deep relationship to the nature of space and the real-world limitations which it imposes on design. By imposing more restrictions (the only ones that we have imposed so far are that lines must end in dots and must not cross), several startling results will follow from this seemingly simple relationship. But first, let's put things in more formal mathematical terms.

4.2   Graphs

The theory of graphs will play a central role in this book since it gives us a way to study spatial structures unencumbered by the details of euclidean geometry. We will go into the subject enough to appreciate the applications to spatial design found in this chapter and the remainder of the book. First we redefine dots, lines, and areas to agree with common mathematical conventions. We call the dots, *vertices, V,* or sometimes nodes, the lines, *edges, E,* and the areas, *faces, F.* Often, we will use the same symbols, *E, V,* and *F,* to mean both the entity and the number of edges, vertices, and faces in the diagram.

The reason for calling the lines *edges* is that we may consider them as the boundary edges of shapes, and the reason for referring to closed areas as *faces* will become clearer when we extend our ideas into three dimensions. (The enclosed areas will become the faces of polyhedra.)

In addition, we are going to call the kind of diagrams we've been drawing *graphs,* or sometimes *networks.* By formal definition, a *graph, G,* is a set of edges and vertices:

$$G = \{V,E\}$$

We are relying here on the reader's naive idea of a *set* as a bunch of things along with a rule of membership that determines whether some object does or does not belong to the set. Sets also have no implied order and there are no duplications.

The definitions of *G, V,* and *E* themselves contain sets. *V* is the vertex set and *E* is the edge set. *E* consists of pairs of vertices taken from the set V. Thus if a and b are vertices, $\{a,b\}$ or just ab is the edge connecting a and b. It should be noted that in this definition:

- $\{a,b\}$ and $\{b,a\}$ are the same thing (order doesn't count in sets) so that we do not give a direction to an edge.
- (a,a) is meaningless (no duplications in sets).

For example:

$$G = \{V,E\}$$
$$V = \{1,2,3,4\}$$
$$E = \{\{1,2\}, \{1,3\}, \{2,4\}\}$$

Given this information, we could draw a diagram showing the vertices and edges with no trouble [see Figure 4.6(a)] However, you should note that the formal definition of what a graph is makes no reference to diagrams; it is a purely abstract idea. As such the graph can be expressed in other ways. For example, we can represent the graph by a matrix in which rows and columns represent vertices and a 1 is placed at each position wherever corresponding vertices are connected by an edge and a 0 is placed in the matrix wherever there's no connection. The matrix is

Vertex number

		1	2	3	4
	1	0	1	1	0
Vertex	2	1	0	0	1
Number	3	1	0	0	0
	4	0	1	0	0

or, with the labels dropped, simply

$$G = \begin{bmatrix} 0 & 1 & 1 & 0 \\ 1 & 0 & 0 & 1 \\ 1 & 0 & 0 & 0 \\ 0 & 1 & 0 & 0 \end{bmatrix} \tag{4.3}$$

We call this the *incidence matrix* and denote it by G to emphasize that the matrix may be considered an abstract representation of the graph.

Armed with either V and E or the matrix, we can also illustrate the graph by Figure 4.6(b). There is no unique way to represent the graph in a diagram, as can be seen by the examples above. However, each of the diagrams unambiguously shows the connections in the graph. We say

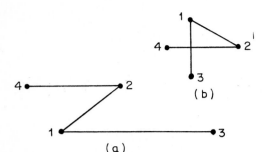

Figure 4.6 A planar graph can be redrawn with no crossovers.

that these graph diagrams are *isomorphic* to one another, meaning that they exhibit the same structure and can be redrawn to look identical. If you think of the graph diagram as being composed of tacks and elastic strings, as we did earlier, you should be able to visualize means of changing one diagram into another. While you should not confuse the diagram of a graph with the graph itself, since a graph is an abstract mathematical concept and the graph diagram is a pictorial representation of the graph, it is common practice to refer to the graph diagrams as graphs, and we will follow that practice in this book.

If two different graphs can be made to have identical matrices by relabeling their vertices, these graphs are *isomorphic*. For example, the two graphs shown in Figure 4.7,

$$G_1 = \{V_1, E_1\} \qquad\qquad G_2 = \{V_2, E_2\}$$

$$V_1 = \{1,2,3,4\} \qquad\qquad V_2 = \{a,b,c,d\}$$

$$E_1 = \{\{1,2\},\{2,4\},\{4,3\},\{3,1\}\} \qquad E_2 = \{\{a,b\},\{b,c\},\{c,d\},\{d,a\}\}$$

are isomorphic even though one has crossing edges while the other does not since by matching up the vertices as follows:

$$1 \leftrightarrow a, \quad 2 \leftrightarrow b, \quad 3 \leftrightarrow d, \quad 4 \leftrightarrow c$$

both can be represented by the matrix

$$\begin{bmatrix} 0 & 1 & 1 & 0 \\ 1 & 0 & 0 & 1 \\ 1 & 0 & 0 & 1 \\ 0 & 1 & 1 & 0 \end{bmatrix}$$

Problem 4.1 Find as many pairs of isomorphic graphs in Figure 4.8 as you can.

The essence of graph theory lies in the fact that two graphs can be visually very different and yet isomorphic, as Figure 4.9 shows. This can be of great use in spatial design when we wish to create a variety

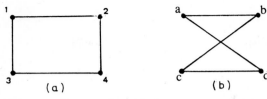

Figure 4.7 Two graphs with the same connectivity; they are graphically identical.

Figure 4.8 Identify the isomorphic graphs in this figure.

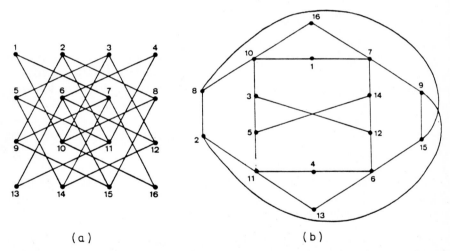

(a)   (b)

Figure 4.9 (a) A graph with many crossovers; (b) an isomorphic copy drawn with no crossovers.

of structures that look quite different but share the same basic plan (see Section 4.17). On the other hand, it is one of the difficult problems of graph theory to determine, in general, if two complicated graphs are isomorphic. They must certainly have the same number of faces, edges, and connected pieces. Each graph must also have the same distribution of the number of edges touching each vertex. However, this is not enough as Figure 4.10 shows. (Why?) Another criterion helpful in deciding whether two graphs are isomorphic is given in Section 4.8.

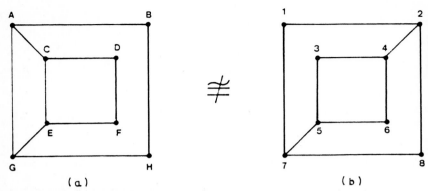

Figure 4.10 Two graphs with the same number of edges and vertices, but they are not isomorphic.

We have seen in Figures 4.6 and 4.7 that the same graph may be represented with crossing edges in one diagram but with noncrossing edges in another diagram. In general, if a graph can be drawn with edges that cross only at the vertices, it is called a *planar* graph, to distinguish it from *nonplanar* graphs that can only be drawn with some edges crossing. Thus the graphs in Figures 4.6 and 4.7 are planar. In Section 4.10, nonplanar graphs will be discussed in greater detail.

As we have noted, the formal definition of graphs does not allow:

- More than one edge between two vertices
- An edge with both ends attached to the same vertex
- Directionality of edges

But when we draw diagrams, it's easy to make sketches of "graphs" having some or all of the above properties. How can they be included in graph theory? The answer is to define new entities:

- *Multigraphs* are graphs with one or more multiple edges, that is, duplicate edges between vertices.
- *Pseudographs* are graphs with loops, that is, one or more vertices have an edge starting and ending at the same vertex.
- *Digraphs* are graphs with directed edges, that is, one or more edges have a specified direction to them.

Figure 4.11 shows examples of a pseudograph and a digraph.

The formal definitions for multigraphs, pseudographs, and digraphs are made by redefining the edge set so that it contains ordered pairs of vertices instead of sets of pairs of vertices. A more illuminating way to

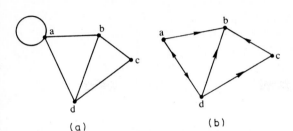

Figure 4.11   (*a*) A pseudograph; (*b*) a directed graph (digraph).

see the differences is to look at incidence matrices for the diagrams of each type. Here are some matrices. The graphical representations of the pseudograph and the digraph shown in Figure 4.11 are

$$
\begin{array}{c}
 & \begin{array}{cccc} a & b & c & d \end{array} \\
\begin{array}{c} a \\ b \\ c \\ d \end{array} &
\left[\begin{array}{cccc}
0 & 1 & 0 & 2 \\
1 & 0 & 1 & 1 \\
0 & 1 & 0 & 1 \\
2 & 1 & 1 & 0
\end{array}\right]
\end{array}
\qquad
\begin{array}{c}
 & \begin{array}{cccc} a & b & c & d \end{array} \\
\begin{array}{c} a \\ b \\ c \\ d \end{array} &
\left[\begin{array}{cccc}
1 & 1 & 0 & 1 \\
1 & 0 & 1 & 1 \\
0 & 1 & 0 & 1 \\
1 & 1 & 1 & 0
\end{array}\right]
\end{array}
\qquad
\begin{array}{c}
 & \begin{array}{cccc} a & b & c & d \end{array} \\
\begin{array}{c} a \\ b \\ c \\ d \end{array} &
\left[\begin{array}{cccc}
0 & 1 & 0 & 1 \\
0 & 0 & 0 & 0 \\
0 & 1 & 0 & 0 \\
1 & 1 & 1 & 0
\end{array}\right]
\end{array}
$$

$$
\qquad\quad \text{Multigraph} \qquad\qquad\quad \text{Pseudograph} \qquad\qquad\quad \text{Digraph}
$$

$$
\begin{array}{c}
 & \begin{array}{cccc} a & b & c & d \end{array} \\
\begin{array}{c} a \\ b \\ c \\ d \end{array} &
\left[\begin{array}{cccc}
1 & 1 & 0 & 0 \\
0 & 0 & 0 & 1 \\
0 & 1 & 0 & 0 \\
2 & 1 & 1 & 0
\end{array}\right]
\end{array}
$$

$$
\text{Combination}
$$

Problem 4.2   Try your hand at drawing the multigraph and the combination graph.

You should compare the matrices and diagrams and see if you can recreate one from the other. There are certain characteristics of the matrices of each type that help in recognizing them:

- A matrix of a graph is symmetric: If you read down a particular column, it will read the same as reading across the corresponding row (the row for the same vertex as the column). In addition, the matrix of a graph always has zeros along the main diagonal—the diagonal line from the upper left of the matrix to the lower right corner.

- The matrix of a multigraph has numbers other than 1s in the matrix (signifying that there are multiple edges).

- The matrix of a pseudograph has 1s on the main diagonal whenever a loop occurs; 1s on the main diagonal signify loops.
- The matrix of a digraph is not, in general, symmetric; corresponding rows and columns are not identical.

4.3 Maps

Everyone is familiar with maps. Figure 4.12 can either be looked at as the map of Europe, or it can be interpreted as an abstract map in a mathematical sense. The mathematical maps that we are going to examine in this section are similar to the cartographer's map in that the faces are countries, the edges are their borders, and the vertices are the corners of the countries. Now let's turn to a formal mathematical definition of a map. A *map M* is a set of edges, vertices, and faces,

$$M = \{V,E,F\}$$

where F is the face set. The edges and vertices satisfy a specified incidence matrix, while each face from the set F consists of a cycle of non-self-intersecting edges (a cycle is defined in Section 4.5) and vertices with no repeats except for the first and last vertex and containing no vertices inside it, i.e., F is a polygon (in Chapter 5 we will consider star polygons whose edges self-intersect). Polygons can have curved edges in this definition, and we will have to extend our notion

Figure 4.12 A map of Europe.

of a polygon to include figures with two edges [see Figure 4.1(a)]. The faces of a map are linked together at edges so that each edge lies in exactly two faces. Finally, in the definition of a map, no country can lie completely within another as Vatican City does within Italy (see Section 4.5). For example in Figure 4.13,

$$f_1 = v_3 e_5 v_5 e_6 v_4 e_3 v_3$$

$$f_2 = v_1 e_1 v_2 e_2 v_3 e_3 v_4 e_4 v_1$$

$$f_3 = v_1 e_1 v_2 e_2 v_3 e_5 v_5 e_6 v_4 e_4 v_1$$

where f_1, f_2, and f_3 are three-, four-, and five-sided polygons, respectively. It may appear strange that the exterior of the map is considered to be a face surrounded by the outer edges, the outside face. An explanation of this is in the next section.

At this point, let us pause to wonder why a concept as natural and familiar as a map needs to be belabored and stated in the technical language of mathematics. Is this language artificial, or is it natural and necessary to convey the meaning? Our intuition about maps serves us well so long as we do not stray too far afield from the concept of a geographical map. Yet in this book, we will consider maps far from this familiar territory. Here, the language of mathematics serves as our only compass. Whether or not this language serves a purpose or is unnecessarily pedantic must await judgment.

By defining the face of a map to be a polygon and each edge of a map to lie in exactly two faces, we have excluded some diagrams that we would like to consider as maps, namely, diagrams with hanging, or *pendant, edges or faces*. In Figure 4.14 the face connecting the hanging triangle lies in only one face, and the enclosed area with the pendant face is called a face even though it is not a polygon. Maps with pendant edges or faces are called *pseudomaps*.

Copies of a map formed by placing the map on a flexible membrane and stretching the membrane without cutting are considered identical

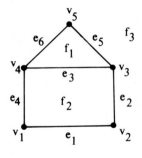

Figure 4.13 A map with three faces (including the outside face), five vertices, and six edges.

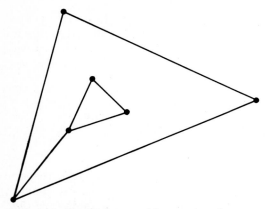

Figure 4.14   A pseudomap with a pendant face.

or *isomorphic*. Edges and faces become distorted but sets *E, F,* and *V* maintain their integrity.

The diagrams of maps and graphs are very much alike. In fact, any map may be considered to be a planar graph or a multigraph by taking into account its vertices and edges and their interconnections but neglecting its set of faces. However, there are subtle differences between graphs and maps:

- A connected planar graph can always be represented as a map or a pseudomap (a connected graph is a graph in one piece).

- A graph can give rise to more than one map (see Section 4.5). Figure 4.15 shows two different maps with the same graph.

- Graphs which are not connected (i.e., which occur in more than one piece) cannot be represented as maps (see Figure 4.16).

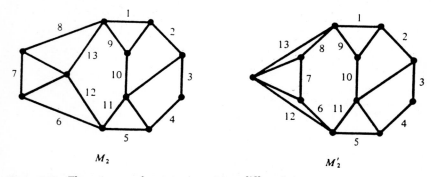

Figure 4.15   The same graph can represent two different maps.

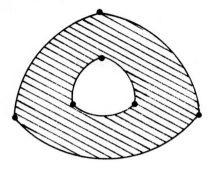

Figure 4.16 A graph which is not a map; the shaded face is not a graphical polygon.

Because of the great similarity in the structures of graphs and maps, we develop their theory together.

4.4   Maps and Graphs on a Sphere

It may seem strange and needless to define the *exterior* of a map to be a face; however, there is a compelling reason for doing this. If the map is drawn on a sphere as shown in Figure 4.17(*a*), it is clear that the outside face is no longer unbounded and should be treated as any other face. In fact, any face could just as well serve as the outer face by imagining the sphere to be an infinitely stretchable membrane pierced at some point within this face as shown in Figure 4.17(*b*) and (*c*). The membrane is then stretched until the puncture point is moved to infinity [see Figure 4.17(*d*)]. A map results in which the punctured face becomes the outside face.

Points on the sphere are now paired with points on the plane. (See Appendix 2.B for another way to pair points on a sphere with points in the plane.) The corresponding maps also share the same sets of faces, edges, and vertices and have the same incidence matrix. Thus, from a mathematical point of view, maps on the plane and maps on the sphere with one point removed are isomorphic. The same holds for graphs.

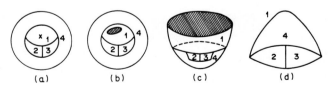

Figure 4.17 Transformation of a map on a sphere to a map on the plane. (*a*) Map on the sphere with face 1 punctured; (*b,c*) the puncture is widened; (*d*) map in the plane with face 1 on the exterior.

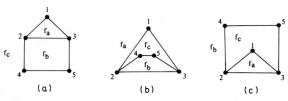

Figure 4.18   The outside face of a map is arbitrary. (*a*) Face *c* is the outside face; (*b*) map redrawn with face *a* as the outside; (*c*) redrawn again with face *b* as the outside face.

A map can always be redrawn with any one of its internal faces serving as the exterior face by first drawing the edges surrounding this face as the outer boundary of the new map as shown in Figure 4.18. The other faces are then redrawn inside this outer boundary, making sure that their connectivities are preserved. It is helpful to label all the vertices of the original map before exchanging faces in order to keep track of the connections between the vertices.

Problem 4.3   Redraw the map shown in Figure 4.19 so that face *f* is the outer face.

Since all the enclosed areas, including one additional outer one, are now considered to be faces, and maps are always considered to be in one piece (connected), we can restate the constraint on space introduced at the beginning of the chapter for the case of maps on the plane or a sphere as

$$F + V - E = 2 \qquad (4.4)$$

This relation, discovered by Swiss mathematician Leonhard Euler and known as *Euler's formula*, is proven for connected planar graphs in the next section.

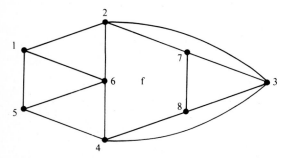

Figure 4.19   Redraw this map with face *f* as the outside face.

Problem 4.4 Show that Euler's formula holds for the graphs in Figures 4.14 and 4.15, but it does not hold for the graph in Figure 4.16. Why is this so?

In Section 4.11 we will generalize this relation to maps drawn on other surfaces.

4.5 Connectivity of Graphs and Maps

Information about the *connectivity* of a graph is given by the incidence matrix. A *path* between two vertices of a graph or map is defined as a succession of adjacent edges in a graph (i.e., a "walk" from vertex to vertex along specified edges). If there is a path connecting each pair of vertices, the map or graph is said to be connected. Since we have specified in our definition of maps that no country can be buried within another, maps are always connected graphs.

A *cycle* is a path from one vertex back to itself excluding the case where the steps are merely retraced. If a graph has no cycles, it is called a *tree*, an example of which is shown in Figure 4.20. The patterns of branching processes shown in Figure 4.2(*c*) are, in a sense, infinite trees.

Any *connected graph* with cycles can be transformed into a tree graph by removing some of its edges and leaving its vertices alone. The tree contains the least number of edges necessary to keep the original graph in one piece (connected). Thus, the result that we found in Equation (4.1*a*) of the introductory exercise holds for trees:

$$E = V - 1 \tag{4.5}$$

As a matter of fact, this equation can be used to prove Euler's theorem, Theorem 4.1.

Theorem 4.1 $F + V - E = 2$ for connected, planar graphs.

proof Transform a connected, planar graph into a tree by removing selected edges. But for each edge that is removed, a face is also eliminated, which pre-

Figure 4.20 A tree graph.

serves the value of $F + V - E$. However, since the tree graph can be viewed as having a single exterior face, Equation (4.5) can be rewritten as

$$F + V - E = 2$$

which must then have been satisfied by the original graph.

A connected graph can be disconnected into more than one segment by removing certain of its vertices and all the edges connected to these vertices. If the graph can be disconnected by removing only one vertex, it is *1-connected*. If two or three vertices must be removed to disconnect it, the graph is *2-connected* or *3-connected* (providing these graphs have more than two or three points, respectively, to start with). Graphs can also be disconnected by removing edges and leaving vertices untouched. However, this is of no interest to us. Examples of 1-, 2-, and 3-connected graphs are shown in Figure 4.21.

All connected planar graphs drawn in the plane can be represented by maps. Theorem 4.2 tells us which of these graphs can be drawn uniquely (except for isomorphic distortions) as maps.

Theorem 4.2 There is only one map corresponding to a 3-connected planar graph. Some 1- and 2-connected planar graphs can be represented by more than one map.

There are two special kinds of connected graphs that we will be referring to in this chapter. The first is a *complete* graph with n vertices, abbreviated K_n, in which each vertex is connected to each of the other vertices. K_5 is illustrated in Figure 4.22(a). The other is the *bipartite* graph, abbreviated $K_{m,n}$, which is defined to be a graph whose vertices are divided into two sets. Each of m vertices of the first set is connected to each of n vertices of the second set as shown in Figure 4.22(b) for $K_{3,2}$.

4.6   Combinatorial Properties

The structure of a graph is determined by its incidence matrix. For example, the sum of the 1s in a row (or column) of this matrix indi-

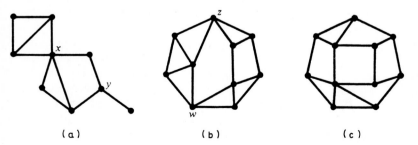

(a)   (b)   (c)

Figure 4.21   (a) 1-connected graph; (b) 2-connected graph; (c) 3-connected graph.

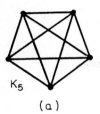
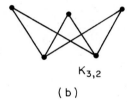

Figure 4.22   (a) A complete graph with five vertices, K_5; (b) a bipartite graph, $K_{3,2}$.

cates the number of edges that touch (are incident to) the vertex corresponding to that row (or column). The number of edges incident to a vertex v is called its *vertex valence*, which is symbolized by $q(v)$ since q, in general, depends on v. We will refer to vertices having odd or even values for their vertex valences as odd or even vertices. Since each edge of a graph contains two vertices, the following relation holds:

$$\sum_V q(v) = 2E \qquad (4.6)$$

where the summation is over all vertices v of set V. In the future, for the sake of brevity, we shall assume that the quantity being summed, e.g., q, depends on the elements of the set indicated under the summation sign and omit the variable, e.g., v. In other words, Equation (4.6) will be rewritten:

$$\sum_V q = 2E$$

Problem 4.5 Show that in a graph with more than one vertex there must be at least two vertices of the same valence. Remember that a graph cannot have more than one edge between vertices.

The same relation holds for maps. In addition, another quantity called the *face valence p* is defined to be the number of edges that surround a given face. Since each edge of a map lies in exactly two faces,

$$\sum_F p = 2E \qquad (4.7)$$

where summation is over all the faces of the map and the same shorthand convention is used as for Equation (4.6).

This equation does not hold for graphs with pendant edges or faces such as the ones shown in Figures 4.14 and 4.20. However, Equation (4.7) holds for these graphs if the pendant edges are counted twice. To make sense out of this seemingly arbitrary counting procedure, imagine that the edges of such graphs are walls and a bug crawls around all the edges of the face with the pendant edge. As the bug crawls along either side of the wall corresponding to the pendant edge, this edge is counted twice.

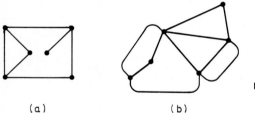

Figure 4.23

(a)   (b)

Problem 4.6 Verify Equations (4.6) and (4.7) for the graphs shown in Figure 4.23.

Many properties of graphs and maps can be deduced from Equations (4.6) and (4.7). One surprising result is expressed by the handshake lemma which states:

At a party, the number of people who shake hands an odd number of times is always even.

proof Draw a graph in which the people at the party are the vertices and an edge connects two vertices if these two people have shaken hands. Divide the graph into two sets of vertices, one set with odd vertex valence q_o and the other with even q_e. From Equation (4.6),

$$\sum_V q_o + \sum_V q_e = 2E \quad \text{or} \quad \sum_V q_o = 2E - \sum_V q_e = \text{even number} \quad (4.8)$$

Since odd + odd = even and odd + even = odd, it follows that the odd numbers on the left-hand side of Equation (4.8) must pair in order to result in an even number. The proof follows since each term on the left-hand side represents a party goer who shook hands an odd number of times.

Problem 4.7 For a hypothetical party with five people, draw the handshake graph and verify the lemma [Baglivo and Graver, 1983].

Another problem in the spirit of the handshake lemma follows.

Problem 4.8 What can you say about the number of people at a party at which everyone knows exactly four other people present, except the host who knows everyone present [Baglivo and Graver, 1983]?

4.7 Regular Maps

There is a family of maps for which each vertex and face is like every other vertex and face. These are called *regular maps*. They could be said to have a "perfect symmetry." If you found yourself placed in a mathematical country defined by such a map, you would experience vistas of sameness in all directions and find yourself hopelessly lost.

In the search for order in a seemingly chaotic world, mathematicians and philosophers since antiquity have been fascinated by structures exhibiting the level of perfection and order of the regular maps. The fact that there are only five of them, as we will show, makes them all the more precious. The next chapter will be devoted to a detailed study of these maps in a different context.

Regular maps are defined as maps whose vertices have identical vertex valence q and whose faces have identical face valence p. For regular maps, Equations (4.6) and (4.7) can be rewritten as

$$qV = 2E \qquad \text{and} \qquad pF = 2E \qquad \text{(4.9) and (4.10)}$$

Figure 4.24(a) and (b) illustrate two infinite families of maps (which we refer to as *trivial maps*) with this property. This leads to Theorem 4.3.

Theorem 4.3 Except for the two trivial families, there are only five regular maps on the plane (or sphere).

proof Solving Equations (4.9) and (4.10) for V and F and replacing in $F + V - E = 2$ yields

$$\frac{2E}{p} + \frac{2E}{q} - E = 2 \qquad \text{(4.11)}$$

After some algebra,

$$pq - 2p - 2q = \frac{-2pq}{E} \qquad \text{(4.12)}$$

Since the right side of this equation is negative, factoring the left side yields

$$(p - 2)(q - 2) - 4 < 0 \qquad \text{or} \qquad (p - 2)(q - 2) < 4$$

The only solutions to this equation, other than the trivial ones, are listed in Table 4.1 along with the number of edges, faces, and vertices of the correspond-

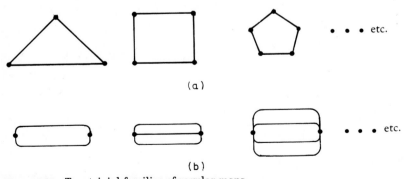

(a)

(b)

Figure 4.24 Two trivial families of regular maps.

TABLE 4.1

q	p	Schlafli notation {p,q}	F	V	E
3	3	{3,3}	4	4	6
3	4	{3,4}	6	8	12
3	5	{3,5}	12	20	30
4	3	{4,3}	8	6	12
5	3	{5,3}	20	12	30

ing maps. F, V, and E are determined by solving Equation (4.12) for E and replacing E in Equations (4.9) and (4.10) to get, after a little algebra,

$$F = \frac{t}{p} \qquad E = \frac{t}{2} \qquad V = \frac{t}{q} \qquad (4.13)$$

where $t = 4pq/(2p + 2q - pq)$

Problem 4.9 Illustrations of the graphs in Table 4.1 are shown in Figure 4.66(a). Before looking at them, try drawing them for yourself. Each of these maps is 3-connected and can, therefore, according to Theorem 4.2, be drawn in only one way.

4.8   New Graphs from Old Ones

Through isomorphism, the visual appearance of a graph can be drastically changed while preserving its underlying structure as we showed in Section 4.2. We now consider ways of transforming the structure of a graph G by adding or subtracting vertices and edges. There are three distinct ways of doing this:

1. Graph G can be augmented to a new graph H by selectively adding additional vertices and edges. Since each vertex and edge of the original graph G is also a vertex and edge of the transformed graph H, we say that G is a *subgraph* of H and that H is a *supergraph* of G. For example, K_5 is a subgraph of Figure 4.25(a) while $K_{3,3}$ is a subgraph of Figure 4.25(b).

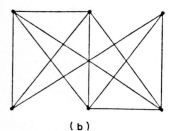 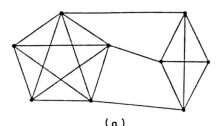

(b)   (a)

Figure 4.25   These graphs contain (a) K_5 as a subgraph; (b) $K_{3,3}$ as a subgraph.

2. Graph H can be reduced to G by selectively erasing certain edges and vertices. However, whenever a vertex is erased, all edges incident to that vertex must also be erased (otherwise G would not be a graph). In this case G is a subgraph of H.

3. An arbitrary number of vertices may be placed within the edges of graph G to obtain graph H, called the *subdivision* of G, i.e., a hole is erased on an edge and a vertex is inserted in the hole. In this case G is usually not a subgraph of H.

In Section 4.2 we showed that it can be tricky to determine whether or not two graphs are isomorphic. Besides the four properties mentioned there, another property preserved by isomorphism is the distribution of subgraphs. That is, if two graphs are isomorphic and you select a subgraph at random from either one, the other will necessarily have an isomorphic subgraph. Hence, you can prove that two graphs are not isomorphic by finding a subgraph of one that is not a subgraph of the other. For example, Figure 4.10(a) and (b) cannot be isomorphic since Figure 4.10(a) has a cycle of eight edges and vertices $A, C, D, F, E, G, H, B, A$ but Figure 4.10(b) does not.

In Section 4.10 these three ways to alter a graph will be used to state an important theorem about nonplanar graphs making use of the obvious fact that supergraphs and subdivisions of nonplanar graphs are also nonplanar.

4.9 Duality

Each map contains the seeds of another map called its *dual*, which is constructed by placing a vertex within each face, including its outside face, and connecting two vertices by an edge if their corresponding faces are adjacent (share an edge). Thus, since each edge of the original map lies in exactly two faces, each edge of the original is paired with one edge of the dual map. An example of a map and its dual is illustrated in Figure 4.26(a) and then redrawn in Figure 4.26(b).

Notice in Figure 4.26(a) that the dual of the dual map is the original. This reciprocal relationship is true of all duals so long as the exterior of the map is considered as a face. In fact, vertex v_1 of the original map corresponds to the outside face f_1 of the dual. However, any face of the dual could equally well serve as the outside face (see Section 4.4). For example, Figure 4.26(c) and (d) shows the dual redrawn so that f_1 is now an inner face and vertex v_4 of the original map corresponds to the outside face f_4 of the dual. (It is possible that for some maps, the redrawn dual may not be isomorphic in a map sense to the

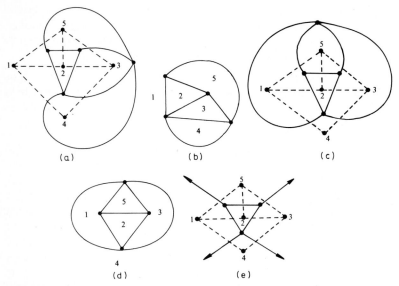

Figure 4.26 Dual maps. (*a*) A map is drawn with dotted lines and its dual map is drawn with solid lines; face 1 is the outside face; (*b*) the dual is redrawn; (*c, d*) the dual map with face 4 as the outside face; (*e*) the vertex corresponding to the outside face of the original map is taken to be the point at infinity.

original dual and so one should probably refer to "a" dual rather than "the" dual.)

Finally in Figure 4.26(*e*), the vertex of the dual corresponding to the outside face of the original map is taken to be the point at infinity in the sense of Section 4.4. In this case the edges of the dual that correspond to the outer edges of the original map are drawn with arrows to indicate that they all meet at the point at infinity.

From a mathematical point of view, a map and its dual may also be considered structurally identical despite their different appearances although they are not isomorphic. To explain what we mean by structurally identical, let's consider a map in terms of its abstract structure. To each face of the original there corresponds a vertex of the dual while edges are paired, i.e.,

$$\text{face} \leftrightarrow \text{vertex}$$

$$\text{edge} \leftrightarrow \text{edge}$$

In this sense, the dual map is encoded in the original, and any statement made about the faces of the original can be translated into an equivalent statement about the vertices of the dual. For example, if a face of the original is surrounded by p incident edges, p incident edges surround the corresponding vertex of the dual. Likewise, if a vertex of the original has q incident edges, the corresponding face of the dual has q edges incident to it. Also, if two vertices are connected by an edge in the original, the corresponding two faces share an edge (are adjacent) in the dual.

Although the original and its dual are structurally identical (but usually not isomorphic), they are visually dissimilar. Duality will play an important role throughout this book as it offers alternative ways of viewing a given structure as we already saw in the duality of Bartók's music (see Section 3.8).

Problem 4.10 Draw the map dual to the one shown in Figure 4.19.

Problem 4.11 For the regular maps shown by looking ahead to Figure 4.66(a), show that one of these maps is self-dual (the map and its dual are isomorphic) while the other maps form dual pairs.

4.10 Planar and Nonplanar Graphs

The development of graph theory has been motivated by the search for solutions to puzzles and games. The bipartite graph $K_{3,3}$, shown in Figure 4.27(a), is the source of one such puzzle for a rainy day.

Problem 4.12 Three people at odds with each other, each represented by one of the upper vertices of the bipartite graph, want to draw water from each of the three wells represented by the lower vertices, but they do not wish to have the possibility of meeting each other during their trips to the well. Can you devise paths from people to wells that meets this condition?

After trying different ways to redraw this graph, you will come to the conclusion that at least one path must cross, i.e., the bipartite graph is *nonplanar*. Figure 4.27(b) shows the graph redrawn with one crossover.

Puzzle solvers have never been put off by the challenge of trying to solve a problem advertised as being impossible; witness the many "solutions" that are still found for squaring the circle or trisecting an angle. However, it is instructive to see how the impossibility of unravel-

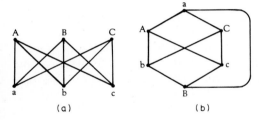

(a)　　　　(b)

Figure 4.27 (a) The bipartite graph $K_{3,3}$; (b) $K_{3,3}$ redrawn with one crossover.

ing a bipartite graph is inherent in the basic constraints on space that we have already accepted.

proof that $K_{3,3}$ is nonplanar If $K_{3,3}$ is planar, it forms a map in the plane for which $V = 6$ and $E = 9$. Replacing these values in Euler's formula for the plane and solving for F yields

$$F = 2 + E - V = 5$$

The smallest cycle of edges on the bipartite graph has four edges. If it is assumed that all faces have face valence $p = 4$, a lower estimate of Equation (4.7) is

$$20 = 4F \leq 2E = 18$$

which gives rise to a contradiction. The conclusion is that the bipartite graph must not have been planar.

It may seem strange that a geometrical property of a graph (edge crossings) has been proven by algebraic means. This is frequently done by mathematicians and it usually endows the proof with an aura of magic. The proof materializes apart from our intuitive understanding, i.e., it is not obvious. When this occurs, there is often a more transparent geometrical proof of the same result. For such a proof we refer you to [Ore, 1963].

Problem 4.13 The complete graph with five vertices, K_5, is another example of a nonplanar graph. Prove this by the same technique that we used to prove that the bipartite graph was nonplanar. Redraw this graph with only one crossing.

Why have we gone to such great lengths to demonstrate and prove that the bipartite graph $K_{3,3}$ and K_5 are nonplanar? It would be reasonable to imagine, and it is most certainly true, that countless other graphs are also nonplanar. However, in 1930 K. Kuratowski made the amazing discovery that all nonplanar graphs must contain within them, in a special sense, either K_5 or the bipartite graph $K_{3,3}$.

Theorem 4.4 (Kuratowski) Every nonplanar graph is a supergraph of a subdivision of $K_{3,3}$ or K_5 (see Section 4.8 for an explanation of supergraph and subdivision).

According to Kuratowski, the graph in Figure 4.9(b) must be nonplanar since it contains a subdivision of $K_{3,3}$ within it (check this!). Although Kuratowski's theorem represents another interesting constraint on space, it is generally not very helpful in spotting nonplanar graphs since the two basic nonplanar graphs are usually well camouflaged within the the graph under consideration. How then does one go about unraveling a complex graph with many crossing edges? A good way is to do what was done for the bipartite graph in Figure

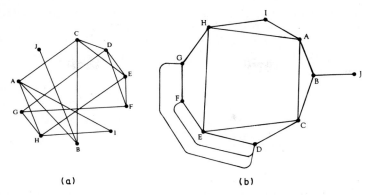

Figure 4.28   Planar graphs can be drawn without edge crossings by reorganiz-
ing its vertices around the periphery of a circle. (*a*) A planar graph with many
crossovers; (*b*) the same graph without crossovers.

4.27(*b*). The vertices are arranged around the periphery of a circle and
numbered so that nearby vertices on the circle connect to each other.
This technique is illustrated in Figure 4.28 for a more complicated
graph.

Of course once you have unraveled the planar graph and made it into
a map, it is easy to generate other maps by reordering the vertices.

Problem 4.14   Apply this technique to drawing another map corresponding to
the graph in Figure 4.28(*a*).

4.11   Maps and Graphs on Other Surfaces

A legend [Tietze, 1965] states that once upon a time in a remote land,
there lived a man with five sons who were to inherit his land after his
death. However, in his will the father made the condition that each of
the five parts into which the land was divided must border on each of
the other four parts. In addition, he required that each of his sons
build a road from his residence to the residence of each of his brothers
and that each of these roads was to run separately without crossings
and without touching the land of a third brother.

When the father died, the five sons worked hard to find a division of
the land to meet the terms of the will—all in vain. The brothers sank
into gloom as it became clear to them that their father's will could not
be fulfilled. Suddenly a traveling wise man appeared and claimed to
have a solution.

Problem 4.15   Before reading on, try to find this solution. Before attempting the
five brothers' problem, solve the same problem assuming that there were only
four brothers (much easier).

The solution to the four brothers' problem is shown in Figure 4.29(a). A solution to the five brothers' problem is shown in Figure 4.29(b). The solution involved building a bridge from the land of brother D to the land of brother E. The moral of this story is that what is possible to do on a surface depends on the nature of the surface.

Up to this point we have limited ourselves to a study of graphs on a plane surface or a punctured sphere, which is equivalent to the plane. In the introduction to this chapter we saw how subtle properties of the plane imposed constraints on graphs. What if we consider graphs and maps on other surfaces? Will there be new constraints? In this section

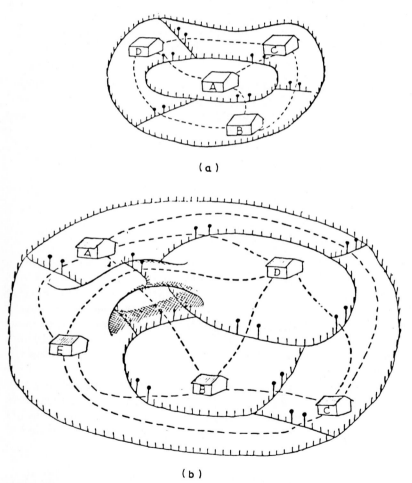

(a)

(b)

Figure 4.29 (a) Solution to the four brothers' problem; (b) solution to the five brothers' problem.

we will explore this question using the intuitive concept of a surface
as a thin membrane rather than the highly abstract mathematical
definition in which a surface has no thickness. The mathematical
point greatly disturbed Buckminster Fuller, who refused to acknowledge its existence since it confounds all practical experience.

The general study of curves and surfaces is carried out in the branch
of mathematics called *topology* [Firby and Gardiner, 1982], [Francis,
1987]. From an intuitive point of view, two surfaces are considered to
be topologically identical, or *homeomorphic* (the topological equivalent of isomorphic), if one surface can be imagined to be constructed of
a flexible membrane capable of being deformed without cutting to
form the other surface. In this deformation, nearby points on the original surface are still nearby on the deformed surface. For example, the
donut and cup of tea in Figure 4.30 are topologically equivalent.

This definition is not very good from a mathematical point of view
since surfaces do not have thickness and cannot be dealt with materi-
ally (despite Fuller's protests). Nevertheless, it gives us an easy way
of visualizing different families of surfaces. For example, by this definition, a cube is certainly homeomorphic to a *sphere*. Also a sphere
with two holes and a bent cylinder, or *handle* as it is called, extending

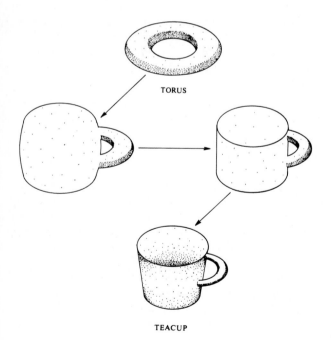

TORUS

TEACUP

Figure 4.30   A torus is shown to be homeomorphic to a teacup.

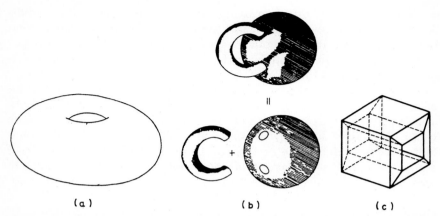

Figure 4.31 The torus in (a) is homeomorphic to a sphere with one handle in (b) and (c) to a cube with a tunnel through it.

from one hole to the other is homeomorphic to a *torus* (inner tube) [see Figure 4.31(a) and (b)] as is the cube with a tunnel pictured in Figure 4.31(c). Spheres with more than one handle are homeomorphic to multitori as Figure 4.32 shows. The sphere is called a *singly connected* surface, which means that any closed curve within it can be shrunk to a point and still remain entirely within the surface. Tori and multitori surfaces are called *multiply connected* surfaces, which means that certain closed curves within the surface (e.g., the curves surrounding the holes) cannot be shrunk to a point without leaving the surface.

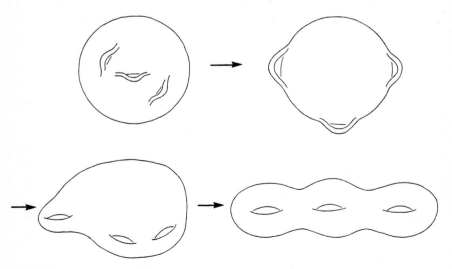

Figure 4.32 A sphere with three handles is homeomorphic to a triple torus.

If a map is drawn on a surface such as a torus or sphere, the value of the *Euler number*, $\chi = F + V - E$, can be computed. For example, $\chi = 2$ for a cube where F, V, and E are the faces, vertices, and edges of the cube while $\chi = 0$ for the map drawn on the cube with a tunnel (picture frame) shown in Figure 4.31(c). (Check this!) It can be proven that maps on homeomorphic surfaces share the same Euler number.

Although surfaces can be extremely complex, as Figure 4.33 shows, this complexity is nicely organized by Theorem 4.5. In it, *closed* surfaces are surfaces that enclose a region of space, while *oriented* surfaces have well-defined notions of "in" and "out" defined at each point on them (see the next section for an example of a nonoriented surface).

Theorem 4.5 Any closed oriented surface in three-dimensional space is homeomorphic to a sphere with h handles.

The number of handles h characteristic of the surface is called its *order*, or *genus*. Thus, a sphere is a surface of order 0 while a torus has order 1. Euler's formula can now be generalized far beyond its original scope.

Theorem 4.6 For any map on a closed oriented surface homeomorphic to a sphere with h handles,

$$F + V - E = 2 - 2h \tag{4.14}$$

An elementary proof of this theorem is suggested in [Beck, 1969]. The method that we used in Section 4.5 to prove the theorem for the plane, i.e., $h = 0$, fails for $h > 0$.

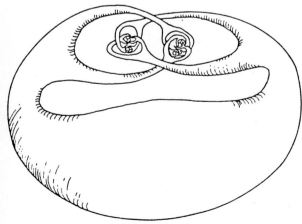

Figure 4.33   Alexander's horned sphere.

4.12   The Torus and the Möbius Strip

Now that we have seen that the nature of graphs depends on the surface upon which they are drawn, perhaps the two nonplanar graphs on the plane, K_5 and $K_{3,3}$, can be drawn without crossovers on some other surface just as the five brothers' problem was solved on a new surface. In fact, we shall show that these graphs can be drawn as planar on a torus.

In order to draw a map on a torus, it is convenient to cut the torus open to a *period rectangle* as shown in Figure 4.34, which is much the same as we did for the cylinder in Figure 2.16. The torus is given two cuts and the edges a and b where the torus was cut are identified. Figure 4.35 shows an example of a map drawn on the torus with $F = 4$, $V = 4$, $E = 8$, and $F + V - E = 0$. Figure 4.36 shows K_5 drawn as a planar graph on the period diagram of a torus.

Problem 4.16   Try your hand at drawing $K_{3,3}$ as a planar graph on a period torus. Cut a cylinder open to a period rectangle with a single crosscut and show that K_5 and $K_{3,3}$ are still nonplanar on the cylinder.

The *Möbius strip* is an example of a nonoriented surface. A rendering of this surface by the artist M. C. Escher is shown in Figure 4.37. A sailboat making one cycle about a Möbius strip with its sail pointed in an "upward" direction, as in Figure 4.38(a), would find its sail pointed in the reverse direction after one cycle about the surface. This indicates that up and down are ill-defined concepts for a Möbius strip. Figure 4.38(a) represents a period rectangle of a Möbius strip. The opposite orientation of the arrows on the left and right sides of the period diagram signifies that the strip is to be given one half twist before gluing the identified edges together.

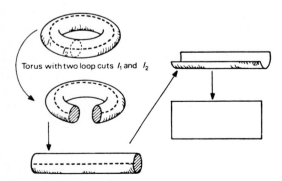

Torus with two loop cuts l_1 and l_2

Figure 4.34   A torus is opened to a period rectangle by cutting two loops on its surface.

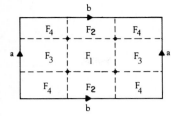

Figure 4.35   A map with four faces, four vertices, and eight edges, i.e., $F = V - E = 0$, drawn on the period rectangle of a torus.

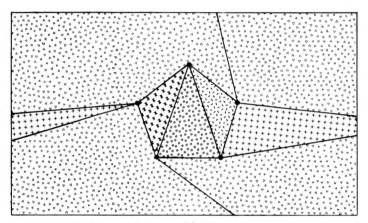

Figure 4.36 K_5 drawn on a torus without crossovers.

Problem 4.17   Build a Möbius strip by folding it up from its period rectangle. Check to see that it is one sided by coloring it with one continuous stroke. Also show that it is bounded by a single closed curve. Enjoy its surprising properties by cutting it parallel to its edge along its centerline. Build another Möbius strip and cut it along a line drawn parallel to the edge but one-quarter of the width of the strip. Finally, show that K_5 and $K_{3,3}$ can be drawn as planar graphs on the Möbius strip [Struble, 1971].

Problem 4.18   Draw a solution to the five brothers' problem on a period torus.

The following general theorem pertains to nonplanar graphs:

Theorem 4.7   Any graph that is nonplanar on the plane or a sphere can be represented with no crossings on some closed oriented surface.

The procedure is simple. Construct a wire model of the graph. "Thicken" the edges and solder them together at the vertices as illustrated in Figure 4.39(a) for a model of K_4. It is more of a challenge to

Figure 4.37 *Möbius Strip II, 1963. (Wood engraving by M. C. Escher. © M. C. Escher Heirs/ Cordon Art-Baarn-Holland.)*

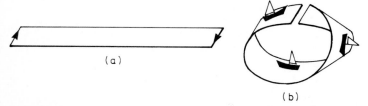

(a)

(b)

Figure 4.38   Folding up a Möbius strip from a period rectangle.

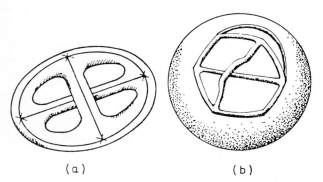

(a)   (b)

Figure 4.39   All graphs are planar on the appropriate surfaces formed by constructing the graph from wires and thickening the wires. (a) K_4 embedded in this way in its wire model; (b) $K_{3,3}$ drawn on a sphere with one handle.

find the surface of lowest order upon which a nonplanar graph is planar. The genus of such a surface is called the *genus of the graph* [Firby, 1982]. Figure 4.39(*b*) shows that $K_{3,3}$ has genus 1 since it is planar on a sphere with one handle.

4.13   Magic Squares

A *magic square* is a square matrix of numbers that add up to the same value if added along the rows, columns, or diagonals. The following is an example of a 3 by 3 magic square:

4	9	2
3	5	7
8	1	6

Magic squares are ancient mathematical structures that some authors feel were used to encode sacred information [Andrews, 1960], [Michell, 1983], [Critchlow, 1976], e.g., the magic square in Dürer's "Melancholia" (see Figure 3.17).

4.13.1   Construction of a magic square

There are many ways to construct n by n magic squares. If n is even, the procedure is far from straightforward [Andrews, 1960]. If n is odd, the following simple procedure can be used:

1. Place a 1 in the square beneath the central square.

2. Place all numbers from 1 to n^2 in the square in order. Any given number is placed one square below and one square to the right of the last number. In this process, think of the square as a period torus. A square beneath the bottom square is equivalent to the top square of the same column. A square to the right of the last square of a row is equivalent to the first square to the left in the same row.

3. During the process of generating numbers, there may be a number already occupying the square for which you are aiming. The rule says to deposit the number two steps below the previous number in the same column of the period torus.

4.13.2   Patterns from magic squares

Once you have generated an n by n magic square, you can use it to generate patterns in many different ways. One way is to place the numbers from 1 to n^2 (the last one) in rows and connect them in the order in which they appear in the magic square. For example, the pat-

Figure 4.40 A magic square pattern.

tern for the 3 by 3 magic square is shown in Figure 4.40. The numbers are connected according to the order in which they appear in the square above.

Problem 4.19 Create a 5 by 5 magic square and draw its pattern, using the above procedure. This square represents the planet Mars in ancient cosmology. According to John Michell [1983], the Litchfield Cathedral in England was dedicated to St. Chad, whose feast day is March 2, i.e., the month of Mars. Its underlying design is suggested by the magic square for Mars. It is built of red brick, the color of Mars. The city of Litchfield was formerly called "Liches from Mars," a remarkable set of coincidences.

Problem 4.20 Ancient cosmology associates the following magic squares with the sun, the moon, and the five planets known at that time (in addition to the Earth): 3 by 3, Saturn; 4 by 4, Jupiter; 5 by 5, Mars; 6 by 6, the sun; 7 by 7, Venus; 8 by 8, Mercury; 9 by 9, the moon.

1. According to Michell, the sun and the moon governed the underlying structure of the New Jerusalum diagram described in Section 1.2. Generate the magic square and patterns for the moon. Since the moon is an odd magic square, you can use the above procedure.

2. The magic square of the sun is

$$
\begin{array}{cccccc}
6 & 32 & 3 & 34 & 35 & 1 \\
7 & 11 & 27 & 28 & 8 & 30 \\
19 & 14 & 16 & 15 & 23 & 24 \\
18 & 20 & 22 & 21 & 17 & 13 \\
25 & 29 & 10 & 9 & 26 & 12 \\
36 & 5 & 33 & 4 & 2 & 31
\end{array}
$$

Draw its pattern. All its numbers sum to 666, a sacred number known in *Revelation* as "the number of the beast." Early Christians associated this number with Rome and pagan rites. According to Michell the chapel at Glastonbury (see Section 1.2) was destroyed during the Reformation because its dimensions embodied this number.

None of this can be taken too seriously from the vantage point of our rational world. Nevertheless, the power that pure numbers held over the ancient mind is fascinating.

4.14 Map Coloring

Exercise 4.2 Take a piece of paper and make a map by drawing five straight or curved lines, each starting and ending at the edge of the paper, and five closed curves. Your map might look like the one in Figure 4.41(a). How many colors do you need to color the map so that two faces that share an edge have different colors?

It is surprising that even though your map might look like a complex work of modern art, you only need two colors [see Figure 4.41(b)]. In fact, you would need only two colors regardless of how many lines you drew in your map. To see this, draw another line in Figure 4.41(a), such as the one straight from one corner to the opposite corner in Figure 4.41(c). Now to see that you only need two colors, all you have to do is to reverse all the colors below or above the line you have drawn. It will look like the drawing in Figure 4.41(d) [Struble, 1971].

Color the maps in Figure 4.21 using the fewest number of colors so that no two faces bordering on each other (i.e., those that share an edge) have the same color. You should find that no more than four colors are needed. Although no one has ever found a map in the plane that needs more than four colors, for nearly 100 years mathematicians tried in vain to prove this easy sounding result for all maps on the plane. Finally, in 1976 two mathematicians, K. Appel and W. Haken, succeeded in proving the four-color problem with the help of a computer, making it the first problem in pure mathematics to use the computer in an essential way.

Although the coloring problem for the plane (and sphere) was solved only recently, it has long been known that seven colors are sometimes needed and always enough to color a map on the torus. Figure 4.42 illustrates two such maps (both isomorphic) called Szilassi Maps on period diagrams of a torus. In each map, seven hexagons are drawn so that each hexagon borders the other six. Figure 4.43(a) is folded up to a torus in Figure 4.43(b) and (c). Another more interesting example of a map requiring seven colors will be given in Section 4.16.

Since the maximum number of colors needed to color a map depends only on the genus or order of the surface, and not on the map or shape

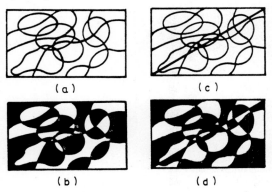

(a)   (c)

(b)   (d)

Figure 4.41 A two-color problem.

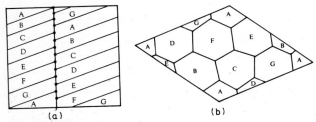

Figure 4.42 A Szilassi map with seven hexagonal faces each bordering on the other six shown on a period parallelogram.

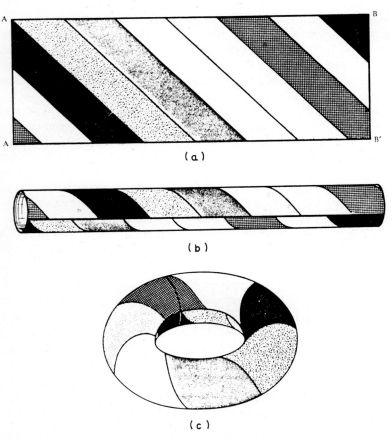

Figure 4.43 The seven-color problem on a torus. (*a*) A period rectangle with seven faces each bordering on the other six is folded up (*b*) to the torus in (*c*).

of the surface, we say that the chromatic number is an invariant of the surface [Firby and Gardner, 1982]. The Euler number is another invariant of the surface (we proved this for a plane surface in Section 4.5).

4.15   Regular Maps on a Torus

Just as there are only *five* regular maps on the sphere (or plane), there are only *three* classes of regular maps that can be created on a torus. As before, we shall call a map regular if each face has identical face valence p and each vertex has identical vertex valence q. One such map was illustrated in Figure 4.42(*b*) in which seven hexagons meet three at each vertex. Let's prove that there are only three classes of regular maps on the torus and find out what they are.

Once again we make use of Equations (4.9) and (4.10) along with Euler's formula for a torus:

$$qV = 2E \qquad pF = 2E \qquad \text{and} \qquad V + F - E = 0$$

Replacing the first two equations in the third we get

$$E \left(\frac{2}{q} + \frac{2}{p} - 1 \right) = 0 \tag{4.15}$$

Since E is not zero, the second factor must equal zero. After a little algebra this leads to

$$(p - 2)(q - 2) = 4$$

The only positive integer solutions to this equation are $p = 3$, $q = 6$; $p = 4$, $q = 4$; and $p = 6$, $q = 3$ or $\{3,6\}$, $\{4,4\}$, and $\{6,3\}$ using Schläfli notation.

There is an important difference between these regular maps and the ones on the sphere in Section 4.4. The regular maps on the sphere are unique in that their numbers of vertices, edges, and faces are fixed. On the other hand, Equation (4.15) shows that the number of edges E is indeterminate and therefore so are V and F. Thus the three regular maps on the torus are actually classes of maps, and there are an infinite number of possibilities in each class. For example, the regular map $\{6,3\}$ with seven hexagonal faces drawn on a period rectangle that has been distorted into a parallelogram was shown in Figure 4.42(*b*). The Hungarian mathematician Lajos Szilassi created a polyhedron based on this map, which will be discussed in the next section.

In fact, an infinite number of all three classes of regular maps can be represented by triangles, parallelograms, and hexagons outlined on the triangular graph paper shown in Figure 4.44. A different map of

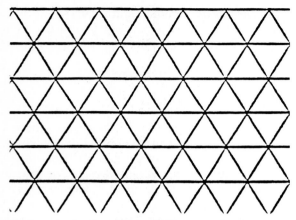

Figure 4.44   A triangular grid illustrates the three regular tilings on a torus {3,6}, {4,4}, and {6,3}.

each class is obtained by adjusting the size of the period parallelogram that frames the map. In Section 5.3, the period parallelogram is permitted to become infinitely large, which leads to three possibilities for regular maps with an infinite number of faces covering the plane.

4.16   Szilassi and Csaszar Maps

We would like to focus now on the Szilassi map since it is interesting from the point of view of design. We showed in Section 4.14 that this map gives an example on the torus of a map that requires seven colors to color it. We can make this map come to life in an interesting way. The faces of the map become polygons with straight-line edges spanned by planar membranes, and these faces combine to form a closed surface that is a distortion (homeomorph) of a torus. Such a surface is an example of a polyhedron, the subject of Chapter 7. One such *Szilassi polyhedron* is illustrated in Figure 4.45 [Gardner, 1978c], [Szilassi, 1986].

> **Construction 4.1**   Using the patterns in Figure 4.46, construct a Szilassi polyhedron. Color each of the faces with a different color to demonstrate the need for seven colors to color some maps on the torus.

A dual map to the Szilassi map, called a *Császár map*, can be drawn by placing a vertex in each of the hexagonal faces of the Szilassi map and connecting vertices by an edge if the corresponding faces share an edge [Gardner, 1975], [Beck, 1969], [Szilassi, 1986]. This dual map {3,6} with all triangular faces has the property that each of its seven vertices are connected to each of the other six, i.e., it is the complete

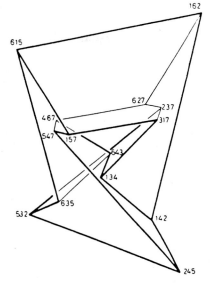

Figure 4.45 A Szilassi polyhedron.

graph K_7. The Császár map can also be materialized in the form of a polyhedron as the following construction indicates.

Construction 4.2 A *Császár polyhedron*, discovered in 1949 by the Hungarian mathematician Ákos Császár, is pictured in Figure 4.47. It can be constructed using the pattern shown in Figure 4.48 where the numbers represent the vertices. Vertices 2, 5, 3, and 4 were selected to form a regular tetrahedron (see Chapter 7). The edge lengths and interfacial or dihedral angles (see Section 7.10) are listed in Table 4.2.

Color the map in the table using the fewest colors. This is equivalent to coloring the vertices of the Szilassi polyhedron so that no two adjacent vertices (connected by an edge) have the same color.

The Császár map gives an example of a map on a torus with the property that each vertex is connected by a single edge to each of the others, and each of its faces is a simple polygon, with the result that no face can have a diagonal. Can other complete graphs be represented as a map on some surface? Certainly the triangle map, K_3, and the regular map, K_4, can be depicted as maps on the plane or sphere.

In order to find still other maps with each vertex connected to all the others, i.e., with faces having no diagonals, we call upon Equations (4.9), (4.10), and (4.14) once again. Since each vertex is connected to the others, $q = V - 1$, and it follows from Equation (4.9) that

$$V(V - 1) = 2E$$

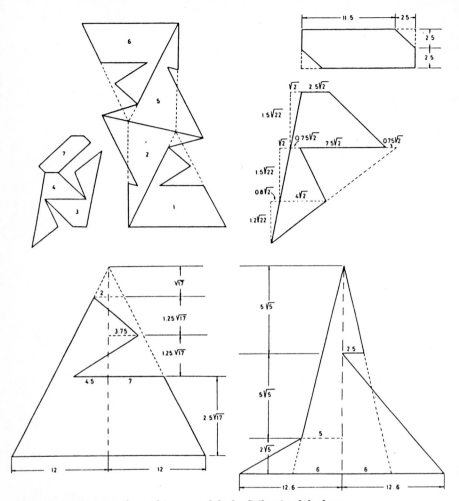

Figure 4.46   Patterns for making a model of a Szilassi polyhedron.

Since no face can have a diagonal, and vertices are connected by a single edge, all faces must have exactly three edges; then $p = 3$, and Equation (4.10) becomes

$$3F = 2E$$

Restating Euler's formula for a surface homeomorphic to a sphere with h handles,

$$F + V - E = 2 - 2h$$

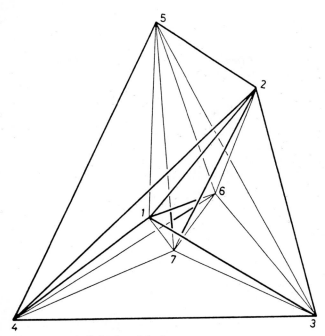

Figure 4.47   A Császár polyhedron.

Inserting the first two of these equations into the third and doing some algebra yields

$$h = \frac{(V - 3)(V - 4)}{12}$$

where h is an integer. There are an infinity of solutions to this equation. Each solution corresponds to a regular map on some surface. The properties of the first three are computed from these equations and listed in Table 4.3.

The case $h = 0$ corresponds to a regular map of K_4 drawn on a sphere, while $h = 1$ represents the Császár map K_7 drawn on a torus. The case of $h = 6$ has yet to be constructed in the form of a polyhedron.

4.17   Floor Plans

4.17.1   Evolution of a floor plan

Up to now we have focused on some mathematical aspects of graph theory. We have shown how graphs follow naturally from the simplest

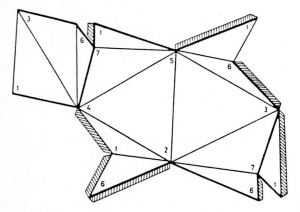

Figure 4.48   Patterns for making a model of a Császár polyhedron.

TABLE 4.2

Edge	Edge length	Dihedral angle (degrees)
(1–6)	10.00	76.133
(2–5)	24.00	70.533
(3–4)	24.00	54.433
(2–4) = (5–3)	24.00	51.050
(2–3) = (5–4)	24.00	52.717
(3–7) = (4–7)	12.89	340.133
(2–7) = (5–7)	17.15	74.417
(1–5) = (6–2)	18.69	339.317
(1–2) = (6–5)	12.55	156.850
(1–4) = (6–3)	12.55	204.467
(1–3) = (6–4)	17.36	41.667
(1–7) = (6–7)	5.86	243.500

TABLE 4.3

h	V	F	E	p	q
0	4	4	6	3	3
1	7	14	21	3	6
6	12	48	66	3	11

notions of placing dots and lines on a piece of paper and lead to complex questions of interest primarily to mathematicians. The remainder of this chapter is devoted to showing some applications of graph theory. In this section we show how graphs provide a natural tool to aid architects in developing a floor plan. In the next section we show that the bipartite graph serves as the appropriate tool with which to study the bracing of structures, and we conclude this chapter with a brief discussion of eulerian and hamiltonian paths through a graph.

There is a stage in the design process that precedes the concrete planning stage. Alexander, whose quote introduces this chapter, has written eloquently on this subject [Alexander, 1964]. In this initial stage, linkages or connections may be drawn between the various components of a design to indicate their relationships to each other in the design of an airplane, an industrial process, or a building. These linkages, or connections, can be understood and manipulated best by using graphs.

Graphs enable the architect to conceive of the relation between the rooms of a building with respect to each other and with respect to the outside environment with no need to specify the details of room shape. In other words, graphs reveal the underlying structure of a floor plan, leaving the details of building planning for the next stage of the design.

In this section we show how graphs can be used in an evolutionary process to design the floor plan of a one-story building.

4.17.2 From floor plan to graph

The *floor plan* of a one-story building can be thought of as a map. This is illustrated by Figure 4.49(*a*). Corresponding to a floor plan, another planar graph can be drawn such as the one shown in Figure 4.49(*b*). This graph, called the *adjacency graph*, places a vertex in each room of the floor plan and the exterior and connects vertices if the corresponding rooms of the floor plan share all or part of a wall.

Another way to indicate the connectivity of the rooms in a house is shown in Figure 4.50(*b*). This *access graph* connects two rooms, represented by vertices of the floor plan in Figure 4.50(*a*), by an edge if there is direct access from one room to the other through a door or partition. Access to the exterior is via a window or a door. The access graph is generally more relevant to the design of floor plans than is

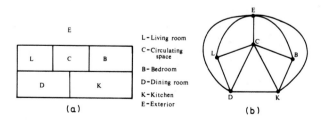

Figure 4.49   (*a*) Floor plan; (*b*) adjacency graph.

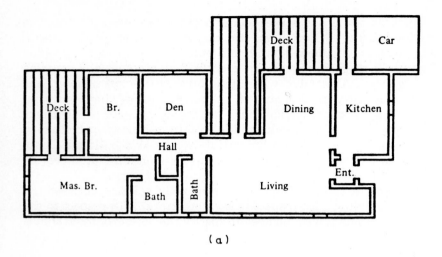

(a)

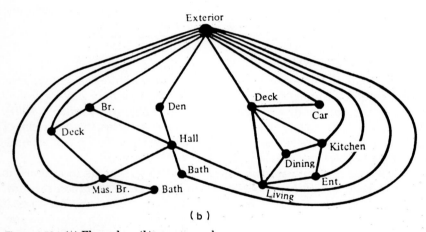

(b)

Figure 4.50   (a) Floor plan; (b) access graph.

the adjacency graph. Both access and adjacency graphs are sometimes referred to as connectivity graphs of the floor plan.

A graphical analysis of three house projects by Frank Lloyd Wright are shown in Figure 4.51(a), (b), and (c) [March and Steadman, 1974]. Although the overall designs of these houses are square, circular, and triangular, respectively, the access graph—identical for all three houses—is shown in Figure 4.51(d). Thus Wright's rich repertoire of designs led to strikingly different final results using the same underlying structure. This also illustrates the potential value of the graphical

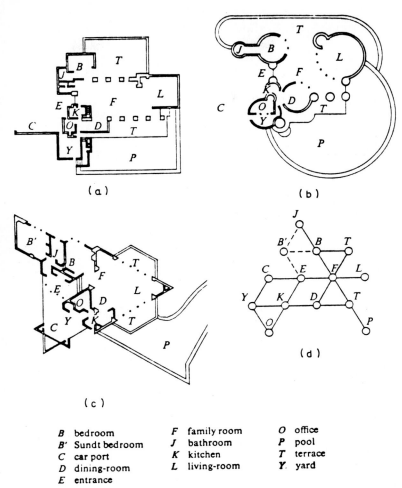

(a)

(b)

(c)

(d)

B	bedroom	F	family room	O	office
B'	Sundt bedroom	J	bathroom	P	pool
C	car port	K	kitchen	T	terrace
D	dining-room	L	living-room	Y	yard
E	entrance				

Figure 4.51 Three houses by Wright: (a) Life house, 1938; (b) Ralph Jester house, 1938; (c) Vigo Sundt house, 1941; (d) access graph for the three projects. The dotted lines refer to the additional bedroom, B, in the Sundt house.

method when used by a skillful designer, although there is no evidence that Wright used this method himself.

4.17.3 From adjacency graph to floor plan

It is more important in designing a building to be able to go from the adjacency or access graph to the floor plan. For simple adjacency graphs this can be done directly.

Problem 4.21 Draw the floor plan of a four-room house in which each room borders on the other three, i.e., K_4 is a subgraph of the adjacency graph.

Not every adjacency graph results in a floor plan on one story, as the next problem shows.

Problem 4.22 Try to draw the floor plan of a five-room house in which each room borders on the other four, i.e., K_5 is a subgraph of the adjacency graph. Why can't this be done?

If the adjacency or access graph is planar, a one-story floor plan can be constructed as the dual to this graph. In this section we outline a procedure for generating a floor plan from partial information about the access graph and apply it to generating the structure of Wright's *Blossom House* shown in Figure 4.52(*b*) [Rowe, 1976].

Step 1. The access graph generally emerges from the first step in the design process. The architect begins by obtaining a partial list of relationships between the rooms of the building from the client, i.e., a partial list of adjacencies is given for a house (↔ means "is connected to"). For example,

a	Conservatory	↔ *c*
b	Rear porch	↔ *f*
c	Dining room	↔ *d, h*
d	Butler's pantry	↔ *c, e*
e	Kitchen pantry	↔ *d, f*
f	Kitchen	↔ *b, e, i*
g	Terrace	↔ *h*
h	Living room	↔ *c, g, i, k*
i	Stairs	↔ *f, h*
j	Library	↔ *k*
k	Hall	↔ *h, j, l, m*
l	Reception room	↔ *k*
m	Porch	↔ *k*

These accesses can be summarized by an incidence matrix (not shown; see Section 4.2), in which the sum of all the 1s in each row represents the number of rooms that access a given room.

An incomplete access graph is drawn [see Figure 4.52(*a*)]. Notice how the accesses correspond to those in the actual floor plan of the Blossom House shown in Figure 4.52(*b*). Also, notice that a vertex corresponding to the exterior of the house has not been included in the

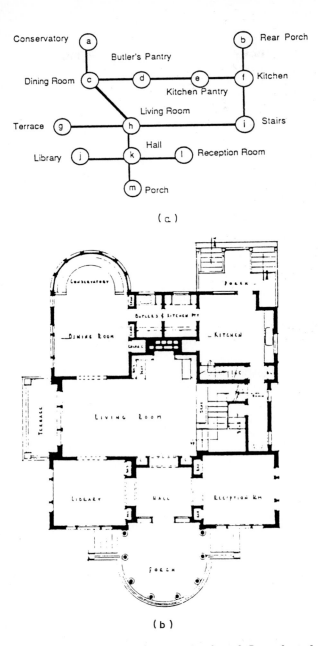

(c)

(b)

Figure 4.52 (a,b) A partial access graph and floor plan of Wright's Blossom House; (c) completed access graph of the Blossom House shown with open vertices and dual map or rough floor plan shown with closed vertices.

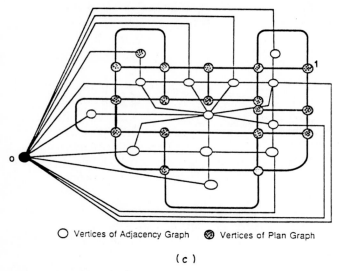

○ Vertices of Adjacency Graph   ⊛ Vertices of Plan Graph

(c)

Figure 4.52   (*Continued*)

access graph. In order to generate the floor plan, it is essential to include the exterior or outside vertex.

The architect must see whether this partial list of connectivities can be realized as the floor plan of a one-story building. This can be done if this partial access graph is a planar graph (as it is in this case). In general, the access graph will have many edge crossings and will have to be redrawn with no crossing edges before proceeding to the next step. The technique for drawing such planar graphs was described in Section 4.10.

Step 2.   The architect completes this partial list of relationships to obtain a complete access graph of the floor plan that satisfies both the client's wishes and the architect's design sense. This is shown by the open dots in Figure 4.52(*c*) where the architect has connected the living room to the butler's and kitchen pantries and to the library in addition to specifying which rooms have openings to the exterior represented by the black dot.

Step 3.   The dual map, whose vertices are shown with closed dots, is superimposed in Figure 4.52(*c*) upon the access graph. The dual of a planar access graph is, in general, a rough floor plan in that the walls encircle spaces representing the rooms in an amorphous way. One trick to drawing a coherent dual is to make sure that the vertex representing the exterior face of the access graph, labeled o in Figure 4.52(c), is on the opposite side of the dual map from the vertex of the dual representing the outside face of the access graph, labeled 1 in Figure 4.52(c).

Step 4.   A problem can arise in generating the floor plan if the exterior face appears as an interior face in the dual. This is not the case in our example, but it can easily happen if the dual is drawn differently [see Figures 4.26(a) and (c)]. If the outside face does appear as an inner face of the dual, the dual must be redrawn with the exterior face on the outside. The method for doing this is to place the map on a sphere, puncture the face o, and stretch it out to the plane. The procedure for drawing a map with a given face on the outside was described in Section 4.4.

Step 5.   In the last step in the evolution of the floor plan, the rough floor plan is changed into a more satisfactory design by distorting the rooms without altering their connectivities as shown by the actual floor plan of the Blossom House in Figure 4.52(b). Notice that the final form of the floor plan is suggested in its rough outlines by the dual map. It is at this final step in the evolution of a floor plan that the design instincts of the architect enter the process. This method is useful to the architect if it suggests possibilities for the design which may not have been obvious to him or her from the start.

Remark 1.   This procedure can also be used to design rooms of a multistory dwelling if the rooms on each story are separated and the rooms on two different levels are connected by a stairwell.

Problem 4.23   It is an interesting design exercise to take an existing floor plan and exchange some centrally connected space with the exterior while maintaining all of the connectivities, i.e., adjacencies or accesses, from room to room. This is done by the method developed in Section 4.4 where the dual map is placed on a sphere and the space to become the exterior is punctured and the dual redrawn with this space as the outside face. Try this idea out for the Blossom House where the living room is to be exchanged with the exterior. Before proceeding, it is best to number the vertices of the dual in order to keep track of the connections of vertex to vertex during the exchange.

Problem 4.24   In a somewhat artificial example, a client has requested an architect to build a house with the following adjacencies:

b bedroom $b \leftrightarrow c, d, e$

c circulation space $c \leftrightarrow b, k, l$

d dining room $d \leftrightarrow b, k, l$

e exterior $e \leftrightarrow b, k, l$

k kitchen $k \leftrightarrow c, d, e$

l living room $l \leftrightarrow c, d, e$

Draw the connectivity graph and construct a floor plan as the dual of the connectivity graph. Show that if the kitchen is required to border on the exterior, the floor plan cannot be realized. Why? Relocate the kitchen so that the floor plan can be realized.

Construction 4.3   Starting with a hypothetical set of client constraints, apply the five-step procedure of this section to create a floor plan of a house, office suite, school, etc., that satisfies these constraints.

4.18   Bracing Structures

In this section we discuss a graphical solution to a problem of bracing an architectural structure. This problem is, for the most part, reproduced from *Incidence and Symmetry in Design and Architecture* by Jenny A. Baglivo and Jack E. Graver [1983], who based their exposition on the work of E. Bolker and H. Crapo [1977]. The structure is a rectangular grid of squares whose edges are steel beams which are pin-jointed at each point. Although each beam in the grid is rigid, the structure itself is not since there is flexibility at the joints. Figure 4.53 shows some possible movements of 1 by 1 and 2 by 3 grids.

We are interested in bracing the structure to make it rigid. The grids in Figure 4.53 can be made rigid by adding cross braces to each square as shown in Figure 4.54. However, considering the high building costs, we wish to make these structures rigid by adding the fewest number of crossbeams possible.

Experiment 4.1.   Many of the ideas discussed here can be tested by using a model of a rectangular grid constructed from cardboard strips and roofing nails. For each beam use a strip of cardboard which is 3 inches by ½ inch; for each crossbeam, use a strip of cardboard which is

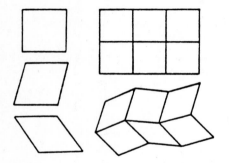

Figure 4.53 Some movements which can occur in a simple square and 2 by 3 grid.

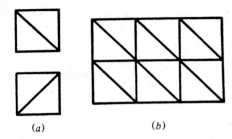

(a)   (b)

Figure 4.54 Bracing a structure to make it rigid by adding a crossbeam.

4 inches by ½ inch. Punch holes in the cardboard strips with centers ¼ inch in from the ends (see Figure 4.55). The holes should be as near to the same diameter of the roofing nails as possible. Thirty nails, 49 strips, and 10 long strips are sufficient for the experiments of this section. Experimental grids are constructed by placing the nails straight up, on a smooth surface, and slipping the short strips over them to form the grid.

1. Construct a simple square. Note that the opposite sides remain parallel however you distort the square. Now place one of the long strips on the diagonal of the square and observe how it becomes rigid.
2. Construct a 3 by 3 grid. Experimentally find the minimum number of crossbeams that will make this structure rigid.

To pursue the discussion further, we need to define certain terms. By a *bracing*, we mean a collection of crossbeams placed in an n by m grid. By a *rigid bracing*, we mean a bracing which makes the grid a rigid structure, that is, in which the only movement possible for the grid is as a single unit. We are interested in characterizing the *minimum rigid bracings*, that is, those which use the smallest possible number of crossbeams. We call this the *bracing problem*.

For a minimum rigid bracing, removal of a single crossbeam destroys the rigidity of the structure. Although it is possible that differ-

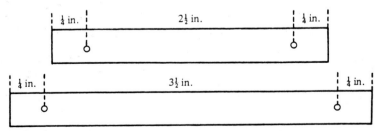

Figure 4.55 Cardboard strips for a bracing experiment.

ent minimum bracings might require different numbers of cross-beams, this turns out not to be the case.

In Figure 4.56, we compare several bracings of a 3 by 3 grid. Bracings (*a*) through (*f*) are all rigid. The others are not. Distortions of bracings (*g*), (*h*), and (*i*) are also included in the figure. Bracings (*d*), (*e*), and (*f*) are minimum rigid bracings.

Experiment 4.2. Construct all of the bracings that are illustrated in Figure 4.56 and check the statements that have been made about them. Can you draw any general conclusions from your experiments?

Consider an *n* by *m* grid. (For the purposes of illustration, we will continue to use a 3 by 3 grid.) The vertical beams along one row of squares of the grid will be called the *elements of the row*. Correspondingly, the horizontal beams down one column of squares of the grid will be called the *elements of the column*. The following lemma gives a very simple, but useful, observation about the movement of elements in any distortion of the grid. Figure 4.57 illustrates this lemma.

Lemma   In any distorted grid all the elements of a row (column) are parallel.

Consider the 3 by 3 grid in Figure 4.58(*a*). By bracing the square in the second row and first column, the first two elements of row 2 are perpendicular to the middle elements of column 1. But then, by the

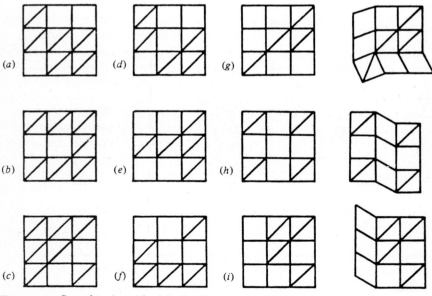

Figure 4.56   Some bracings of a 3 by 3 grid.

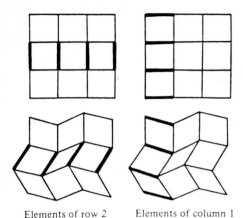

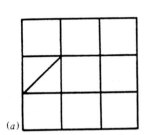

Elements of row 2   Elements of column 1

Figure 4.57 In any flexing of a grid the elements of each row and each column remain parallel.

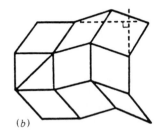

(a)   (b)

Figure 4.58 By the lemma, the elements of row 2 are perpendicular to the elements of column 1.

lemma, all of the elements of row 2 will be perpendicular to all the elements of column 1 under any distortion. This is illustrated in Figure 4.58(b). (Using your model of the 3 by 3 grid, verify these statements.)

A rigid bracing has the properties that the elements in a fixed row are parallel to the elements in every other row; the elements in any column are parallel to the elements in every other column; and the elements of each row are perpendicular to those of each column. This sounds as though there are very many constraints on the structure, but let us take the analysis of the lemma one step further. Consider a 3 by 3 grid as braced in Figure 4.59(a), and place a second crossbeam on the second row and the third column position of our grid [see Figure 4.59(b)]. As before, the elements of row 2 will always be perpendicular to the elements of column 3. Since the first crossbeam assured us that the elements of row 2 would be perpendicular to those of column 1, we can conclude that the elements of columns 1 and 3 will be parallel no matter how we distort the grid. This is illustrated in Fig-

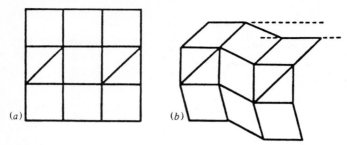

Figure 4.59 By the lemma the elements of row 2 are perpendicular to the elements of columns 1 and 3.

ure 4.59(b). Thus, we have satisfied three of the constraints using only two crossbeams. By properly placing additional crossbeams we will be able to control all of the constraints. But the analysis is delicate. Theorem 4.8 provides a quick method for knowing where to place the crossbeams to ensure a rigid bracing and how many crossbeams are needed for a minimum rigid bracing.

We now have the basis for a theoretical method of dealing with the bracing problem. Consider an n by m grid. Represent the rows by n vertices labeled $r_1, r_2, \ldots, r_n$, and represent the columns by m vertices labeled $c_1, c_2, \ldots, c_m$. If the square which is in row i and column j is braced, we place an edge between vertices r_i and c_j. A bracing of the grid can then be represented by a subgraph of the complete bipartite graph $K_{n,m}$; we call this the *bracing subgraph* of that bracing of the grid. The bracing subgraphs of three of the bracings in Figure 4.56 are pictured in Figure 4.60. A careful study of these subgraphs leads to the following observations: first, the bracing subgraph for the minimally braced grid, Figure 4.56(d), is a tree; second, the bracing subgraph for the nonrigid grid, Figure 4.56(h), is not connected; third, the bracing subgraph for the overbraced grid, Figure 4.56(a), is connected but contains a circuit. These observations, along with observations from other examples, lead to Theorem 4.8.

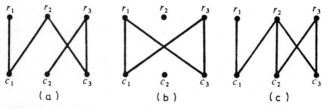

Figure 4.60 Bracing subgraphs for three bracings in Figure 4.56 illustrating Theorem 4.8: (a) minimally braced grid; (b) nonrigid grid; (c) overbraced grid.

Theorem 4.8

1. A bracing of an n by m grid is rigid if and only if the corresponding bracing subgraph is connected.
2. A bracing of an n by m grid is a minimum rigid bracing if and only if the bracing subgraph is a tree.

Since we know the relationship between the number of vertices and edges for a tree graph is given by $E = V - 1$, we see that every rigid bracing of an n by m grid contains at least $n + m - 1$ crossbeams. Furthermore, if a rigid bracing contains more than $n + m - 1$ crossbeams, it is always possible to find a set of beams to delete without affecting the rigidity of the structure.

proof The proof is presented here only in its most general outlines. If the bracing subgraph is connected, there is a connected path leading from any row r to any column c, e.g., $r = r_h, c_i, r_j, c_k = c$. By applying the lemma, it can be shown that each element of row r_h and r_j is perpendicular to each element of column c_i and that each element of columns c_i and c_k is perpendicular to each element of row r_j. From this we infer that each element of r_h is perpendicular to each element of c_k and in particular the element of the grid in the hth row and kth column must be a square. Thus bracing the square in the row r_h and column c_k does not alter the rigidity of the grid. By the same argument each square of the grid can be shown to be effectively braced. Part (2) of the proof follows from the fact that a tree is a connected graph with the least number of edges.

4.19 Eulerian Paths

It is thought that graph theory had its origin in a paper written by Euler in 1736. In this paper Euler used graph theory to solve several popular puzzles of the time, such as the *bridges of Koenigsberg* [Ore, 1963], [Euler, 1979].

The different parts of the city of Koenigsberg, today known as Kalingrad, lay on either bank of the river Pregel, between a fork in the river and on the island of Kneiphof as shown in Figure 4.61(a). Seven bridges connected the various parts of the city, and people had always wondered whether it was possible to walk across all seven bridges without retracing a bridge.

Euler saw that the problem could more easily be studied by reducing island and banks to points and drawing a graph (or, as we will sometimes say, a *network*), in which two points are connected by an edge whenever there is a bridge connecting the corresponding land masses, as shown in Figure 4.61(b).

In this way Euler was able to abstract the problem so that only information essential to solving the problem was highlighted and he could dispense with all other aspects of the problem. It is for this reason that graphs find great utility as a conceptual tool in many different disciplines.

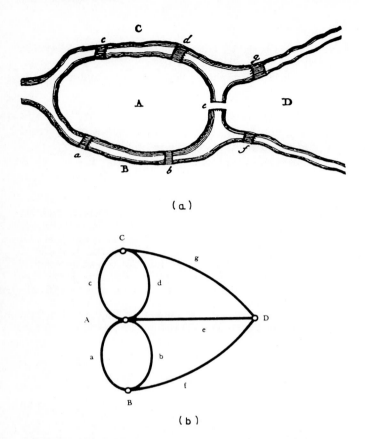

Figure 4.61 The seven bridges of Koenigsberg (*a*) as originally drawn by Euler in the Proceedings of the St. Petersburg Academy of Sciences in 1736; (*b*) as a graphical representation.

The Koenigsberg bridge problem is an example of a class of similar problems concerning graphs that can be stated as follows:

Given a connected graph, find a path that traverses each edge of the graph without retracing an edge.

Such a path is called an *eulerian path*, or E path as we shall refer to it. If the beginning and end point of the Euler path are the same, the eulerian path is called an *eulerian circuit*. Restating this problem leads to a famous rainy day recreation. Can you draw a given graph without taking your pencil off the page and without retracing edges?

Problem 4.25 Which of the graphs shown in Figure 4.62 can this be done for?

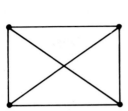 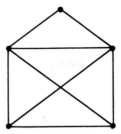

Figure 4.62   Which of these graphs have Euler paths?

After attempting to find E paths in several graphs, it is natural to ask whether there is some simple criterion by which you can predict, a priori, whether a graph contains such a path. Some experimentation and application of logic should convince you that in order to have an E path, whenever the path enters a vertex, there must be another path that leaves it. The only exceptions to this rule occur for the beginning and ending point of the path, if these points are different. Restating:

> A necessary condition for the existence of an eulerian path through a con- nected graph is that all vertices be even, i.e., have even vertex valence with the possible exception of two.

What is somewhat surprising is that this simple condition also guaran- tees that the graph contains an E path. However, we will not prove this.

Problem 4.26   The graph shown in Figure 4.63 represents the hallways of a mu- seum. Pictures are to be hung on one side of each hall. If possible, design a tour that will enable a person to see each exhibit exactly once. Indicate where the entrance and exit should be built. Number the edges and represent the tour as a sequence of edges.

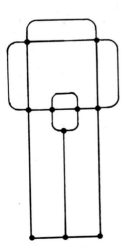

Figure 4.63   A schematic of the circulation space of a museum. Does this graph have an Euler path? If so, find it.

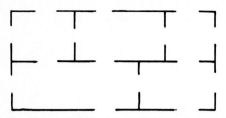

Figure 4.64 Find a path through each of these floor plans that goes through each door exactly once.

Problem 4.27 A floor plan is shown in Figure 4.64 for a five-room house. All doorways are shown. If possible, design a walk that takes a person through each doorway once only.

Along with his discovery of graph theory, Euler set forth the general properties of networks in a set of four rules [1979]:

1. The number of odd nodes must be even or zero (handshake lemma of Section 4.6).
2. If a network has no odd nodes, it can be traveled along a path using all the edges without repeating an edge, beginning and ending at any node (i.e., there exists an Euler cycle).
3. If a network has only two odd nodes, it can be traveled along an E path that begins at one of them and ends at the other one. Any route (a path with no repeating edges) that begins at an even node, however, cannot traverse the network on an E path.
4. Any network that has more than two odd nodes can be fully explored by several disconnected routes without traveling over a branch more than once. If it has $2n$ odd nodes, it can be fully explored in n routes, each traveled on an E path.

Problem 4.28 This is an old puzzle that asks whether you can draw the diagram in Figure 4.65 with three strokes of the pencil. You are not permitted to go over any line twice. Use Euler's rules to analyze this problem.

Jearl Walker [1986] shows how networks can be used to solve mazes. There are procedures guaranteeing that one can find a path through a maze, if such a path exists, even when no map is explicitly

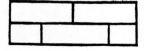

Figure 4.65 Can you draw this with three strokes of your pen?

given. Other procedures enable people who are lost to retrace their steps to the beginning of a labyrinth. Also, some of the procedures useful in exploring a maze have applications to problems of computer processing, traffic control, electrical engineering, and many other fields.

4.20   Hamiltonian Paths

As simple as it is to find a necessary and sufficient condition for a graph to have an eulerian path, the problem of finding those conditions that predict when a graph possesses a path containing each vertex of the graph once only has yet to be solved. Such a path is called a *hamiltonian path*, or H path, after the mathematician William Rowan Hamilton, who first studied this problem.

Problem 4.29   Find a hamiltonian path through each of the regular maps shown in Figure 4.66(*a*). Which of them does not possess an eulerian path? Show that the graph in Figure 4.66(*b*) has no hamiltonian path.

Donald Crowe shows how H paths of two-, three-, and higher-dimensional cubes can be incorporated into a strategy for solving an old puzzle known as the *Towers of Hanoi* [Beck et al., 1969]. For this puzzle, N circular discs with holes in their centers, each with a different radius, are piled on one of three posts in order of decreasing radii as shown in Figure 4.67. The object of the puzzle is to transfer the discs to the last post so that they appear, once again, in order of decreasing radii. The middle post can be used for intermediate transfers, but at no time in the transfers can a disc of larger radius sit atop one of smaller radius. The total number of transfers to transfer N rings is $2^N - 1$. A legend surrounding this puzzle has the priests at the high temple of Benares working day and night to transfer 64 rings from one

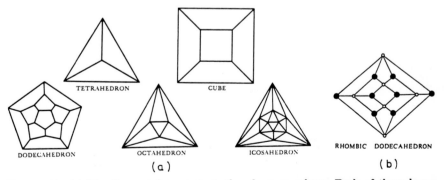

Figure 4.66   (*a*) The five regular maps on the plane or sphere. Each of them has a hamiltonian path. (*b*) A graph with no hamiltonian path (the correspondence of these graphs to polyhedra will be explained in Section 7.5).

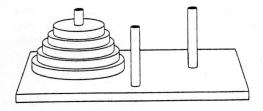

Figure 4.67   The Tower of Hanoi.

diamond needle to another at the center of the world after which "tower, temple, and Brahmins alike will crumble into dust, and with a thunderclap the world will vanish." However, undue worry is not called for since it takes 18,446,744,073,709,551,615 moves to carry this out.

We find that the order of the vertices in the H path of an n-dimensional cube (hypercube) yields a strategy for carrying out the transfers of N rings. For example, a two-dimensional cube, or square, and its H path is shown in Figure 4.68. If movements to the left or

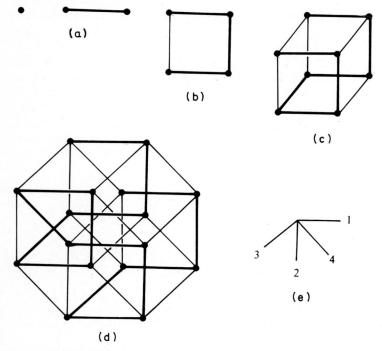

Figure 4.68   Hamiltonian paths on a (a) line segment; (b) square; (c) cube; (d) tesseract (four-dimensional cube); (e) directions of the one-, two-, three-, four-cubes.

right are considered as movements in the first dimension, while movements up and down are movements in the second dimension, the sequence of moves constituting the H path of the two-dimensional cube, using the vectors in Figure 4.68(e), is 121 as shown in Figure 4.68(b). But this is also the strategy for transferring two discs from post 1 to post 3, where disc 1 is the smaller of the two discs, i.e., transfer disc 1 to post 2, then disc 2 to post 3, then disc 1 to post 3.

Likewise, the strategy for transferring three rings is 1213121, which corresponds to the H path of the three-dimensional cube where 3 refers to a movement in the third dimension, as shown in Figure 4.68(c). (Check to see that this strategy succeeds in transferring the three rings.)

This line of thinking to an H path for a four-dimensional cube yielding the strategy for transferring four rings: 121312141213121. But what do we mean by a four-dimensional cube? Strictly speaking, we are unable to represent such a cube in three-dimensional space, but we can depict its three-dimensional projection by taking each of the eight vertices of the cube and translating them one unit in a given direction as shown in Figure 4.68(d), similar to the way the three-dimensional cube was generated from the two-dimensional cube.

Problem 4.30 Another three-dimensional projection of a four-dimensional cube is shown in Figure 4.69(a). You can see that it divides space into eight compartments, C, counting the exterior of the cube as a compartment. Count edges, faces, and vertices and show that they satisfy Ludwig Schläfli's generalization of Euler's formula:

$$F - E + V - C = 0 \qquad (4.16)$$

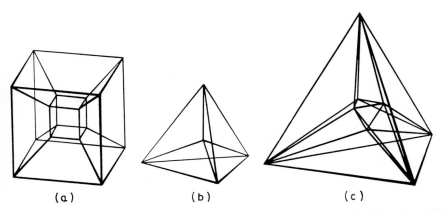

     (a)   (b)   (c)

Figure 4.69 (a) Cube in four-dimensional space; (b) Tetrahedron in four-dimensional space; (c) octahedron in four-dimensional space.

Verify Schläfli's equation for the projection of a four-dimensional tetrahedron, shown in Figure 4.69(b) and a four-dimensional octrahedron shown in Figure 4.69(c). Notice that each vertex in the 4-cube and 4-tetrahedron is 4-valent (q = 4); each edge is in common to three cells, while each face is in common to two cells. Graphs with these valencies or greater possess the necessary condition of a family of graphs known as *4-polytopal* graphs since they are graphical representations of polyhedra in four-dimensional space [Coxeter, 1973]. Such graphs will play an important role in determining the rigidity of three-dimensional structures in Section 7.8.

Four-dimensional and higher-dimensional cubes are now being used as optimal networks for the flow of information in parallel processing computers [Hillis, 1987]. It should also be mentioned in passing that the old Chinese rings puzzle [Ball, 1967] is essentially the same as the Tower of Hanoi. That is, the solution to the Chinese rings—suitably interpreted—gives the same hamiltonian circuit on the n-cube as the Tower of Hanoi. We can say that the Tower of Hanoi, the Chinese rings, and the n-dimensional cube have isomorphic structures. In Sections 10.13 and 10.14 we shall see that n-dimensional cubes play an important role in characterizing polyhedra.

5

Tilings with Polygons

*Pattern is born when one reproduces the
intuitively perceived essence.*
　　　SOETSU YANAKI, UNKNOWN CRAFTSMAN

5.1　Introduction

Something basic in the human mind has led us to create repeating
patterns of geometric shapes. Such patterns have been woven into fabrics or carved and painted on the walls of temples and buildings since
the dawn of civilization. In nature, the surface of the skin or the stalks
of a plant reveal intricate patterns of geometric shapes. Artists and
architects also work at subdividing space in ways that are pleasing to
the eye. From the point of view of design, the possibilities for creating
geometric patterns that cover the entire plane or a limited region of
the plane are endless. In this chapter we will examine several of these
patterns with an eye to understanding their underlying structures.
Once a simple pattern is generated, it can serve as the source of countless other patterns which are transformations of it and widen the repertoire of interesting possibilities.

In Chapter 4 the edges of a graph were shown to be completely
amorphous and to have the function of indicating connections between
pairs of vertices. The faces of a map were shown to be equally mallea-
ble and were defined by cycles of vertices and edges. Now we consider
the edges to be straight and of definite lengths and the cycles of edges
and vertices to define polygons in the usual geometrical sense. The
polygons are arranged to fill up the plane without gaps. Such tilings,
also known as tesselations, pavings, or mosaics, have appeared in human activities for millennia.

The geometry of tiling played a central role in the art, science, and
culture of Islam. The first mathematical investigations of tilings were
carried out by Kepler three and a half centuries ago. Much of what is

presently known about this ever-growing subject can be found in B. Grünbaum and G. C. Shephard's *Tilings and Patterns* [1987]. In the tilings that we will study, two tiles will be disjoint, will share a single vertex, or will share an entire edge. These, so-called, edge-to-edge tilings eliminate many possibilities but enable us to consider tilings as extensions of the maps of Chapter 4. Try the following exercise before reading further.

Exercise 5.1   Get a piece of triangular graph paper like the kind shown in Figure 4.44 and draw a few designs. The triangular grid is an extremely versatile design medium. Many of these designs have the appearance of patterns found in Islamic art (see Figure 5.1).

Exercise 5.2   Get some marshmallows and toothpicks. Find as many patterns as you can with the restriction that all marshmallows are surrounded by identical patterns of polygons.

In this chapter, we first focus on tilings in which the same number and order of a single kind of polygon surrounds each vertex of the tiling. We refer to these as *regular tilings*. They are regular maps in the sense of Section 4.15. Generally the polygons we will consider will have equal angles and edges, although occasionally we will deviate from this restriction. We will refer to such polygons as *regular polygons*, which should not be confused with regular tiling.

Next we look for tilings known as *semiregular tilings*, in which more than one kind of polygon surrounds each vertex. We then consider several ways in which tilings can be transformed to develop interesting designs based on regular and semiregular tilings including the parquet deformations of William Huff, the movable tilings of William Varney, and the shadow-and-light transformations of Janusz

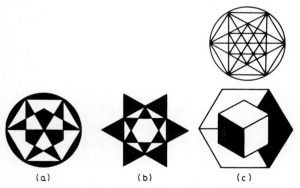

(a)   (b)   (c)

Figure 5.1   The versatility of the triangular grid. These are created by shading a portion of the grid enclosed by the circle.

Kapusta. Some designs based on a special class of tilings with penta-gons are presented, as is a unified approach to origami based on tilings by modular units discovered by Peter Engel. The chapter con-cludes with a brief discussion of Islamic art. First let's find out a few things about polygons.

5.2   Polygons

5.2.1   Convex polygons

The sum of the internal angles of a polygon with n sides is

$$\sum_V \theta = 180 \, (n - 2)$$

An easy way to see this result is to recognize that any polygon can be triangulated—that is divided into triangles as shown in Figure 5.2. In each case the number of triangles is two less than the number of sides.

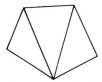

Figure 5.2   A pentagon divided by diagonals into three triangles.

The average angle of a polygon is the sum of their angles divided by the number of angles, which equals n (the number of sides). That is

$$\theta_{\text{avg}} = \frac{180 \, (n - 2)}{n} \tag{5.1}$$

For regular polygons (not to be confused with regular tilings) with n sides, denoted by $\{n\}$, the internal angles are all identical (and so are the edge lengths), and so each interior angle equals the average value. Some internal angles of regular polygons are listed in Table 5.1. These are the only polygons which arise in the tilings of this chapter and as the faces of the polyhedra of Chapters 7 through 10.

5.2.2   Star polygons

If the edges of a triangle are extended, they do not envelop any new regions of space as shown in Figure 5.3(a). The same goes for a square [see Figure 5.3(b)]. However, the sides of a pentagon intersect to pro-duce the star-shaped figure shown in Figure 5.3(c). *Star polygons* can

TABLE 5.1

n	θ (degrees)
3	60
4	90
5	108
6	120
8	135
10	144
12	150

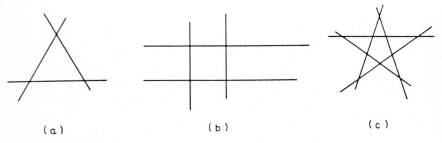

(a)   (b)   (c)

Figure 5.3 Star polygon formed as a convex polygon.

serve both as tiles to tile the plane (see Section 5.9) and as the faces of polyhedra (see Section 7.14). They are also very interesting objects from a mathematical point of view.

In general, a star polygon is obtained by drawing a circle with a compass [Davis and Chinn, 1969]. Readjust the opening of the compass but make sure that it is not greater than the diameter of the circle. Place the compass point anywhere on the circle, say at point P_1, and allow the pencil to intersect the circle at P_2. Place the compass point at P_2 and intersect the circumference at P_3. Proceed in this manner always in one direction either clockwise or counterclockwise. This yields a sequence of points, $P_1, P_2, P_3, \ldots$ and chords $P_1 P_2, P_2 P_3, \ldots$. The question is, will the points ever come back and fall on the first point? Or, said another way, will the polygons ever close? The answer depends on the ratio of the circumference of the circle to the length of arc marked out by the compass setting. If this ratio is an integer n, the points return after one revolution and result in the regular n-gon, $\{n\}$. If the ratio is a rational number, m/n, the points return after m revolutions and result in a star polygon, $\{n/m\}$. If the ratio is irrational, the points never return but become dense on the circumference of the circle, as we saw in Section 3.7.2, when laying down stalks around the periphery of a plant at irrational angles based on the golden mean.

In this way, several species of n-gons can be obtained by placing n

evenly spaced points on a circle and by connecting every third point on the circumference or every fourth point, etc. If this is done for a seven-sided figure, three distinct species of heptagon are obtained, as shown in Figure 5.4. One is a regular heptagon, {7}. One closes after two turns, {7/2}. The other closes after three cycles, {7/3}. Contrast this with the polygons arising from circles with eight points shown in Figure 5.5. There are only two species, the regular octagon, {8}, and the star octagon, {8/3}. Instead of {8/2} the octagon breaks into two squares, and in place of {8/4}, the polygon degenerates to an intersecting set of line segments. In general, a polygon or star polygon {n/m} with n sides is obtained by connecting every mth point on the circumference of a circle whenever n and m have no common factors, i.e., they are relatively prime (see Appendix 1.A).

The preceding relationship between the geometry of star polygons and the theory of numbers was discovered by Louis Poinsot, a French mathematician [Davis and Chinn, 1969]. Gauss discovered that polygons with a prime number of sides could be constructed using only a compass and straightedge if and only if the number of sides was figured by the formula

$$N = 2^{2^{n}} + 1$$

Since $n = 0, 1, 2, 3, 4$ leads to the primes 3, 5, 17, 257, 65,537, these polygons can be constructed with compass and straightedge. However, $n = 5$ leads to a composite number (not prime) and so it cannot be constructed, showing a close relationship between geometry and the theory of numbers.

One note of caution about star polygons. There is sometimes confusion between star polygons and polygons with the shape of stars (see Section 5.9). For star polygons, the points at which the edges intersect are not vertices, whereas these points are vertices of the star-shaped polygons. These latter polygons are examples of nonconvex curves. *Convex* curves such as the one shown in Figure 5.6(a) are curves that

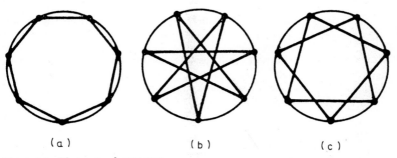

(a)   (b)   (c)

Figure 5.4   Three star heptagons.

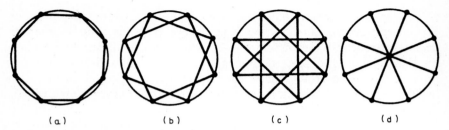

Figure 5.5 Four star octagons (one is not connected and another is degenerate).

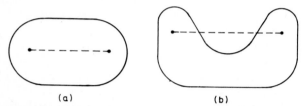

Figure 5.6 (a) A convex curve; (b) a nonconvex curve.

have no indentations while *nonconvex* curves have bulges or depressions such the one in Figure 5.6(b). A closed convex curve is defined to be one such that any two points placed within it can be connected by a straight line also lying within the curve as shown in Figure 5.6(a). If part of the connecting line lies outside of the curve for some pair of internal points, as shown in Figure 5.6(b), the curve is nonconvex.

Star polygons have also been used as mystical symbols, and they have been incorporated in mandalas such as the sacred Sri Yantra diagram shown in Figure 5.7. According to John Michell [1988], the star heptagon makes a surprise appearance in the New Jerusalem pattern shown in Figure 1.3. As Figure 5.8. illustrates, four star heptagons fit exactly into the pattern of 12 spheres, marking off 28 equal intervals of the lunar month. In this way the solar and lunar cycles are combined in a single geometric construction. Michell also feels that this unusual coherence of an irregular 12-gon with a star 7-gon is at the very foundation of the New Jerusalem as it is described in *Revelation* 21:

> Then one of the seven angels that held the seven bowls full of the seven plagues came and spoke to me and said, "Come, and I will show you the bride, the wife of the Lamb." So in the spirit he carried me away to a high mountain, and showed me the holy city of Jerusalem coming down out of heaven from God—It had a great high wall, with twelve gates, at which were twelve angels; and on the gates were inscribed the twelve tribes of Israel. There were three gates to the east, three to the north, three to the

Figure 5.7 The Sri Yantra is drawn from 9 triangles, 4 pointed downward and 5 pointed upward, thus forming 42 triangular fragments around a central triangle.

south, and three to the west. The city wall had twelve foundation-stones, and on them were the names of the twelve apostles of the Lamb.

5.3   Regular Tilings of the Plane

Buried in the triangular grid of Figures 4.44 and 5.1 are three regular tilings of the plane by congruent polygons: a tiling with triangles, six surrounding each vertex, or {3,6}; parallelograms, four surrounding each vertex, or {4,4}; and hexagons, three surrounding each vertex, or {6,3}, pictured in the top row of Figure 5.12. Here we use the Schläfli notation {p,q} where p is the face valence and q is the vertex valence of the map.

Each of these tilings is a regular map in the sense of Section 4.7 with the property that each vertex is surrounded identically by congruent faces. If we restrict ourselves to regular tilings with regular polygons, we can prove one of the oldest results from the theory of tilings.

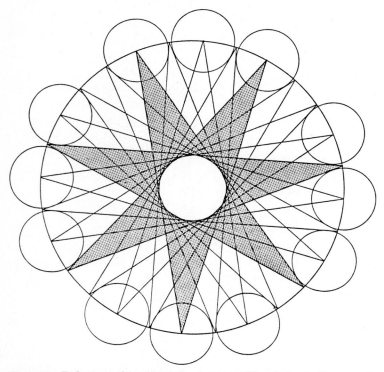

Figure 5.8 References throughout the chapters of *Revelation* to the geometry of the New Jerusalem repeatedly demand that the number 12 be combined with the number 7 to symbolize the union of body and spirit. This union is achieved through the New Jerusalem ring of 12 lunar circles. Accommodated by this ring is a figure made up of four star heptagrams having 28 horns, the number of phases in the lunar cycle. The regularly spaced horns fit neatly between the circles, touching their sides, or terminate at their centers.

Theorem 5.1 The only regular tilings on the plane are {3,6}, {4,4}, and {6,3}.

proof for the case of tilings by regular polygons Consider a p-sided regular polygon, $\{p\}$. From Equation (5.1) each internal angle of a regular polygon is

$$\theta = \frac{180\,(p-2)}{p} \text{ degrees}$$

Surround a typical vertex of the tiling by q regular p-sided polygons. The sum of the internal angles around the vertex is

$$\frac{180q\,(p-2)}{p} = 360 \text{ degrees}$$

After a little algebra we can rewrite this equation as

$$(p - 2)(q - 2) = 4$$

This has positive integer solutions: $\{q,p\} = \{3,6\}, \{4,4\}, \{6,3\}$.

We shall make several remarks about this result.

Remark 1. Tilings with regular pentagons are impossible, although Kepler obtained some very interesting tilings with pentagons as a result of trying to tile the plane regularly with pentagons. More will be said about such tilings in Section 5.11. For now, try the following exercise.

Exercise 5.3 Cut a regular pentagon with edges of about ¾ inch out of cardboard and see what kind of tilings you can get by replicating the pentagon on a sheet of 8 by 10 paper. Try to arrange your pattern so that the leftover space assumes interesting shapes.

Remark 2. Although it is harder to prove, these are the only regular edge-to-edge tilings possible with congruent (not necessarily regular) polygons of any sort. In fact it is easy to see that:

Any triangle can tile the plane as $\{3,6\}$. Just rotate the triangle around the midpoint of one of its sides to form a parallelogram.

Any four-sided polygon tiles the plane as $\{4,4\}$ [see Figure 5.9(a)]. Here, each quadrilateral is rotated about the midpoint of its side to form the adjacent quadrilateral. The quadrilaterals need not be convex as Figure 5.9(b) shows.

Any hexagon with opposite sides parallel and equal tiles the plane as $\{6,3\}$.

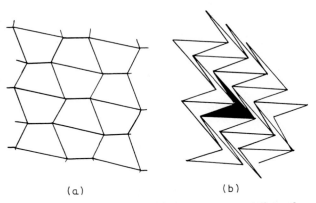

(a)   (b)

Figure 5.9 Tiling the plane with (a) convex quadrilaterals; (b) nonconvex quadrilaterals.

Remark 3. We mentioned in Section 4.15 that regular maps on the torus can be interpreted as tilings on the plane. Since a tiling on the period rectangle (or parallelogram) diagram of a torus must match at the opposite edges of the rectangle (or parallelogram), the period rectangle (or parallelogram) can be replicated in the directions of its edges to fill up the plane with a regular tiling. The three regular maps on the torus derived in Section 4.15 are in fact the three regular tilings of the plane in a topological sense.

The regular maps on a plane unwrapped from a torus satisfy $(p - 2)(q - 2) = 4$, while the five regular maps on a plane derived from a punctured sphere (see Section 4.7) satisfy $(p - 2)(q - 2) < 4$. Another class of regular maps on what is known as the *hyperbolic plane* are discussed in Section 12.10 and Appendix 2.B. They satisfy, $(p - 2)(q - 2) > 4$. One such mapping of $\{7,3\}$ onto the euclidean plane is shown in Figure 5.10. Such regular tilings of the plane can be eliminated from consideration if we impose the restriction that all tiles of a tiling enclose circles with diameters no smaller than a preassigned di-

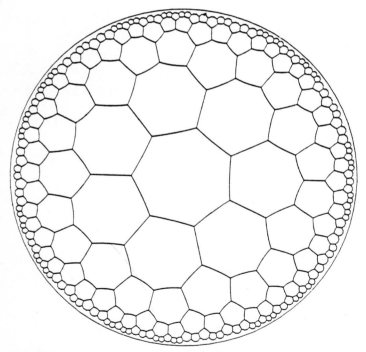

Figure 5.10 Tiling the plane with noncongruent heptagons with three heptagons surrounding each vertex, i.e., $\{7,3\}$.

ameter and can be enclosed by circles of diameters no larger than a preassigned diameter.

5.4 Duality

Each of the regular tilings by regular polygons has another tiling associated with it. Place a dot at the centroid of each polygon of the tiling and connect dots with a straight line if the polygons share an edge. What emerge are the following *dual tilings* in the sense of Section 4.9:

$$\{3,6\} \leftrightarrow \{6,3\}$$
$$\{4,4\} \leftrightarrow \{4,4\}$$
$$\{6,3\} \leftrightarrow \{3,6\}$$

Thus the dual tilings to the regular tilings with congruent tiles remain within the family of regular tilings. The duality of hexagons and triangles is illustrated in Figure 5.11.

5.5 Semiregular Tilings

Now that we have found the three regular tilings of the plane with regular polygons, let's relax the condition that only one kind of polygon be used but still require that each vertex be surrounded identically (see Exercise 5.2). We will start focusing on triangles grouped around a single vertex and successively remove polygons from around this vertex and replace them by regular polygons that fit evenly into

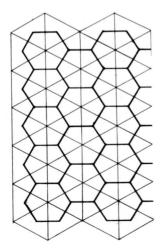

Figure 5.11 The regular tiling {3,6} is dual to {6,3}.

the gap using Table 5.1 as an aid. For example, if two triangles of $\{3,6\}$ are removed, this leaves an angle of 120 degrees, just large enough to fit a hexagon according to Table 5.1. The resulting sequence of four triangles and one hexagon surrounding the vertex is referred to by the Schläfli symbol, 3.3.3.3.6, or $3^4.6$ for short. If three triangles are removed, this leaves room for two squares. The resulting sequence of polygons surrounding the vertex now has two distinct possibilities, $3^3.4^2$ or $3^2.4.3.4$.

Problem 5.1   There are 21 ways to arrange regular polygons around a vertex. See how many of the 18 possible kinds of vertices with more than one kind of polygon you can find by successively removing regular polygons from the regular tilings and replacing them with different species of polygons.

Once all 18 possible ways are found to surround a vertex by regular polygons, the question arises as to whether the tiling near the vertex can be extended to a tiling of the entire plane. The end result of the search for tilings with two or more regular polygons surrounding each vertex leads to the 8 possibilities shown in Figure 5.12 known as *archimedean*, or *semiregular*, tilings. The other 10 tilings cannot be extended from around the single vertex to a tiling of the entire plane. The nature of space prevents them from tiling the plane, each for its own reason. For example, Figure 5.13 shows that any sequence of polygons surrounding a vertex of the form $3.x.y$ can tile the plane only if $x = y$. As a result a number of possibilities that fit locally such as 3.9.18, 3.10.15, 3.7.42, and 3.8.24 cannot be continued to tile the entire plane whereas 3.12.12 can.

We should be clear at this point that in each of the semiregular tilings not only does the same species of regular polygon surround each vertex, but also all vertices are surrounded by polygons in the same cyclic order. If order is not a requirement, there are an infinite number of different ways to tile the plane. For example, in Figure 5.14 a zigzag strip of tiles from the $3^2.4.3.4$ tiling are shifted to a new relative position to get a nonregular tiling with two squares and three triangles around each vertex [Grünbaum and Shephard, 1977]. In a similar way, an unlimited number of tilings can be gotten by altering other rows.

Another way to picture the regular and semiregular tilings is shown in Figure 5.15 where each vertex is replaced by identical circles of arbitrary radius.

5.6   Symmetry

What makes these archimedean tilings so aesthetically pleasing is their high degree of symmetry. By the symmetry of a pattern, we

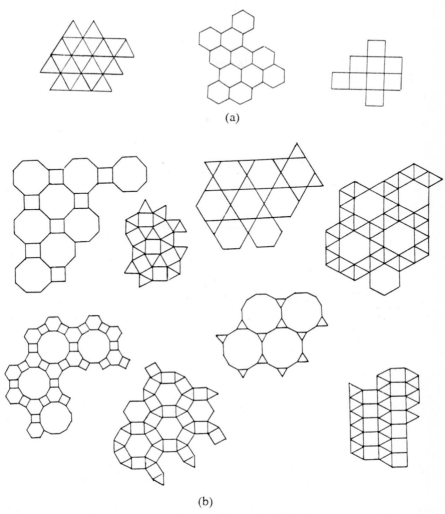

(a)

(b)

Figure 5.12 The three regular tilings and eight semiregular tilings of the plane. The tiling $3^4\,6$ exists in two mirror-symmetric (enantiomorphic) forms.

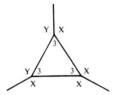

Figure 5.13 Tilings of the form $3.x.y$ require that $x = y$.

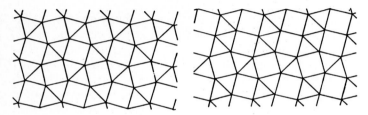

Figure 5.14 Many distinct tilings that have only vertices of species $3^2.4.3.4$ may be obtained by changing the relative positions of horizontal zigzag strips in the tiling at the left.

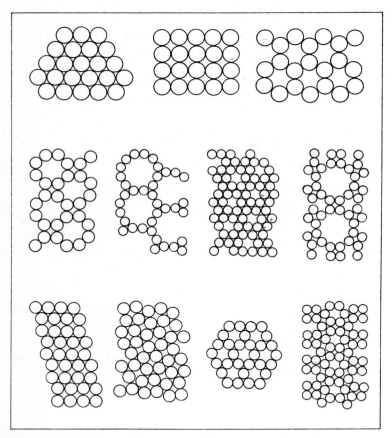

Figure 5.15 Regular and semiregular tilings drawn with circles.

mean that the pattern possesses organized repetitions of some motif. As you progress through this book, this notion will be made more precise. First of all, with the exception of $3^4.6$, each tiling has lines of symmetry. By a *line of symmetry* we mean that the entire tiling is obtained by reflecting half of it in a mirror placed along the line. This also means that if a tiling has a line of symmetry, it looks the same when viewed in a mirror. Only $3^4.6$ has a distinct mirror image, or enantiomorphic form. Each tiling also has *centers of symmetry*. This means that to each point of the tiling there corresponds another point diametrically opposite it with respect to the center (the entire tiling is reproduced by rotating it by a half-turn about this center). Not only does each tiling look alike at the local level of a single vertex, but if the tiling is reproduced on tracing paper, any vertex of the traced tiling can be superimposed on an arbitrary vertex of the original tiling in such a way that the two tilings coincide after a possible mirror reflection. Such tilings are called *uniform* by mathematicians. These tilings also have the property that if they are translated in suitable directions a certain distance, they once again match up. Such tilings are called *periodic*.

5.7 Duality of Semiregular Tilings

Again, duality offers alternative images of the semiregular tilings. Place a vertex at the center of symmetry of each of the tiles of a semiregular tiling and form a dual tiling by connecting two of these vertices if their corresponding faces share an edge of the original tiling. Since all vertices of the original are surrounded identically, all tiles of this dual must be congruent. An example is shown of $3^2.4.3.4$ and its dual in Figure 5.16. Notice that the dual tiles the plane with congruent pentagons. However, this does not contradict the impossibility of regular pentagonal tilings since it is apparent that some of the vertices have three incident edges while others have four.

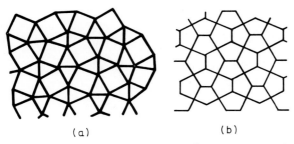

(a)   (b)

Figure 5.16   The archimedean tiling $3^2.4.3.4$ and its dual.

The concept of a dual tiling is generally problematic except for symmetric tilings like the regular and semiregular ones [Grünbaum and Shephard, 1988]. For graphs, any point on a face can be taken as the vertex of the corresponding dual, and its dual is truly reciprocal in that the dual of the dual is isomorphic to the original. On the other hand, to define a dual tiling we must specify a particular point on each face of the tiling to serve as a vertex of the dual, for example, the center of symmetry. The problem is that, in general, no point distinguishes itself. However, the archimedean tilings can be recovered from their duals by placing a vertex at the meeting point of the angle bisectors of each tile of the dual.

5.8   The Module of a Semiregular Tiling

A manufacturer wishing to produce a set of tiles that cover the plane in a semiregular fashion does not have to create all the tiles individually. Each tiling has a basic *module* which can be rigidly moved to stamp out the entire tiling. Let's determine this module for a typical tiling, 3.6.3.6. Several elements of this tiling are shown in Figure 5.17(a) along with the dual tiling. As you can see, a typical tile of the dual is made up of ⅙ of each of two of the original's hexagons and ⅓ of each of two of the original's triangles. Thus, since all tiles of the dual are congruent, the tiling must have hexagons and triangles in the ratio

$$\frac{2}{6}\text{ hexagon}: \frac{2}{3}\text{ triangle} \qquad \text{or} \qquad 1 \text{ hexagon}: 2 \text{ triangles}$$

Figure 5.17(b) shows two such modules. You may check to see that this module can be translated to generate the entire 3.6.3.6 tiling.

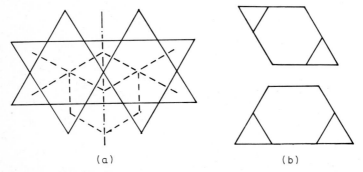

(a)   (b)

Figure 5.17   (a) A portion of 3.6.3.6 with its dual superimposed; (b) two modules of the 3.6.3.6 tiling.

Problem 5.2 Find another module for 3.6.3.6. By similar construction, determine a module for each of the other semiregular tilings.

5.9 Other Tilings with Regular Polygons

Grünbaum and Shephard have catalogued many interesting classes of tilings [1987]. For example, more than one kind of vertex may be permitted. O. Krötenheerdt has discovered that there are exactly 135 n-uniform tilings where n takes values no greater than 7. A 7-uniform tiling is shown in Figure 5.18. Figure 5.19 shows two of the seven families of semiregular tilings that are not edge to edge while Figure 5.20 shows one of the four semiregular tilings that include star-shaped polygons.

5.10 Transformations of Regular Tiling

Starting with a tiling of the plane and applying a set of rules of transformation to the tiles, there are several ways in which new tilings can be generated. Tilings as ordinary as the regular tilings can then serve as the starting point of tilings that are quite complex and interesting. In this section and the following ones we consider four kinds of trans-

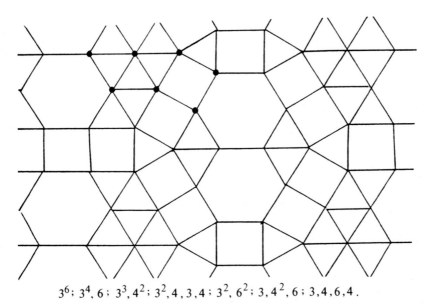

$$3^6; \; 3^4, 6; \; 3^3, 4^2; \; 3^2, 4, 3, 4; \; 3^2, 6^2; \; 3, 4^2, 6; \; 3, 4, 6, 4.$$

Figure 5.18 A 7-uniform tiling.

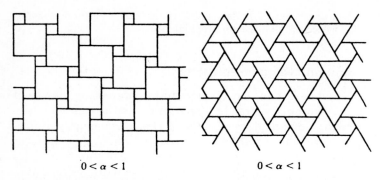

$$0 < \alpha < 1 \qquad\qquad 0 < \alpha < 1$$

Figure 5.19   Two of the seven families of uniform tilings that are not edge to edge.

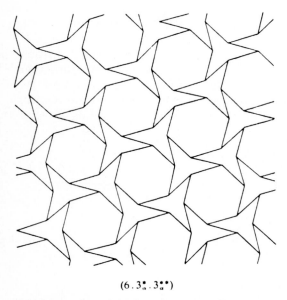

$$(6 . 3^*_\circ . 3^{**}_\circ)$$

Figure 5.20   One of the four 1-uniform tilings of star polygons.

formations: vertex motion, distortions, augmentation-deletion, and one-dimensional parquet deformations.

5.10.1   Vertex motion

The regular tilings inherent in the triangular grid shown in Figure 4.44 can be made dynamic by considering the tiling by rhombuses

{4,4} formed by joining adjacent triangles. A bunch of rhombuses are cut out of cardboard. Pairs of rhombuses are attached by hinging them according to the pattern shown in Figure 5.21. This has the effect of splitting apart the vertices in the tiling and making the tiling movable. If the dual tiling to the triangular grid is drawn on the opposite side of the rhombuses, movement of the tiling gives rise to a transformable tiling of irregular hexagons which are completed in the open portion of the tiling.

Varney, an architectural designer, used this idea to create the geometric design of panels for a 68-foot radar dome built by ESSCO, Inc. of Concord, MA. In addition to regular hexagons, he used three kinds of irregularly shaped hexagon panels to construct his dome in order to prevent interference with the incoming signals [Varney, 1988]. The structural design for this *radome* was made by William Ahern. The geometric design for the radome is shown in Figure 5.22. The fact that 12 pentagons appear along with the hexagons is a necessary consequence of the tiling of a sphere by hexagons and pentagons and will be

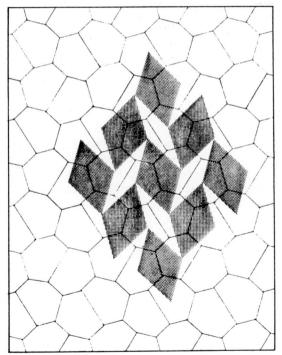

Figure 5.21   The movable triangular grid of William Varney.

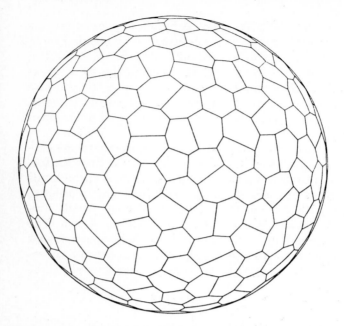

Figure 5.22   Geometric design for a 68-foot radome.

discussed in Section 9.8. Ron Resch has illustrated the act of creating movable tilings in his fascinating film, *Paper and Sticks* [1989].

5.10.2   The K-dron

Kapusta is a Polish architect and designer with an interest in philosophy. He discovered a way to make the regular tilings with squares {4,4} dynamic by lifting each square into the third dimension as an 11-faced polyhedron which he patented in 1987 and calls the *K-dron* (see Figure 5.23) [Kapusta, 1989]. The existence of so many facets

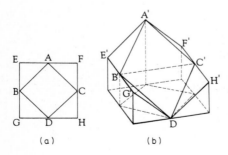

Figure 5.23   The K-dron. (*a*) Top view is a tiling with squares; (*b*) the squares are lifted from the plane to form the 11-faced K-dron.

causes entirely new patterns to be created by the interplay of light and shadow as it impinges on the K-dron from different directions. In Figure 5.24, three entirely different patterns are created as a tiling with K-drons reflect morning, noon, and late afternoon light.

The K-dron itself has a remarkably simple structure. It is created by lifting one panel of the square tiling, shown in Figure 5.23(b), into the third dimension by raising point A to A' [see Figure 5.23(b)] an arbitrary perpendicular height h from the base plane and points B, C, E, F, G, H a distance half this height or $h/2$ to points B', C', E', F', G', H'. Point D remains anchored to the base plane. This results in a sphinx-like structure in which half of a pyramid sits upon a rectangular parallelopiped base. The top surface of the K-dron is a diamond with outwardly folding triangles reflected from each quarter of the diamond as shown in Figure 5.25(a). The five faces that make up this diamond configuration are essential to the K-dron since they reflect the light. The other six faces make up the base and are shown in Figure 5.25(b). Figure 5.25(c) shows how the faces of a K-dron tile a rectangle.

If the half-pyramid is properly hinged along $B'C'$ [as in Figure 5.23(b)], it collapses into the base to form a parallelopiped of height $h/2$. Also two congruent K-drons fit together to form a rectangular parallelopiped of height h. The dimensions of the diamond depend on the height h. When h equals the width of the base, it has its diagonals in the proportion $\sqrt{2}:1$. In this case the two K-drons form a cube. With small values of h, the resulting shallow K-drons can be used for dynamically changing wall decorations or acoustical tiles. For larger values of h, the resulting polyhedra have a great deal of versatility and can be used for packaging, modular furniture, artistic sculptures,

Figure 5.24   The effect of light on a tiling with K-drons. (a) Morning; (b) noon; (c) afternoon.

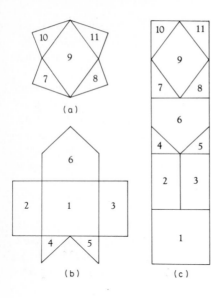

Figure 5.25 Folding-up the K-dron based on a cube from the plane. (*a*) The sectors of a diamond with diagonals $\sqrt{2}:1$ reflect outward; (*b*) the K-dron base; (*c*) the 11 faces of the K-dron tile a rectangle.

or toys. The structure of the K-dron is closely related to the symmetry of a cube. We will have more to say about this in Section 7.13.4.

5.10.3 Distortions

The distortion operation consists of expanding, contracting, twisting, flattening, and stretching polygons either in isolation or in aggregation. One special type of distortion operation involves *n*-zonogons [Baracs et al., 1979], [Williams, 1972]. An *n-zonogon* is a $2n$-sided polygon where pairs of opposite sides are parallel and equal. For example, the parallelograms and hexagons that combine to tile the plane regularly are 2-zonogons and 3-zonogons. Adjacent tiles are related by being translations of each other.

An *n*-zonogon can be constructed by specifying a star of *n* directed line segments (vectors) representing the direction and length of its sides all emanating from a common origin. This is referred to as an *n*-vector star. The vectors are numbered according to the sequence of edges in the resulting polygon. For example, a star of three vectors and the resulting 3-zonogon is shown in the cartesian coordinate system in Figure 5.26(*a*). The vectors are named by the points in the grid that the tips of the vectors intercept when the vectors are anchored at the origin, e.g., the three vectors (5,2), (2,5), (−2,2). Figure 5.26(*b*) shows a convex zonogon defined by these vectors while Figure 5.26(*c*) illustrates another nonconvex zonogon defined by the same vectors in a different order.

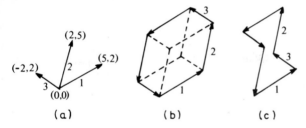

Figure 5.26 Zonagons. (*a*) A three-vector star; (*b*) 3-zonogon with a sequence of vectors 1-2-3; (*c*) 3-zonogon with the sequence 1-3-2.

All zonogons have a *center of symmetry*. In addition, an *n*-zonogon can always be decomposed into $n(n - 1)/2$ parallelograms (the number of ways in which two vectors can be chosen from a set of *n* vectors) in a number of ways N that increases rapidly with *n*. For a 3-zonogon, $N = 2$, and Figure 5.26(*b*) shows the 3-zonogon subdivided into two sets of three parallelograms. An exact formula for N is given in Section 10.13 where the notion of zonogon is generalized to three-dimensional space.

In a practical construction, it is easy to lengthen or contract linear elements; however, it is difficult to modify the complex joining mechanism where two edges meet. What is important about zonogons is that they can be contracted or expanded in a direction parallel to any pair of opposite sides, as shown in Figure 5.27, without altering the angles between adjacent sides (the internal angles). This is done by merely lengthening or contracting one of the vectors in the vector star without changing its direction. In this way, if the angles surrounding a vertex sum to 360 degrees before a transformation, they continue to do so after the deformation. Thus any space-filling aggregate of zonogons will remain space filling after distorting an individual zonogon in this way and then adjusting adjacent zonogons of the tiling accordingly as shown in Figure 5.28.

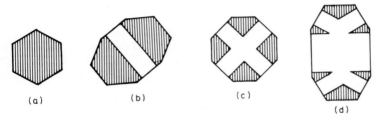

Figure 5.27   Examples of stretching an individual zonogon.

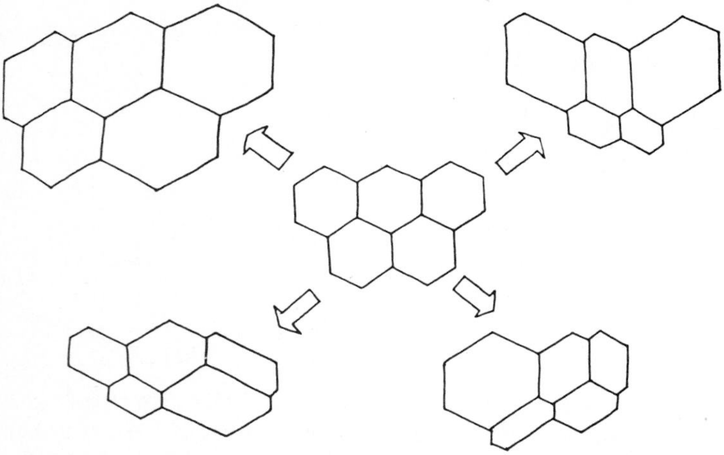

Figure 5.28   Examples of stretching aggregated zonogons.

Construction 5.1   Construct an interesting aggregate of 3-zonogons starting with a 3-vector star of your choosing. Your aggregate should illustrate the capability of 3-zonogons to fit together in distorted forms.

5.10.4   Augmentation-deletion

We are all familiar with how dramatically the scene changes in the fall when leaves fall off of the trees or in the spring when nature blossoms forth again. The *augmentation-deletion* operation of a tiling can also result in profound changes in appearances. This method of transformation involves either the addition or subtraction of vertices, edges, and faces on existing entities [Williams, 1972]. This may be done either symmetrically or randomly. For example, Figure 5.29(a) shows two transformations of {4,4} with certain edges removed, while Figure 5.29(b) shows $3^2.4.3.4$ with augmented and deleted edges and vertices.

5.10.5   One-dimensional parquet deformations

Perhaps the most interesting and versatile family of deformed tilings is the one-dimensional *parquet deformations* developed by Huff, a professor of architecture at the State University of New York at Buffalo. I first learned about Huff's work by reading an article in *Scientific American* by Douglas Hofstadter [1983]. This section is, in large part, excerpted from that article. One-dimensional parquet deformations produce tilings that deform in a single direction, for example, from left to right. Thus it produces a visual effect akin to what music does to

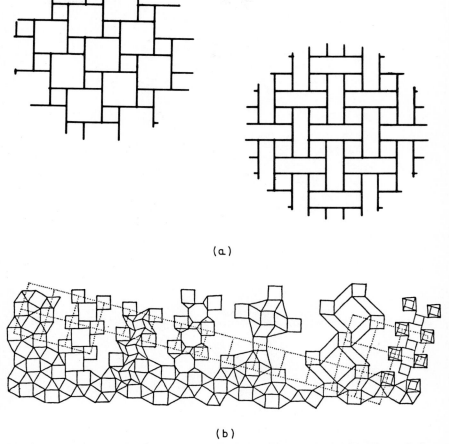

(a)

(b)

Figure 5.29   (a) Tesselations derived by deletion of certain edges and vertices in the {4,4} tiling; (b) the $3^2.4.3.4$ tiling with augmented and deleted edges and vertices to generate new tilings.

the ear. While music transforms sound through the single dimension of time, parquet deformations vary along a single spatial dimension. The tilings that Professor Huff's students have made are reminiscent of M. C. Escher's famous woodcut, *Liberation*, shown in Figure 5.30 and of D'Arcy Thompson's continuous deformations [1966].

In "Consternation," shown in Figure 5.31, the regular triangular tiling {3,6} falls apart at first chaotically, then it reforms into a tiling in which hexagons and cubes vie for perceptual supremacy. Once again, the triangles group together to form hexagons that maintain

Figure 5.30 *Liberation* by M. C. Escher. (© *M. C. Escher Heirs/Cordon Art - Baarn - Holland*)

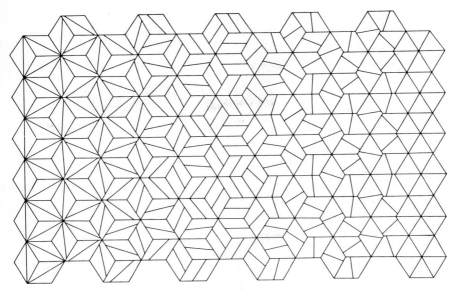

Figure 5.31 *Consternation.*

their integrity throughout the transformation. The vertices of these hexagons maintain their relative positions while three of the six internal partitions are continuously rearranged in an orderly fashion and the other three remain fixed.

What are the basic elements of a parquet deformation? First of all Huff begins with a single tile. According to Hofstadter,

> Typical devices in his repertoire of transformations are lengthening or shortening a line; rotating a line; introducing a "hinge" somewhere inside a line segment so that it can "flex"; introducing a "bump" or "pimple" or "tooth" (a small protrusion or extrusion having a simple shape) in the middle of a line or at a vertex; shifting, rotating, expanding, or contracting a group of lines that form a natural subunit; and variations on these themes. To understand these tilings you must realize that a reference to "a line" or "a vertex" actually refers to a line or a vertex inside the smallest repeating element, or unit cell (the hexagon unit in "Consternation"), and therefore when one such line or vertex is altered, all the corresponding lines or vertices that play the same role in the copies of that cell undergo the same change. Since some of those copies may be at 90 degrees (or some other angle) with respect to the master cell, one locally innocent-looking change may induce changes at corresponding spots resulting in unexpected interactions whose visual consequences can be quite exciting. After a line is deformed and all the other lines so respond, the tiles in the new zone of figures remain congruent with one another. Huff feels that it is this congruence of tiles that makes them appealing both from the standpoint of design and mathematics.

Many unexpected patterns emerge in parquet tilings. It is a useful intellectual exercise to attempt to read the spatial patterns and try to understand the intricate and subtle transformations that take place. It is also fun to try your hand at constructing one of Huff's tilings.

5.11   Nonperiodic Tilings

Although regular pentagons cannot tile the plane, two geometric figures called a *kite* and a *dart*, which can be formed by dissecting a regular pentagon and reassembling its parts, can be used to tile the plane in strikingly beautiful ways [Gardner, 1978*b*; 1989], [Penrose, 1979]. These so-called *nonperiodic* tilings provide a simple mathematical model for describing a new class of quasicrystals (to be discussed in Section 6.10) whose approximate pentagonal symmetry defy the traditional tenets of crystallography which require crystals to be periodic and forbid pentagonal symmetry.

A *periodic* tiling is one in which the entire configuration can be translated (without rotation) to a new position which reproduces the original tiling. We say that such a tiling is invariant under translation. Both regular and semiregular tilings are periodic. Until recently it was thought that any set of forms that tile the plane nonperiodically can tile periodically as well. For example, the polygonal forms called *enneagons* shown in Figure 5.32(*a*) tile the plane both periodically and nonperiodically. On the one hand, the enneagons stack to fill space; on the other hand, the spiral form in Figure 5.32(*b*) cannot be translated without also moving its center.

Therefore great interest met Robert Berger's discovery in 1964 that there is a set of tiles that tiles nonperiodically but for which there is no way of tiling periodically. To carry out this tiling Berger needed more than 20,000 kinds of tiles. Sometime later Raphael Robinson reduced the required set of tiles to six. This enables us to better appreciate Roger Penrose's discovery of two tiles, the *kite* and *dart* shown in Figure 5.33, which are guaranteed to tile the plane nonperiodically if certain rules are followed stating how the pieces are to be combined. (Note that each tile separately or both together tile periodically if no other restriction is imposed.) The kite is constructed from two of one type of golden triangle while the dart is constructed from two of the other kind of golden triangle (see Section 3.5). During the tiling process, the blue curve drawn on the kite and dart is allowed to meet only the blue curve of another kite or dart to form a continuous curve that winds through the tiling. The same holds for matching the tiles so as to ensure a continuous red curve wafting through the tiling. One such tiling is shown in Figure 5.34.

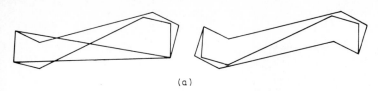

(a)

Spiral Tiling

(b)

Figure 5.32 (a) A pair of enneagons forming an octagon that tiles periodically; (b) a nonperidic tiling with congruent shapes: a spiral tiling by Heinz Voderberg.

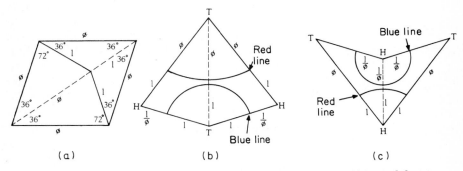

Figure 5.33 (a) Construction of a kite and a dart; (b), (c) a coloring of kite and dart to force periodicity.

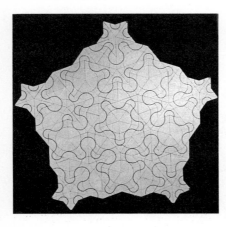

Figure 5.34 A cartwheel pattern with kites and darts.

Most noteworthy about these tilings is their either exact or approximate pentagonal symmetry. As a matter of fact, regular decagons appear throughout the tilings. This is not so surprising once we realize that the tiles were constructed by dissecting a regular pentagon and reassembling its parts. Try to find a kite and dart in the star pentagon shown in Figure 3.18(a). The relation of these tiles to a pentagon also accounts for their golden mean measurements (see Section 3.6). Penrose first came upon his discovery by attempting to tile the plane with regular pentagons, an impossible task as we saw in Exercise 5.3. However, in the process he discovered that when certain arrangements of the tiles are disallowed, the·gaps left over from the pentagon tilings coalesced into four tiles which could be further reduced to two by using rules of combination. Penrose also·constructed other tiles that were equivalent to the kites and darts, including a pair of chickens. In Section 6.10, we will show how a pair of Penrose rhombuses lead to nonperiodic tilings and suggest a model for the phenomenon of quasicrystals. All Penrose tilings can be obtained by specific markings on the pair of Penrose rhombuses [Penrose, 1979]. These markings define a special grid. In Section 12.18 a similar concept will be illustrated for generating Islamic patterns.

Although these tilings have no region that replicates itself by translation, they always seem to be striving to do so but never quite succeeding. Wherever we look, we see a configuration that looks familiar in the sense that we have seen something just like it at one or another point of the tilings. We can make this statement more precise by stating a remarkable theorem developed by Conway. In colloquial language, the theorem can be described as follows: Let's say that you are residing in a finite region of a Penrose tiling (or universe) of diameter *d*. Call this region your town. If you are suddenly transported to another universe (a different tiling) and there are as many of such tilings as there are real numbers in the number system, how far must you wander to find an exact replica of your town? Conway proved that you need not wander more than a distance of 2*d* from your new position, although the exact distance is unpredictable.

Many of the interesting properties of Penrose tilings come about from the property that any one of the tilings can be reconfigured so that a new tiling is obtained with kites and darts scaled up in size, or *inflated*. This is done by splitting each of the darts along their lines of symmetry and attaching all short edges of the original tiling to each other, leaving the long edges as the boundaries of the new tiling as shown in Figure 5.35.

Construction 5.2 [Gardner, 1978*b*] Construct a pattern of at least 60 kites and 100 darts. In any nonperiodic tiling, you will need exactly 1.618... times as many kites as darts. A Penrose tiling can be made by starting with darts and kites and expanding around one vertex. Each time that you add a piece to an edge, you must choose between a kite and a dart. Sometimes the move is forced; sometimes it is not. Sometimes either piece fits, but later you may encounter a contradiction and you will have to go back and make the other choice. The more that you play with the pieces, the more you will become aware of the forcing rules. The discussion by Martin Gardner goes into more details about the practical aspects of construction.

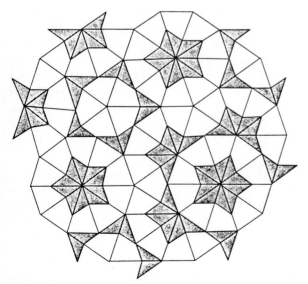

Figure 5.35 A new tiling is obtained by inflating the tiles of the old one.

5.12   Origami Patterns

The Chinese invented the paper folding art of *origami* over 1000 years ago, and they endowed it with the aesthetic principles that are at the heart of their culture. As Peter Engel, an American master of the art, says, "the success of a completed figure depends on the creator's eye for form—is it a mere likeness of the original, or does it delve deeper into the form's essential character? [Engel, 1988]" Engel points out that origami has been taken up in this country by mathematicians rather than artists. He says: "To the mathematician, the beauty of origami is its simple geometry. Latent in every pristine piece of paper are undisclosed geometric patterns, combinations of angles, and ratios that permit the paper to assume interesting and symmetrical shapes."

An origami figure always begins with a single square piece of paper. Only folding, with no cutting or pasting, is permitted. Traditional origami uses four basic folded bases: the *kite, fish, bird,* and *frog* shown in Figure 5.36(*a*). Engel's contribution to this craft has been to show that when these bases are unfolded, as they are in Figure 5.36(*b*), they reveal a sequence of geometric patterns based on a single module. The basic module is represented in the kite pattern by the shaded region. It is reflected about the diagonal of the square to produce the entire pattern. When the same pattern is replicated four times, it results in the fish base. Eight replicas makes the bird base while 16 repeats give rise to the frog base.

The kite base is constructed by folding the square on its diagonal to form a right triangle. Two additional folds produce the kite pattern. On the other hand, the fish base is constructed by folding the right triangle in half to form two right triangles. If each of these right triangles is folded into a kite base, the fish base pattern appears when the paper is opened up. Repeating this procedure by folding the triangle into four and

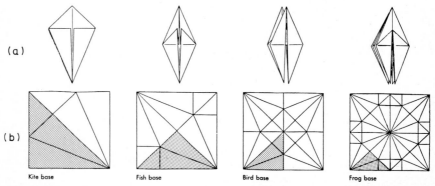

(a)

(b)

Kite base   Fish base   Bird base   Frog base

Figure 5.36   (*a*) The fundamental bases of traditional origami: kite; fish; bird; frog; (*b*) patterns formed by unfolding the bases.

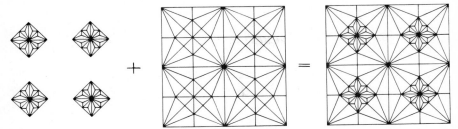

Figure 5.37 Grafting four frog bases onto four bird bases produces the folding pattern for an octopus.

eight right triangles produces the bird and frog base patterns. Once the patterns are obtained, the base easily folds up into itself.

Engel looks at the basic module as being a self-similar component, or *fractal*, of the entire pattern. He was able to break out of the restrictive four-base mold of Japanese origami by extending these fundamental patterns to additional stages of development and by grafting one base upon another. For example, Figure 5.37 shows that grafting four frog bases onto four bird bases produces a complex folding pattern which the author used to make the octopus shown in Figure 5.38.

One word of caution. Production of the underlying pattern is only the first step in creating the final work of art, which requires much work, ingenuity, and patience. I refer the reader to Engel's book on origami [1989].

Figure 5.38   An origami octopus by Engel.

5.13 Islamic Art

5.13.1 The Temple of Ka'ba and the Dome of the Rock

Islamic culture succeeded in creating art, architecture, science, and mathematics entirely integrated within a spiritual realm. Although we have only a sketchy record of the nature of the traditional Islamic consciousness, we can piece together some idea of how and why this integration was achieved by studying modern commentators. Most notable of these is Titus Burckhardt, a well-known scholar of the art and culture of Islam, but we must be aware that much of Burckhardt's discussion of Islamic art is based on his own understanding of its culture and history and not on commentary by the Islamic artisans or artists themselves. Burckhardt [1976, 1987] and other authors [Chorbachi, 1988] disagree about the spiritual interpretations.

Burckhardt feels that in the structure of the sacred Ka'ba in Mecca lie the philosophical underpinnings of Islamic religion and art. This temple, which claims its origin to the time of Abraham, is approximately cubic (actually 10 by 12 by 16 meters). The four corners of the base point approximately to the four cardinal directions of the earth, with the vertical axis of the zenith defined by the top and bottom faces. The Ka'ba itself is considered the "navel" of the earth, toward which all Muslims must direct their prayers.

By nature, the Islamic religion is both static and dynamic. The static is symbolized by the fact that all locations of prayer are considered equivalent with respect to the unity of the center (at Ka'ba) while the dynamic is manifested by the requirement that all Muslims carry out a pilgrimage once in their lifetimes to the Ka'ba where they must circumambulate the temple in a symbolic circle. The *cube* or *square* symbolizes the *earthly* with its dualities of hot and cold, moist and dry, and axes of spatial orientation. The *circle* symbolizes the realm of the *celestial* surrounding the source of all being and dominated by the element of time in the form of the zodiac (see Section 1.2).

One of the oldest surviving Muslim monuments is the Dome of the Rock in Jerusalem which encloses the rock forming the summit of Mt. Moriah. This mountain is the supposed location of the Great Temple of Solomon, the site at which Abraham is said to have performed the sacrifice of his son and the place where Mohammed is said to have ascended into heaven. This structure was built in 688 by Abn Al Malik to serve as a substitute for the Ka'ba at a time when Mecca had fallen into the hands of a rival caliph. The Dome of the Rock is designed to shelter this sacred site beneath a central cupola and an octagonal deambulatory [shown in Figure 5.39(a)] in a style that can be traced back to the architecture of Byzantine times and is found in many of

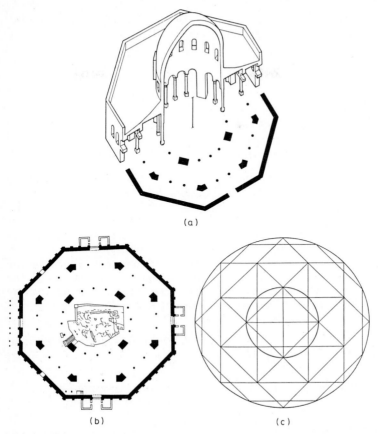

Figure 5.39 The Dome of the Rock. (*a*) Perspective view; (*b*) plan view; (*c*) geometric scheme.

the sanctuaries of that period. The dome is supported by 12 columns and 4 pillars shown by the central area of Figure 5.39(*b*). The 12 columns are arranged with 3 each to the north, south, east, and west as in the New Jerusalem diagram (see Figure 1.3). Surrounding this circle is a second series of 8 pillars and 16 columns arranged octagonally. The columns of the second set are spaced with relation to the inner ones in such a way that they radiate into the center through the intersection points of two squares inscribed in the inner circle that form the star octagon {8/2} shown in Figure 5.39(*c*). The octagonal columns themselves form another star octagon {8/3}. The complete set of 12 columns connects to form a grid of rectangles and squares. The circular cupola again represents the celestial domain contrasted with the earthly crystal of the octagon. Burckhardt explains that the 40 supporting columns and pillars correspond to the number of saints who, according to Mohammed, constitute the spiritual pillars of the world in every age.

5.13.2 Islamic tiling

Unlike Christian sacred art, the art of Islam contains no graven images. Islamic art is best known for its arabesque and polygonal forms bordered by undulating woven strips as shown in Figure 5.1. Here is how Burckhardt [1976] describes the spirit behind Islamic art:

> A sacred art is not necessarily made of images...it may be no more than the quite silent exteriorization of a contemplative state.... It reflects no ideas but transforms the surroundings by having them share an equilibrium whose center of gravity is unseen.... Ornamentation with abstract forms enhances contemplation through its unbroken rhythm and endless interweaving.... Continuity of interlacement invites the eye to follow it, and vision is transformed into rhythmic experience accompanied by the intellectual satisfaction given by the geometric regularity of the whole.... Study of Islamic art, or any other sacred art, can lead to a profound understanding of the spiritual realities that lie at the root of a whole cosmic and human world.

J. Bourgoin published an extensive collection of Islamic patterns in 1879 [1973]. Underlying each pattern, Bourgoin shows a grid from which the pattern is developed. Many of these grids, such as the one shown in Figure 5.40, are regular tilings by triangles; others are developed from regular tilings by squares (not shown). Unlike Burckhardt, Bourgoin lists his tilings with no commentary. Keith Critchlow feels that the triangular tilings were used because of their platonic symbolism through the form of the tetraktys (see Section 1.2), and tilings based on square patterns may have been suggested by hidden symmetries in the number relations of magic squares (see Section 4.13). Figure 5.41, from Critchlow's book, *Islamic Patterns* [1976], shows the interest of Islamic artists in pentagonal tilings, which Critchlow feels can be traced to the sacred properties exhibited by the golden mean. There may even be some foreshadowing of the nonperiodic Penrose tilings, discussed in Section 5.11, in ancient Islamic tilings [Chorbachi, 1988]. In Section 12.18, we will describe these patterns by a more refined method developed by H. Lalvani based on their symmetry [1982], [1990].

A. K. Dewdney recently described a practical method of creating homemade Islamic tilings [1988]. A set of intersecting and self-intersecting lines weave through the tilings, as shown in Figure 5.42. These lines are unrestricted except for the fact that each must originate and end at the boundary. If each crossing is alternately designated as either an overpass or an underpass, whenever one arrives at

Figure 5.40 An Islamic pattern by J. Bourgoin with underlying triangular grid.

a previously designated crossing, it has the required structure. Why is this? This series of crossing lines is exactly the class of lines that we encountered in the two-colored map of Figure 4.41. That the assignments are always correctly made follows from a two-coloring of the regions in Figure 5.42 (see Section 4.14). Say one travels along the road bordered on the right by a region of some color. After the crossing, the color on the right changes. Thus one can say that an overpass always leads to, say, the color red (on the right) while an underpass leads to, say, blue. It follows that the road crossings must be assigned correctly after a cycle.

5.13.3 Islamic art and mathematics

Although one can read spiritual meanings into the art of Islam, we are still left with a profound silence on the matter by the artisans and art-

Figure 5.41 An Islamic pattern with tenfold symmetry.

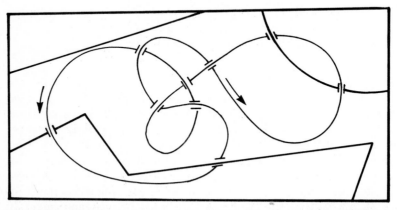

Figure 5.42 Dewdney's "over-under" rule for the construction of Islamic tilings.

ists themselves. Wasma Chorbachi, a specialist in Islamic art, states in a recent article that not once in the hundreds of manuscripts and folios she has examined in libraries throughout the world is there a practitioner's comment on the spiritual meaning behind the art [1988]. In fact, quite to the contrary, she has unearthed volumes from the thirteenth and fourteenth centuries that are totally preoccupied with practical and geometrical concerns, as exemplified by one book with the title *What the Artisan Needs of Geometric Problems*.

Figure 5.43(*a*) shows one panel that Chorbachi has studied (disregard the dotted lines and surrounding dodecagon). If the kite shapes are divided into right triangles with sides *a, b, c* as in Figure 5.43(*b*), an ancient proof of the pythagorean theorem attributed to Bhaskara

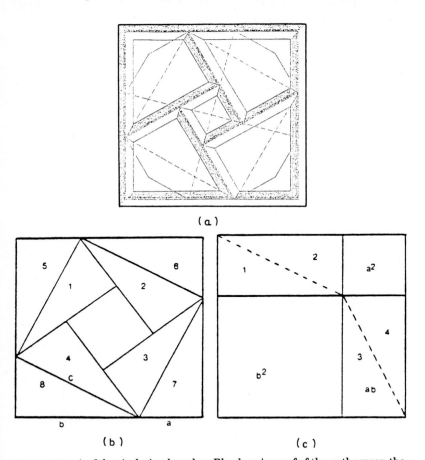

Figure 5.43   An Islamic design based on Bhaskara's proof of the pythagoren theorem. (*a*) The design as given by Critchlow; (*b*), (*c*) Bhaskara's proof as given by Chorbachi.

follows from the fact that the inner square has side $b - a$ and the area of the square made up of the inner square and triangles 1, 2, 3, 4 is

$$c^2 = (b - a)^2 + 4\frac{ab}{2} = b^2 + a^2$$

Also, as Figure 5.43(c) shows, the very outer square has side $a + b$ and illustrates the relation

$$(a + b)^2 = a^2 + b^2 + 4\frac{ab}{2}$$

Another object of Chorbachi's research is the tiling shown in Figure 5.44. The square is divided into four congruent sectors by two perpendicular lines and each sector is divided, in turn, into polygons of three kinds including a symmetric kite shape as shown in Figure 5.44(a).

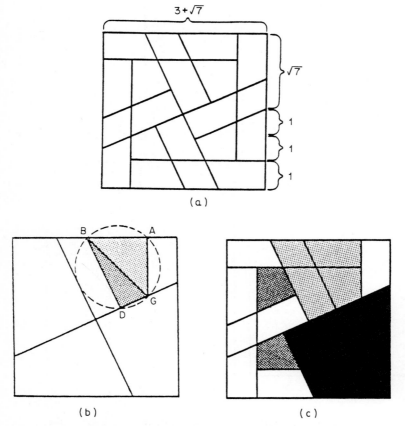

(a)

(b)   (c)

Figure 5.44   Chorbachi's analysis of an Islamic pattern with fourfold symmetry based on a geometric problem.

The entire tiling has a fourfold symmetry, which means that a quarter turn about the center brings all the tiles of one sector onto the tiles of another. The key to understanding this tiling is an asymmetric quadrilateral $ABDG$ with proportions $1:2:2:\sqrt{7}$ inscribed in a circle as shown in Figure 5.44(b). The sector of the square is obtained by adding two gnomons of unit widths to two sides of the quadrilateral as shown in Figure 5.44(a). Figure 5.44(c) shows that similar kites at three different scales can also be found within the tiling. Just as we saw in the last section for origami, in Section 5.11 for Penrose tilings, in Section 1.7 for the Modulor, and in Section 2.12 for fractals, this gives another example of how good design is the result of the repetition of a limited number of congruent modules along with the reproduction of these elements at varying scales. In her article, Chorbachi has generated many of her own tilings based on this asymmetric quadrilateral and its geometric properties.

Exercise 5.4 Subdivide a square into four congruent sectors as in Figure 5.44(a). It makes a good puzzle, for persons not aware of the origin of the sectors, to put the pieces together to re-form the square. It makes an even better puzzle to put the pieces together to form two squares such as in Figure 5.43(a). This can always be done.

It is interesting that the same panel that Chorbachi sees in strictly geometric terms, Critchlow prefers to think of in spiritual terms. For example, the solid dodecagon and dotted square in Figure 5.43(a) is Critchlow's doing. His interpretation is

> The coincidence of twelve and four suggests the zodiacal symbolism controlling or embracing the fourfold axial kite shapes which can be taken to symbolize the four seasons, the four elements, and the four qualities of hot and cold, moist and dry.

So we have been thrown back to the sacred architecture of the Ka'ba and the Dome of the Rock. Perhaps future research will be able to show these two visions of Islamic art to be of one cloth.

6

Two-Dimensional Networks and Lattices

Everything that we can see, everything that we can understand is related to structure—perception is in patterns not fragments.   Cyril S. Smith

6.1   Introduction

Tilings of the plane arise naturally in both the artificial and natural worlds. Whether we observe the structure of soap films, the structure of cellular elements of living organisms, the growth of plants, the structure of crystals, the organization of rural markets, the optimal layout of cities, the equilibrium of forces within frameworks of cables, or the geometrical possibilities in a design, we find that a simple geometry of networks and lattices lies beneath the surface. Beyond the physical, biological, and sociological mechanisms involved in these complex systems, much can be learned about them from studying their geometry. In this chapter we shall study some of the geometric constraints that underlie some of these phenomena.

6.2   Planar Soap Films

Have you ever watched a drop form on the faucet over your sink? Look at it more carefully. Notice how the drop forms, grows slowly, and suddenly falls. Every time this happens the drop is always the same size and shape at the time of its plunge [Boys, 1959]. Why does the drop remain clinging to the faucet instead of immediately falling under the force of gravity?

From these observations it is reasonable to conjecture that water develops a surface skin that responds to the weight contained within by stretching. The tensile force developed by this skin is known as *surface tension*. The surface tension comes about from forces between the molecules of liquid. The forces on the molecules within the interior are balanced by those on their neighbors. Those molecules at the surface, however, have unbalanced forces acting on them. In response to the forces exerted by the molecules below the surface, the surface molecules are continually pulled under the surface. In this way the surface tends to be a shape with minimal area. If there were no competing nonmolecular forces such as gravity, the surface area would be the exact minimum possible within its geometric constraints. This is borne out by the shape of the water surface in very narrow capillary tubes or in small droplets of mist and also in minute organisms or small cellular elements of larger organisms [Thompson, 1966].

In Sections 8.9 and 10.11 we will examine some geometric constraints imposed on three-dimensional soap films by the requirement of minimal surface area. Here, we consider the simpler case of soap films constrained to grow along minimal networks in the plane.

Exercise 6.1   Place three thumbtacks between two sheets of glass as shown in Figure 6.1. Submerge this sandwich of glass and thumbtacks in a soap solution and observe the soap films that cling to the tacks. Can you predict how the films will lie across the tacks, without carrying out the experiment? Problem 6.1, posed by Jakob Steiner [Courant and Robbins, 1941], [Bern and Graham, 1989], [Stevens, 1974], answers this question. Try to solve it before reading on.

Problem 6.1 (Steiner's Problem)   Three villages, *A*, *B*, and *C* are to be joined by a system of roads of minimal total length. Mathematically, three points *A*, *B*, *C* are given in the plane and a fourth point *P* in the plane is sought so that $AP + BP + CP$ is a minimum. In other words, connect points *A*, *B*, *C* with the shortest set of line segments.

Two possible solutions to Problem 6.1 are shown in Figure 6.2:

1. Measure the sum of the network of lengths in Figure 6.2(*a*) and compare it with the result of Figure 6.2(*b*) where *P* is taken to be

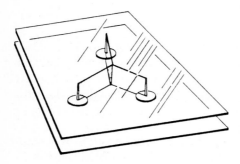

Figure 6.1   A soap film solution to Steiner's problem.

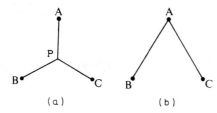

Figure 6.2 Minimal networks for three points.
(*a*)Minimal network; (*b*) a nonminimal network.

point *B*. Steiner proved that the configuration of Figure 6.2(*a*) is the shortest possible linkage of points *A*, *B*, *C* if these points form a triangle with no angle greater than 120 degrees. The three angles surrounding *P* are 120 degrees.

2. For three arbitrary points *A*, *B*, *C* forming a triangle with no angle greater than 120 degrees, locate point *P* with compass and straightedge. Hint: Construct an equilateral triangle on each edge of triangle *ABC*. Construct a circle circumscribing each of these equilateral triangles (a method for doing this is described in Section 6.5) and use Theorem 1.3 (that the central angle of a circle is twice the inscribed angle that intercepts the same arc) to construct angles of 120 degrees, at the point of intersection of these three circles.

3. Where is point *P* if the triangle has an angle greater than or equal to 120 degrees? Hint: Note in Figure 6.3 how the films transform as ∡*ABC* is moved along the line between its original position and the junction point, *P* when ∡*ABC* = 120 degrees.

It is clear from Steiner's problem that if the thumbtacks are placed at *A*, *B*, *C*, three soap films will join at point *P*. The angle between the planar faces of the films, known as the dihedral angle (see Section 7.10), or angle at which the planar surfaces intersect in edge view, is 120 degrees. If an angle of triangle *ABC* is greater than or equal to 120 degrees, point *P* must coincide with the vertex incident to that angle. Soap films have the property that the tensile forces they exert within the surface of the film are the same in all directions and at all points. For this reason a soap bubble never has regions of concentrated stress, but rather distributes stress evenly across its entire surface. Move the tack to a new position and the whole configuration adjusts itself almost instantaneously so that once again the tension in the bubble is the same at every point. This supplies another justification for the configuration of films given by Steiner's problem since, as shown in Figure 6.4, three forces of the same magnitude are in equilibrium if they are symmetrically placed around a point. In three di-

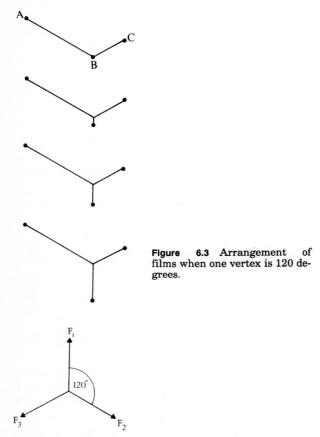

Figure   6.3 Arrangement   of films when one vertex is 120 degrees.

Figure 6.4 A soap film exerts equal tensions in three symmetric directions around the center of a soap film.

mensions, as we will see in Section 10.11, four edges meet at each vertex, each pulling with equal force symmetrically around the vertex.

What happens if we add a fourth thumbtack? How does the configuration of Steiner's problem generalize? First consider four points located as in Figure 6.5. Notice that Figure 6.5(c) and (d) yields line segments whose total lengths are less than those in Figure 6.5(a) and (b). Again, we find that soap films spanning four thumbtacks assume the positions of either Figure 6.5(c) or (d) but cannot remain for long in the configuration of Figure 6.5(a) or (b).

How can we connect the four points of Figure 6.5 to form a stable network of soap bubbles? We find that the films drawn in Figure 6.5(c) and (d) are the only stable configurations. However, the configuration

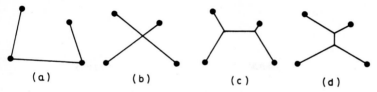

Figure 6.5 Possible networks formed by four points. Networks (a) and (b) are unstable while networks (c) and (d) are stable.

in Figure 6.5 (d) is shorter than the one in Figure 6.5(c). Doesn't this contradict our restriction to minimal surfaces? Why do soap bubbles sometimes make the "mistake" of choosing a nonminimal arrangement? This error in judgment can be explained by considering the analogy of a ball rolling down a mountain (shown in Figure 6.6) seeking the lowest position A at which to come to rest. However, it may come to rest in a mountain valley located still up in the mountains at point B. Points A and B are both local minima of potential energy and stable resting positions for the ball. However, if the ball is displaced from position B by rolling it uphill a bit, it may move down to position A. In the same way, the soap film in Figure 6.5(c) may be transformed to the arrangement shown in Figure 6.5(d) by gently blowing on it.

Thus we see how Steiner's problem generalizes. N points are, in general, connected by a tree graph (see Section 4.5) in which three soap films surround each junction point with angles of 120 degrees. Furthermore, a theorem of topology developed by Leonhard Euler states that there are at most $N - 2$ vertices with three incident edges ($q = 3$) in any polygonal linking up of N points. For example, three different ways to connect the six points lying at the vertices of the regular hexagon are shown in Figure 6.7 where one of the incident edges degenerates in Figure 6.7(c).

Although this problem is easy to state and it is easy to construct solutions for small N, there is no practical algorithm to solve the problem for large N, say $N = 100$. All known algorithms require an exponential number of operations as N grows large, or, as it is said, they

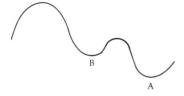

Figure 6.6 Point B is a metastable point; it is a local minimum of potential energy. Point A is a stable point; it is an absolute minimum.

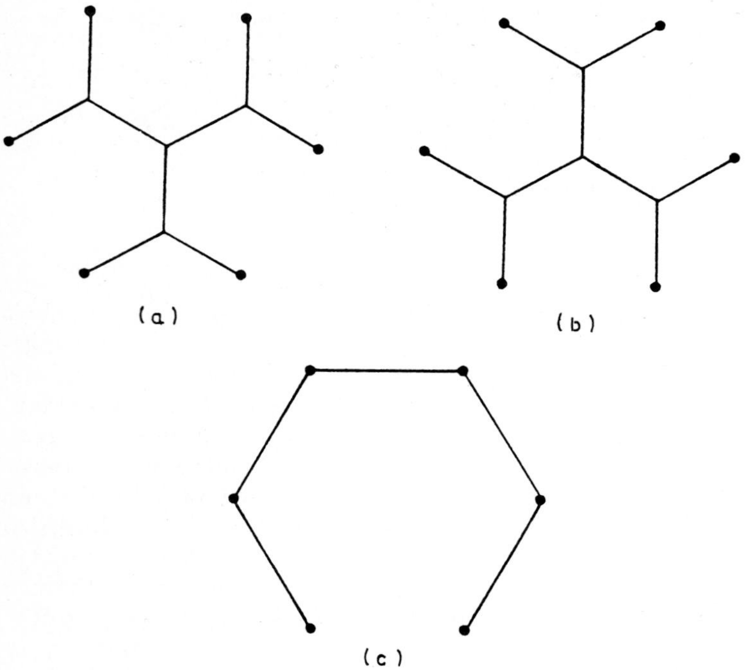

Figure 6.7 Three stable networks with six vertices.

can be solved only in exponential time. Problems of the Steiner type have been used to construct telephone, pipeline, and roadway networks and, most recently, to design electronic integrated circuits in which the networks are rectilinear. An unusual application, developed by David Sankoff, uses Steiner trees to determine plausible phylogenetic trees in which edges correspond to a relation between organism and ancestor that assumes the fewest mutations [Bern and Graham, 1989].

6.3 Random Cellular Networks

Random soap bubble froths are representative of cellular patterns of all kinds in two- and three-dimensional space. The random soap bubble pattern of cells with three edges incident to each vertex occurs in many diverse contexts: the granular patterns on the surface of metals [Rivier and Weaire, 1984], [Smith, 1965], the structure of biological tissues [Dormer, 1980], the cracking patterns of dried mud (see Figure

4.2); and the organization of rural market patterns of agricultural societies discussed in the next section.

The irregular shapes of the disordered boundaries of these froths reveal nothing of the inner order of the structure within. The shape of the boundaries results from the vicissitudes of time, but the internal order is immutable. This situation arises in crystalline materials where nucleation of a crystal occurs at some local inhomogeneity of the medium. Thereafter, the crystal grows in strict accordance with the geometry of lattice structures (see Section 6.7). Grain boundaries are produced when regions of crystal growth of different origins impinge upon each other. As Cyril S. Smith, a metallurgist, says [1965],

> In the space-filling aggregate, the individuals limit each other. They may be arranged randomly or regularly, but however undetermined the shape of an individual, the conditions of joining at the points where three or more meet are defined. Structure on one level, by its imperfections or variations, always gives rise to a new kind of structure on a larger scale. A local configuration will always have some connection to neighboring ones. In ever-decreasing degree, every part is dependent on the whole and vice versa.

In the frontispiece of his book (not shown), *Fundamental Tissue Geometry for Biologists,* K. J. Dormer illustrates the geometrical similarity of the inner tissue from the shaft of a bird feather and the fruit flesh of a crab apple. Although these cells differ both biologically and chemically, considered as geometric patterns they are almost interchangeable. How is it that physical and biological systems that are influenced by such different external forces, nevertheless end up with similar patterns? Smith feels that at the scale of these phenomena, it is the geometric constraints on space that are the controlling factor rather than external forces that determines form [1954], [Dormer, 1980]. One such geometric constraint is given by Theorem 6.1.

Theorem 6.1 For an infinite tiling in which vertex valence $q = 3$ at each vertex and each face of the tiling contains or is surrounded by a sphere no smaller or larger than some preset diameter, $\langle p \rangle = 6$, where $\langle p \rangle$ is the average number of edges per face, i.e.,

$$\langle p \rangle = \sum_F \frac{p_i F_i}{F} \tag{6.1}$$

where summation is over all the faces.

proof For tilings on the infinite plane,

$$F + V - E = 2 \tag{6.2}$$

$$3V = 2E \qquad (6.3)$$

$$\sum_F p_i F_i = 2E \qquad \text{or} \qquad F\sum_F \frac{p_i F_i}{F} = 2E \qquad (6.4)$$

Replacing Equations (6.1), (6.3), and (6.4) in (6.2) yields

$$\left(\frac{2}{\langle p \rangle} - \frac{1}{3}\right)E = 2$$

It follows that $\langle p \rangle = 6$ when $E \to \infty$.

Dormer has studied the geometry of cellular structures in very general terms [1980]. He has isolated three primary transformations that cell structures can undergo. They are shown in Figure 6.8. In T_1 the edges are merely rearranged as we saw in Figure 6.5 for soap bubbles. In T_2 a three-sided cell disappears eliminating one face and six edges from itself and the surrounding cells. In T_3 a cell undergoes mitosis in which one cell splits in two. Notice that an n-gon parent cell gives rise to two daughter cells having a total of $n + 4$ edges. This enables us to deduce that the dividing cells must be 7-gons in order for all the cells of the network to have an average of six edges per cell. The calculation goes like this: if the parent cells have an average of m edges per cell, the average of parents and daughters must be six edges per cell, or

$$\tfrac{1}{3}[m + (m + 4)] = 6 \qquad \text{or} \qquad m = 7$$

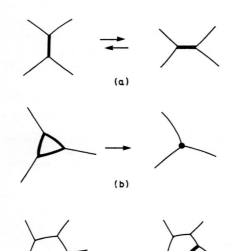

(a)

(b)

(c)

Figure 6.8 Elementary cell transformations. (a) T_1, or neighbor exchange; (b) T_2, or cell disappearance; (c) T_3 or mitosis.

Figure 6.8(*c*) shows that each dividing cell releases two edges into the nondividing population. Unless there is some mechanism to eliminate the excess edges, these edges will build up, destroying the hexagonal equilibrium. In fact Dormer shows how this results in an elaborate bookkeeping system that enables the ensemble of cells to maintain its equilibrium. A hexagon borrows an edge from another cell, depleting its edges; the division of the new 7-gon releases two edges to the surrounding cells; one of these edges is used to repay the donor while the other is loaned to yet another cell in order to prepare it for mitosis. Dormer extends this analysis to cells in bounded domains and to three-dimensional froths of cells (see Section 10.11).

6.4 Rural Market Networks

If one views the patterns of random soap froths described in the last two subsections anthropomorphically, they betray a kind of social ethos. As Smith says,

> The freedom of a structural unit inflicts and suffers constraints whenever its closer interaction with some neighbors makes cooperation with others less easy. Social order intensifies the interfacial tension against a differently ordered group.

It is this tension between marketing requirements of population settlements which results in the patterns observed in rural market networks of agricultural societies [Plattner, 1975].

Central-place theory was developed in the 1930s by the German geographer Walter Cristaller and elaborated by another German economic geographer August Losch to describe the organization of rural markets. Although this theory is highly idealized, it has been extremely successful in describing the dynamics of the marketing practices of these societies.

Christaller's model is predicated on the existence of a featureless landscape with population settlements spread equidistantly from each other and interconnected by a grid of pathways that can be traveled with equal ease. Population settlements are represented by a triangular grid, as shown in Figure 6.9(*a*). Markets for high-value goods, e.g., clothing or medical or legal services, called *A markets* are established at another triangular grid, a subset of the first, marked with open dots, as shown in Figure 6.9(*b*). Surrounding these grid points are circular boundaries demarcating the maximum distance an individual must travel to purchase an A market product. Overlap between circles is replaced by a line segment in Figure 6.9(*c*) to obtain hexagonal A market domains.

Markets for low-valued products, e.g., fresh vegetables or incidental

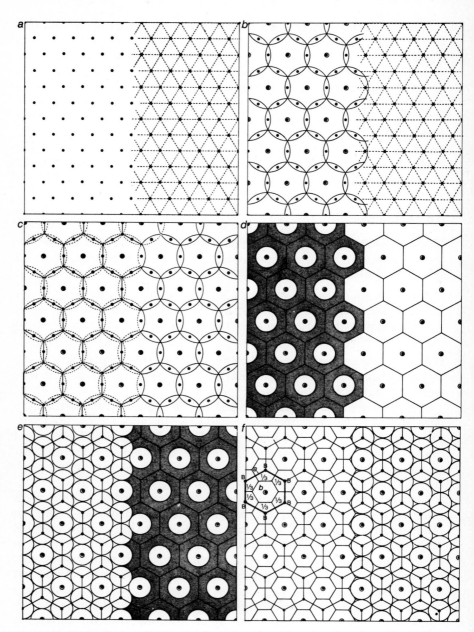

Figure 6.9 Christaller's model for the placement of rural markets based on a featureless landscape. (*From "Rural Market Networks" by Stuart Plattner. Copyright © 1975 by Scientific American, Inc. All rights reserved.*)

household items, called *B markets*, are established at each A market grid point. These are indicated in Figure 6.9(*d*) by solid dots. The regions of demand for these products also lie within circles surrounding the grid points, but these circles are naturally of smaller radius than the A market circles since a person will travel a smaller distance to obtain a commonplace item than for something of great value. But, as you can see from Figure 6.9(*d*), this leaves large regions of unmet demand on the part of most of the settlements for the lower-valued products. This demand is met by establishing B markets at the six settlements surrounding the A markets as shown in Figure 6.9(*e*). Points of tangency between adjacent B markets are replaced by line segments to form hexagonal B market domains.

The entire pattern is illustrated in Figure 6.9(*f*) in which each B market lies at a vertex of one of the space-filling A market hexagonal domains and equidistant from three A markets. An A market hexagon is composed of one entire B market and six ⅓ sectors of the surrounding B market hinterlands, or the equivalent of three B market regions. In the same spirit, a series of different-valued markets sets up an elaborated market hierarchy with self-similar structure, so that a single A market gives rise to 3 B markets, 9 C markets, 81 D markets, etc. Figure 6.10 shows how this elaborated hierarchy works for three different-valued markets. The highest order A markets are represented by large open dots, the B markets by smaller open dots at the vertices of the A market hexagons in addition to the A market sites, while C markets are established at the vertices of the B market hexagons in addition to all the A and B market sites represented by solid dots.

The actual networks have highly irregular market domains with boundaries more like the irregular shapes of random soap froths since these domains are determined by many social and geographical idiosyncracies in both time and space. Nevertheless, as Stuart Plattner reports, the anthropologist G. William Skinner has found that the dynamics of the market systems of the Chinese province of Szechwan are governed quite well by Christaller's model. The model can also be modified to take into consideration other geometrical and social circumstances such as a nonhomogeneous landscape in which communication in certain directions is hampered by such constraints as a mountain range while in other directions it is enhanced by such advantages as a navigable river.

Another example of the interaction between the geometry of cellular patterns and social context is Bill Hillier's analysis of the arrangements of building clusters and roads within towns and villages by a geometrical language that he calls *space syntax* [Hillier and Hanson, 1984]. He has developed this geometrical language to study the way in

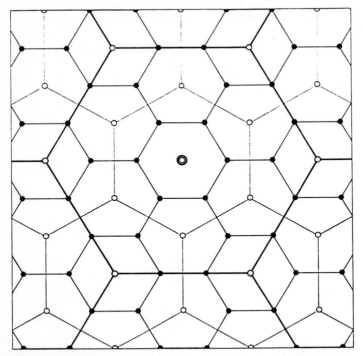

Figure 6.10   Elaborated market hierarchy. (*Copyright © 1975 by Scientific American, Inc. All rights reserved.*)

which the plan of a community addresses the tensions between the needs of neighborhood residents for security and social intercourse and the need to allow outsiders to obtain access to portions of the community. Hillier has been able to use his geometry to study why some communities have been successful in their planning while others have not.

6.5   Dirichlet Domains

A map of Cambridge, MA, school districts is shown in Figure 6.11. The black markers represent the schools. The map is drawn so that each point of a school district is nearer to the school in that district than to any other school. Check to see that this criterion holds. The school districts are called the *Dirichlet domains* of the set of points represented by the schools, where a Dirichlet domain of a point from a set of points is defined to be the points of space nearer to that point than to any of the other points of the set [Loeb, 1976]. The points whose Dirichlet domains border the Dirichlet domain of another point are said to be its *neighbors*. You will notice that all but one vertex of the map is the

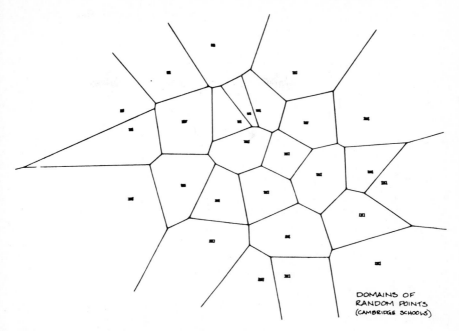

Figure 6.11   D domains of Cambridge, MA, schools.

meeting point of three districts (Dirichlet domains). Thus each of the three schools from these districts is equidistant from a common vertex. Why does it usually occur that a vertex of the map is surrounded by exactly three Dirichlet domains? The one exception to this rule is the vertex surrounded by four domains. We shall see why in a moment.

We would like to find a way to construct the Dirichlet domains, or D domains as we will call them, of any set of points, and thus be able to draw a map similar to Figure 6.11. Let's first consider the D domains corresponding to two points A and B shown in Figure 6.12. The boundary of the D domains is clearly the perpendicular bisector of line segment AB.

Now let's consider three points A, B, and C shown in Figure 6.13. Clearly points on the perpendicular bisectors of BC, CA, and AB are

Figure 6.12   D domains for two points.

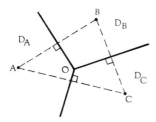

Figure 6.13   D domains for three points.

equidistant from B and C, C and A, and A and B, respectively. Also, it is well-known that the perpendicular bisector of the sides of any triangle meet at a common point O. In fact, this point O is the center of the unique circle, the circumscribed circle, that passes through A, B, and C. (Prove this!) Therefore A, B, and C each lies in a region formed by the perpendicular bisectors of the line segments incident to that point from the other two points as indicated in Figure 6.13 by domains D_A, D_B, and D_C.

Also, since three points uniquely determine a circle, the points of a complicated situation such as that of the Cambridge school districts can be expected to form groups of three points on a circle about a common boundary point of the D domains. Four or more points can also be found around a common boundary point, but this is an exception since three points determine the circle while the fourth point is unlikely to lie on that circle. In a physical manifestation of D domains, boundary points with four incident edges are not structurally stable. A small perturbation causes the domains to lapse into a pattern with trivalent edges just as for the soap bubble patterns in Section 6.2.

Problem 6.2   Although the triangle is one of the simplest of geometric shapes, it is a rich source of mathematical ideas. In fact, any triangle determines many unique points, including the following five, all of which can be constructed with compass and straightedge: (1) the meeting points of the perpendicular bisectors of the sides—the center of the circumscribed circle, (2) the meeting point of the angle bisectors—the center of the inscribed circle, (3) the meeting point of the medians (lines drawn from a vertex to the midpoint of the opposite side)—the centroid or balance point of the triangle, (4) the meeting point of the altitudes drawn to each side from the opposite vertex, and (5) the center of a remarkable circle known as the *nine-point circle*. On the circumference of this circle lie nine special points; they are the three intersection points of the altitudes with the opposite sides, the midpoints of each side, and three additional points which are identified in [Coxeter, 1961]. Choose a triangle, and construct these five points with compass and straightedge.

Problem 6.3   Prove that any point lying in one of the D domains defined above for the case of three points is nearer to its corresponding point than to the other two points.

If we can extend the method of constructing Dirichlet domains from two and three points to the case of four points, we can use the same procedure to find the D domains of any number of points, e.g., the Cambridge school districts. Consider four points *A, B, C,* and *D* in Figure 6.14. If the D domain of *A* borders on the D domains of either *B, C,* or *D*, the boundary of D_A must include a segment of the perpendicular bisector of *AB, AC,* or *AD*. However, the perpendicular bisector of *AD* lies outside of the domain defined by the perpendicular bisectors of *AB* and *AC*. Thus, the D domains of *A* and *D* do not border on each other. The D domain of *A* is then seen to be the innermost envelope formed by the perpendicular bisectors of the line segments joining *A* to each of the other points. The D domains of *B, C,* and *D* are determined in the same manner and are illustrated in Figure 6.14. This procedure can just as well be applied to find the D domains of any number of points.

Given a regular tiling of the plane by congruent polygons, we can ask whether the tiles are D domains of some set of points, one of which lies within each tile. Although this question has no simple answer, we can show that regular tilings with triangular faces, such as the one in Figure 6.15, are also D domains if all angles are less than 90 degrees. The restriction on the angles ensures that the meeting point of the perpendicular bisectors lies within each triangle. However, if the triangles have angles greater than 90 degrees, they may also be D domains [Grünbaum, 1989]. Furthermore, the centers of the triangular D domains in Figure 6.15 are the vertices of a dual tiling to the triangular domains, i.e., a tiling with hexagons. These hexagons can be observed to have opposite edges equal and parallel, i.e., they are zonogons, with vertices that lie on a common circle whose center is the center of the domain. These partic-

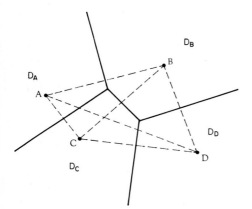

Figure 6.14   D domains for four points.

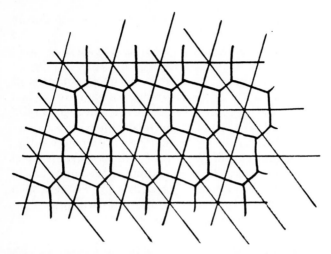

Figure 6.15 The D domains of a tiling with triangles constitute another tiling with hexagons.

ular hexagons are themselves D domains of the set of points at the vertices of the triangles.

Problem 6.4 Prove that the tiles of an infinite tiling by directly congruent quadrilaterals are D domains if the vertices of each quadrilateral lie on a circle.

6.6   Spider Webs, Dirichlet Domains, and Rigidity

A group of architects and mathematicians including Janos Baracs, Henry Crapo, Ethan Bolker, Walter Whiteley, and others based at the University of Montreal have revived work done by nineteenth-century mathematicians and engineers to determine the conditions under which frameworks built of iron bars and pins are rigid [Crapo, 1978]. Their studies led them to the work of James Clerk Maxwell, a physicist who, in 1864, discovered a geometric tool for studying the static equilibrium of forces on a plane framework: a planar graph called the *reciprocal figure*. As stated in Ash et al. [1988]:

> This figure was a kind of dual graph to the original framework with the dual edges perpendicular to the original edges and forces. Maxwell built his reciprocal by piecing together the polygons of forces expressing the vector equilibrium at each joint. He then observed that this construction yields a polyhedron in space which projects onto the framework. These results belong to the field of graphical statics, which withered around the

turn of the century, along with much of projective geometry. [See Appendix 6.A for a brief discussion of projective geometry.]

Recent work on the statics of frameworks grows from these roots. Perhaps the best way to understand these ideas is to examine a structure known as a *spider web*: a framework with no crossing edges and some edges going to infinity which has an internal static equilibrium formed entirely with tension in the members (an *internal equilibrium* of forces in a framework is a set of tensions and compressions in its members in the absence of external loads). Figure 6.16(*a*) shows such a spider web. Those edges not attached to other members are taken to be the ones going to infinity, and these are considered to be pinned to the ground. Figure 6.16(*b*) shows the reciprocal diagram of this spider web, which will be explained below. Figure 6.17(*a*) shows another simple spider web. According to Ash et al. [1988]:

> A tiling of the plane is called a spider web if it supports a spider web stress: a set of nonzero tensions which leads to mechanical equilibrium at each vertex. More specifically, a spider web stress is a nonzero force F_{VE} in each edge E at a vertex V, directed from V along the edge, such that:
>
> 1. For a finite edge E joining V and V', the forces at the two ends are equal in size and in opposite directions: $F_{VE} = -F_{V'E}$.
> 2. For each vertex V, the vector sum of forces on the edges leaving V is zero.

Spider webs are interesting and important. If they are built with cables, and pinned to the ground on the infinite edges [as in Figure 6.16(*a*)], they are rigid in the plane. At the other extreme, if a plane bar-and-joint framework has the minimum number of bars needed to restrain V joints

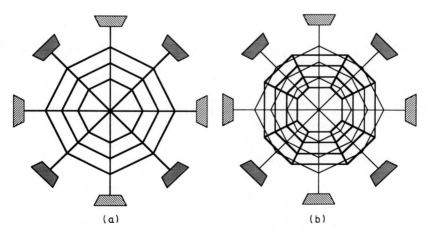

(a)   (b)

Figure 6.16 A plane spider web (*a*) has an internal static equilibrium with tension in all members and (*b*) a convex reciprocal figure derived from this equilibrium.

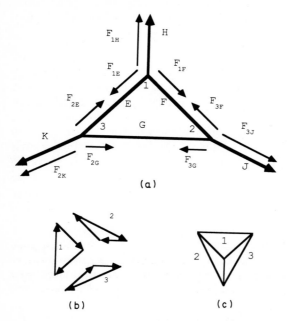

(a)

(b)   (c)

Figure 6.17 The arrows in (a) show the tensions of a spider web stress on a cell decomposition. The polygons of forces for the equilibria at the vertices (b) are pieced together and rotated 90 degrees to form a convex reciprocal figure (c).

[we will show this to be $E = 2V - 3$ in Section 7.8], the appearance of a spider web signals that it is shaky.

The three cycles of vectors in Figure 6.17(b) represent an equilibrium of forces at the three vertices of Figure 6.17(a). In Figure 6.17(c) these cycles are joined together and rotated by 90 degrees to form the reciprocal diagram of the spider web.

In general, any edge-to-edge tiling, or *cell decomposition* as [Ash et al., 1988] refers to it, of the plane has a reciprocal diagram associated with it. To each edge of the cell decomposition, for example, the light lines in Figure 6.18, there is an edge of the reciprocal diagram at right angles to it, for example, the dark lines in this figure. The reciprocal diagram will also have as many vertices as the cell decomposition has faces. It is a kind of "dual tiling"; however, unlike an actual dual, the edges of the reciprocal need not intersect the corresponding edges of the parent tiling, and vertices of the reciprocal need not lie within the faces of the original. If the reciprocal has only convex cells as in Figure 6.18(a) and (c), it is called a *convex reciprocal*. However, a decomposition may also have a reciprocal with nonconvex cells as in Figure

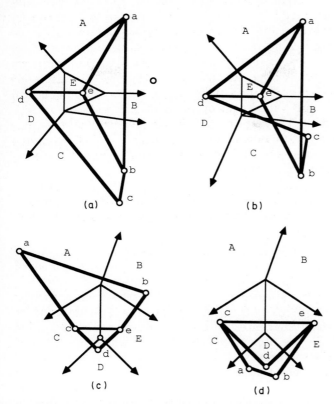

Figure 6.18 Some cell decompositions (light lines) with reciprocal figures (heavy lines). The cell decomposition in (*a*) has only convex reciprocals, that in (*b*) has only noncovex reciprocals. A single cell decomposition may have both convex (*c*) and nonconvex (*d*) reciprocals.

6.18(*d*) or with cells that are not well defined as in Figure 6.18(*b*). The D domains of a set of points form a cell decomposition, and the set of interconnections between the centers of the D domains always forms a reciprocal as shown in Figure 6.19. However, we must be careful here since not every cell decomposition comprises the D domains of some set of points.

We make three important remarks about the relationship of reciprocals to spider webs:

1. Spider webs always have convex reciprocals because of their polygons of force, and conversely, any cell decomposition with a convex reciprocal can be realized as a spider web.

2. Reciprocals are, in general, not unique since the tension in the

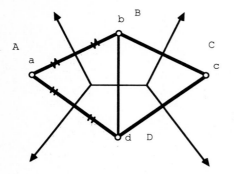

Figure 6.19 Some cell decompositions are the D domains of the vertices that make up its reciprocals.

cables of a spider web is not uniquely determined. When reciprocals are not unique, the structure is said by structural engineers to be *statically indeterminate*.

3. It was discovered by Whiteley, Ash, and Bolker [Ash et al., 1988] that any spider web is the plane section of the D domains of a set of points in three-dimensional space (three-dimensional D domains, which are natural generalizations of two-dimensional ones, will be discussed in Section 10.6). The reciprocal diagram is also obtained by orthogonally projecting the centers of the D domains cut by the plane onto the cutting plane (see Appendix 6.A). (An *orthogonal projection* is one in which an object is projected by parallel lines perpendicular to an image plane from a point at infinity, e.g., like the projection of an object to its shadow on the ground by the sun shining directly overhead.) It is a little difficult to draw a picture of this for three dimensions, so we illustrate it for the case of two-dimensional D domains sectioned by a line (see Figure 6.20). The black dots represent the boundaries of the sectioned D domains while the open circles represent the vertices of the reciprocal.

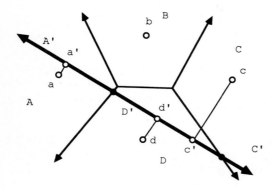

Figure 6.20 Any section of a plane Dirichlet tessellation creates a sectional Dirichlet tessellation on the line (the black dots on the heavy line), with a convex reciprocal (the circles) given by the orthogonal projection of the plane centers.

Some of the most interesting work of the Montreal group has to do with the relation between the rigidity of two- and three-dimensional frameworks and the projection of polyhedra. Although we are anticipating a bit by talking about polyhedra at this point (see Chapter 7), nevertheless, the connection of spider webs and reciprocal diagrams to projected polyhedra is easy to see. Consider a set of connected plane faces in the form of a bowl with no top, called a *polyhedral bowl*. Two adjacent faces of this bowl always join at an edge (see Figure 6.21). A cell decomposition is obtained by orthogonally projecting the edges onto a plane. The reciprocal tiling is obtained by piercing the plane by normal lines to the plane faces of the bowl from a point within the bowl. The piercing points are the vertices of the reciprocal tiling, while the lines connecting the points corresponding to two adjacent faces of the bowl, which intersect the projection of their common edge at right angles, must be an edge of the reciprocal.

The point of projection of both the edges and the normals can be taken to be on the top face to the bowl (or, for that matter, on any face of a convex polyhedron parallel to the plane of projection). This is illustrated in Figure 6.22 for the case of a two-dimensional polygonal bowl. The polygonal bowl projects to a cell decomposition of a line (given by the black dots), and the normals to the edges produce a convex reciprocal represented by the open circles.

It has also been proven by K. Q. Brown that each finite Dirichlet

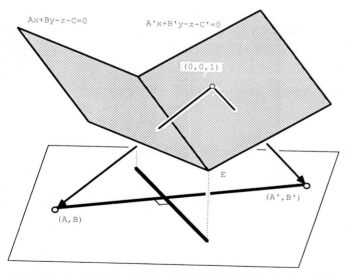

Figure 6.21 A cell decomposition is obtained by orthogonally projecting a polygonal bowl and its normals onto a plane.

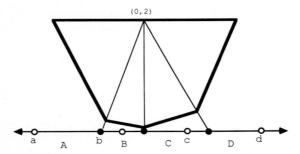

Figure 6.22 The cell decomposition is the projection of a closed polygon from a point on the edge.

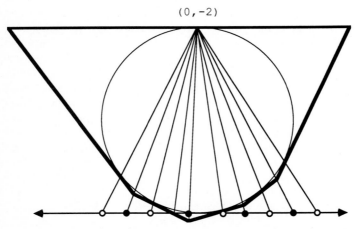

Figure 6.23 Each Dirichlet tessellation on the line is the projection of a convex polygon with an inscribed circle, from the point of contact of one edge. The centers are the projections of the other points of contact.

tiling of the plane is the projection of a polyhedron all of whose faces touch a common sphere from a point on one of its faces onto a plane parallel to this face. This is illustrated in Figure 6.23 for the case of the Dirichlet tesselation of a line projected from a convex polygon with an inscribed circle touching each edge. Some of these ideas will be discussed further in Section 7.8 with regard to the rigidity of polyhedral frameworks (also see Appendix 6.A).

6.7 Lattices

If the lines of the triangular graph paper in Figure 4.44 are removed, leaving only the vertices of the triangles, an orderly set of points called a *triangular lattice* remains. Lattices provide models for under-standing the structure of crystals where atoms of the crystal lie at points of the lattice. In general, lattices, and therefore crystals, pos-sess two kinds of *long-range order: orientational* and *translational* [Nelson, 1987]. These can be seen in the triangular lattice where the lattice points assume the configuration of billiard balls when they are racked up at the start of the game.

In this two-dimensional lattice, the atoms sit in hexagonal cages, the D domains of the lattice points. Neighbors of a lattice point are defined as the centers of the bordering D domains to that point. Thus, each lattice point of the triangular lattice has six neighbors. The crys-tal can be broken down into a repeating pattern of hexagons, as shown in Figure 6.24. Because all the hexagons have the same orientation—that is, because the sides of each hexagon are parallel to the sides of all the others—the crystal is said to exhibit long-range orientational

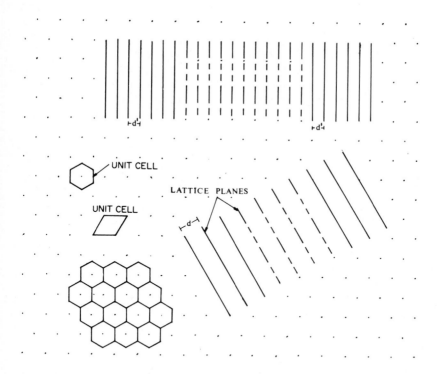

Figure 6.24   A periodic lattice illustrates two kinds of order that are inherent in conventional crystals. Long-range translational order is demonstrated by the two families of parallel lines. Long-range orientational order is demonstrated by the two kinds of unit cells, hexagons and parallelograms, that tile the lattice without change in orientation.

order. The hexagons are called unit cells, where a *unit cell* of a lattice is the smallest unit of the lattice that replicates the whole by translation.

The other kind of long-range order present in a lattice or crystal can be demonstrated by drawing a family of parallel lines on the lattice, as shown in Figure 6.24. When the lines are drawn so that every atom lies on one line or another and every line contains more than one point, the lines will be spaced exactly evenly across the crystal. If the lattice points and the family of parallel lines are drawn on an overlay and this overlay is moved by translation without rotation in a direction perpendicular to the lines, there is some new location, shown by the arrow in Figure 6.24, at which the lattice points and lines of the overlay coincide; the lattice is *invariant under translation*.

In a lattice or a conventional crystal there are many families of parallel lines (another set is shown across the top of Figure 6.24); thus there are many different directions in which the lattice is invariant under translation. However, it can be shown that any two nonparallel directions, such as the ones specified by the two vectors in Figure 6.24, are sufficient to translate the lattice to a new position so that any point is made to coincide with any other point. Thus the environment of any one point of a lattice or crystal is identical to any other point.

Everything that has been said for the triangular lattice continues to hold for more general so-called skew lattices, as is shown in Figure 6.25. They have translational invariance in two nonparallel directions perpendicular to parallel sets of lattice lines. The unit cells are 3-zonogons (hexagons with opposite sides parallel and equal), which are also D domains of each lattice point. Thus, each lattice point can be said to have six neighbors corresponding to the points of the bordering domains. For the special case where two of the directions of translation are at right angles, the two points A and B coincide, and the hexagonal cells degenerate to rectangles having only four neighbors.

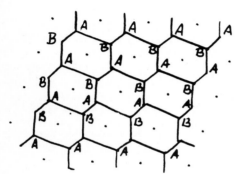

Figure 6.25 D domains of a lattice.

The unit cell of a lattice can also be defined as the parallelogram formed by two nonparallel line segments with lattice points at the vertices but with no lattice point within it, as shown in Figure 6.24. These two nonparallel line segments define two vectors that characterize the lattice. The entire lattice is obtained by making a rubber stamp in the form of a unit cell and stamping it out successively in each of the two nonparallel directions.

Problem 6.5 Find areas of the polygons shown in Figure 6.26 by counting unit squares. Check your results against the general formula given by *Pick's law* [Coxeter, 1961],

$$A = \frac{C}{2} + I - 1$$

where C denotes the number of lattice points lying on the boundary and I refers to the number of lattice points inside the boundary.

Three-dimensional lattices serve as models for three-dimensional crystals. The family of parallel lines for two-dimensional lattices become planes known as *lattice planes* in the three-dimensional case. When beams of x-rays are directed at a crystal, they are reflected and scattered by the lattice planes. By studying the directions in which the beams are scattered and the intensity of each scattered beam, in-

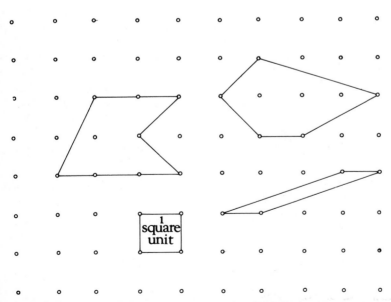

Figure 6.26 Use Pick's law to find the areas of these lattice polygons. Note that parallelograms with no lattice points within them are unit cells.

vestigators can determine which families of lattice planes must exist in the crystal; often they can deduce the location of the atoms.

We will discuss three-dimensional crystals in a little more detail in Chapter 10. In the last section of this chapter, we introduce a new class of quasicrystals that appear to have fivefold symmetry. The inability of conventional crystals to have fivefold symmetry is related to the fact that unit cells or D domains of lattices cannot be pentagons, i.e., there are no regular tilings of the plane with regular pentagons.

6.8   Pattern Generation with Lattices

Patterns that tile the plane by translation in two nonparallel directions are said to have the symmetry of a lattice. Of the regular tilings of the plane, only the hexagon and parallelogram, both zonogons, can tile with lattice symmetry. The tiling by a triangle or a general quadrilateral requires the tile to be rotated by 180 degrees about the midpoint of its edges in the case of directly congruent tiles or by some other combination of rotation or reflection when the tiles are not all directly congruent (see Figure 5.9) to obtain an adjacent tile. In Section 10.13, we shall see that the only tilings of three-dimensional space with lattice symmetry are by polyhedra, which are generalizations of the zonogon. Can we generate more interesting two-dimensional patterns with lattice symmetry than these regular ones?

Consider the tile shown in Figure 6.27. It tiles the plane since it is merely a rectangle that has been transformed by adding a triangle and a semicircle to it while removing an identical triangle and semicircle from it. The new tile has the same area as the original rectangle from which it was derived, and its tiling of the plane is simply a reallocation of space from the old tiling with rectangles. This transformed tile is called the *fundamental pattern*, or *motif*, of its tiling. The fundamental pattern is the smallest element of the total pattern that,

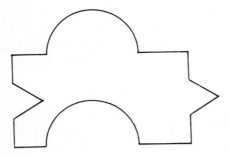

Figure 6.27 This modification of a rectangle is a space filler.

when acted upon by all of the symmetries of the tiling (in this case lattice symmetry), succeeds in tiling the plane. Any tiling with lattice symmetry can be transformed in this way to give new tilings. Escher has used variations on this theme to produce many fanciful tilings such as the one shown in Figure 6.28.

Mathematician William J. Gilbert from the University of Waterloo has come up with an easy way to create attractive tilings of the plane with lattice symmetry [1983]. His procedure is described as follows:

1. Place the origin of a cartesian coordinate system at a point of the lattice, and let neighboring points of the lattice be displaced from the origin at the points $(k,0)$ and (h,l) for integer values of h, k, and l as shown in Figure 6.29. In other words the two nonparallel directions that characterize the lattice are the vectors $(k,0)$ and (h,l). Also, the unit cell of the lattice is the parallelogram formed by these two vectors and has area kl. Unless $h = 0$, the lattice is said to be skew, i.e., the lattice points are not arranged in a rectangular pattern. All the points of the lattice lie at the grid points $(ak + bh, bl)$ for all integer values of

Figure 6.28 *Black and White Knights* by M. C. Escher. (© M. C. Escher Heirs/Cordon Art-Baarn-Holland.)

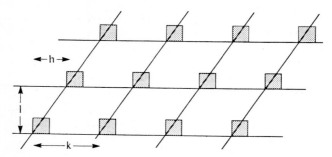

Figure 6.29 Two vectors determine the size and shape of the lattice.

a and b. The location of these points is determined from a linear combination of the lattice vectors in the sense that

$$a(k,0) + b(h,l) = (ak + bh, bl)$$

2. The rectangles of area $kl = n$ which are k units long and l units wide, and whose lower left-hand corner lies at one of the lattice points, fill the plane as shown in Figure 6.30. Each of these rectangular bricks encloses n grid squares which are numbered from 1 to n.

3. A fundamental pattern is formed by selecting the numbers from 1 to n with no repeats from one or more bricks. Of course if all numbers are chosen from the same brick, the resulting pattern will be the bricks themselves. However, more interesting patterns can be formed by allocating the numbers to several bricks as is done in Figure 6.31(a) for a 4 by 3 rectangle. This pattern must have the same symmetry as the skew lattice from which it was derived, i.e., it is invariant in the two nonparallel directions of specified vectors. To distinguish one pattern in a tiling from another you can add color or shading or use some other distinguishing design idea. One such pattern is shown in Figure 6.31(b) and a more interesting design is shown in Figure 6.33(a).

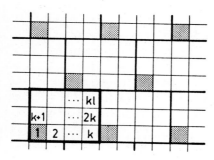

Figure 6.30 The lattice is "squared off" and all squares are labeled.

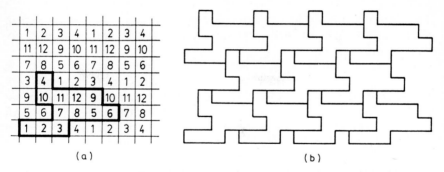

1	2	3	4	1	2	3	4
11	12	9	10	11	12	9	10
7	8	5	6	7	8	5	6
3	4	1	2	3	4	1	2
9	10	11	12	9	10	11	12
5	6	7	8	5	6	7	8
1	2	3	4	1	2	3	4

(a) (b)

Figure 6.31 (a) The shape of the fundamental pattern is determined; (b) the fundamental pattern is replicated.

3	2	1	1	4	7	3	2	1	1	4	7
9	5	7	2	5	8	9	5	7	2	5	8
6	8	7	3	6	6	6	8	7	3	6	6
9	6	3	7	8	9	9	6	3	7	8	9
8	5	2	4	5	6	8	5	2	4	5	6
7	4	1	1	2	3	7	4	1	1	2	3	7
3	2	1	1	4	7	3	2	1	1	4	7
9	5	7	2	5	8	9	5	7	2	5	8
6	8	7	3	6	6	6	8	7	3	6	6
9	6	3	7	8	9	9	6	3	7	8	9
8	5	2	4	5	6	8	5	2	4	5	6
7	4	1	1	2	3	7	4	1	1	2	3

Figure 6.32 A tiling of the plane with fourfold rotational symmetry.

4. The method can be extended to include 90-degree rotations as well as translations. Here, the unit cell of the lattice is a square which is subdivided into four subsquares each containing n^2 grid squares as shown in Figure 6.32. The subsquare is called the *fundamental domain* of the tiling since it is the smallest element of the plane in which we are permitted to create a pattern. The grid squares in one of these fundamental domains are numbered from 1 to 9. These numbers are then recopied in the other rotated squares in the appropriate rotated positions. We are then free to choose any pattern made up of the numbers from 1 to 9, with no repeats from one or more subsquares, to form a fundamental pattern. The entire pattern can be formed by rotating

the fundamental pattern to the other subsquares and, once again, stamping out the unit cell successively in the two nonparallel directions of the lattice.

If we place a pin at the center of the unit cell, perpendicular to its plane, rotation of the pattern through 90 degrees around the pin replicates the entire pattern. Four successive rotations around the pin reproduce the same pattern four times; the pin is therefore called an axis of *fourfold rotational symmetry*. An example is shown in Figure 6.33(*b*). Notice that the midpoints of the edges of the large square that make up the unit cell are the sites of twofold rotations (i.e., the pattern matches up after a half turn). The constraints on space that force these additional symmetries will be discussed in Section 12.14.

Problem 6.6 Create a lattice design with mirror symmetry by subdividing a square into two half-squares by perpendicular mirror lines, or lines of symmetry as they are called. Grid squares in the right half-square (the fundamental domain) are reflected in the mirror to the left half-square. The fundamental pattern is formed as above. Gilbert's method can also be used to create lattice designs in three dimensions [1983].

6.9   Dirichlet Domains of Lattices and Their Relation to Plant Growth

In Section 3.7 we showed how the stalks of a plant are placed successively around its periphery beginning with some initial stalk. We can follow the lead of the geometer H. Coxeter [1953] and model the process of laying down stalks by picturing it to take place on the surface of a semiinfinite cylinder with a 1-unit radius that has been cut open along a line on the surface parallel to its axis as shown in Figure 6.34 for the case of a pineapple. Thus the cylinder now looks like a rectangular strip 2π units wide pictured on an x,y cartesian coordinate system stretching from $x = 0$ to $x = 2\pi$. In this figure, the first stalk is laid down at the origin of the coordinate system. Successive stalks are displaced from their predecessors by the divergence angle, $2\pi/\phi^2$ radians, or 137.6 degrees, in the example shown in Figure 6.34. In addition, the stalks rise along the surface of the cylinder by h units for each new stalk laid down. This rise is called the *pitch*. The stalks are represented by the hexagonal D domains of the lattice points. Sunflower-like plants can be represented by tiling a polar coordinate system with D domains.

Notice in Figure 6.34 that the numbers of the F series alternate on opposite sides of the y axis. For example, stalk 5 occurs to the right after two turns around the cylinder, 8 occurs to the left after three turns about the cylinder, while 13 occurs to the right again after five turns. Marzec and Kappraff [1983] showed that this is related to the

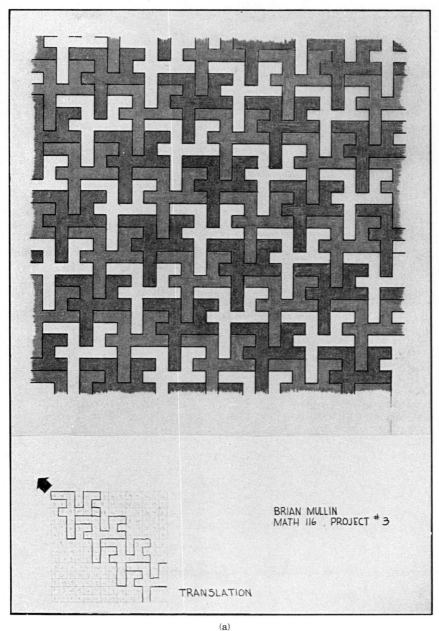

(a)

Figure 6.33   Two patterns illustrating Gilbert's method.

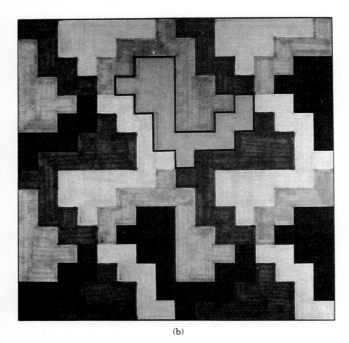

(b)

Figure 6.33 (*Continued*)

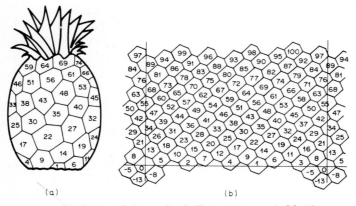

(a)   (b)

Figure 6.34 Relation of pineapple phyllotaxis to a period lattice.

fact that the continued fraction expansion (see Section 1.6.3) of the divergence angle mod 2π, or $1/\phi^2$, has convergents

$$\tfrac{2}{5}, \tfrac{3}{8}, \tfrac{5}{13}, \dots$$

In this model of a pineapple, the initial stalk is adjacent to the fifth, eighth, and thirteenth stalks, and these stalks line up with the initial stalk along a series of 5, 8, and 13 diagonal lines of hexagons which represent the numbers of clockwise and counterclockwise spirals evident on the surface of the plant, i.e., 5, 8, 13 phyllotaxis. However, if the pitch h were a smaller value, later stalks—for example, the thirteenth and twenty-first stalks—would be nearest neighbors of the initial one and would correspond to higher phyllotaxis numbers. Also, there is some transition value of h at which the pattern of growth changes from 8,13 to 13,21 phyllotaxis, and at this transition point the D domains become rectangles.

Other models lead to roughly similar conclusions. For example, R. O. Erickson presents a model in which the stalks are circles which pack together to fill the lattice. A careful analysis shows how the phyllotaxis number, pitch, and divergence angles are related [1983].

N. Rivier et al. [1984] have also developed a crystallography on a circular disc. They define the stalks as the D domains of a sequence of computer-generated growth centers given by the algorithm

$$r(\ell) = a\sqrt{\ell}$$

$$\theta(\ell) = 2\pi\lambda\ell$$

where r and θ are the polar coordinates of the disc, ℓ labels individual cells, λ is the divergence angle, and a is the typical cell's linear dimension. By representing stalks on the computer as D domains and studying the constraints on space imposed by Euler's theorem, they have simulated the growth of plants and shown that golden mean growth (λ related to ϕ) results in a homogeneous and self-similar pattern of nearly isotropic (identical) cells. For example, when $\lambda = 1/\phi$, the daisy-like structure shown in Figure 6.35 results, whereas when $\lambda = {}^{13}\!/_{21}$, the result is a rational approximation to $1/\phi$—the spider web with highly nonisotropic cells shown in Figure 6.36.

The D domains for the golden mean growth patterns are regular tilings with hexagons except for a few pentagonal and heptagonal cells which are defects in the regular tilings (see Section 6.3). In a regular tiling with hexagons all cells have the same orientation; the pentagons and heptagons are sources of positive and negative curvature, respectively, in the growth pattern. Their geometry is based on the following consequence of Euler's theorem for the tiling of a disc ($F + V - E = 1$) in which each vertex has exactly three incident edges (any other vertex valence is structurally unstable for reasons mentioned in Section 6.5):

$$\sum_{n=1}^{x} F_n (6 - n) = 6 \qquad (6.5)$$

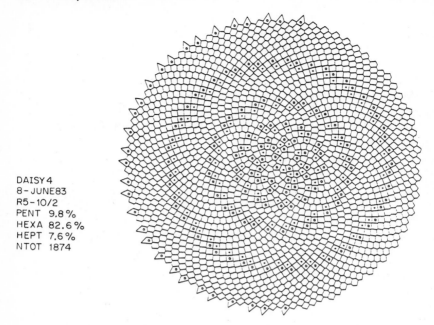

DAISY 4
8 - JUNE 83
R5 - 10/2
PENT 9.8 %
HEXA 82.6 %
HEPT 7.6 %
NTOT 1874

Figure 6.35 Daisy structures. Notice the self-similarity and the glide circles.

where F_n is the number of cells with n sides. Thus it is consistent with this equation that any finite set of cells that tile a finite domain are hexagonal except for an unlimited number of pairs of pentagonal and heptagonal cells in addition to six isolated cells of positive curvature (pentagons) as shown in Figure 6.37. This is also consistent with the description of the cell structure of the infinite random soap bubble patterns given by Theorem 6.1. Of course, Equation 6.5 does not forbid octagonal or other shaped tiles.

In Rivier's model of plant growth, pentagon-heptagon pairs of dislocations in an otherwise homogeneous pattern of hexagonal cells are located on concentric fault circles, and they screen the strain energy caused by the isolated pentagonal faults which are forced by the geometric constraints. The dislocations also prevent the hexagonal cells from gliding over themselves because of shearing forces. It turns out that concentric circles of dislocations, forming boundaries between defect-free (hexagonal) grains which can glide on each other, are seen in large daisies and sunflowers as shown in Figure 6.35. What is even more interesting about Rivier's simulation is that it appears to model the growth of cellular patterns that occur in Benard-Marangoni convection (see Figure 6.37), whereby a fluid heated from below exhibits convective motion above a certain temperature threshhold. It is be-

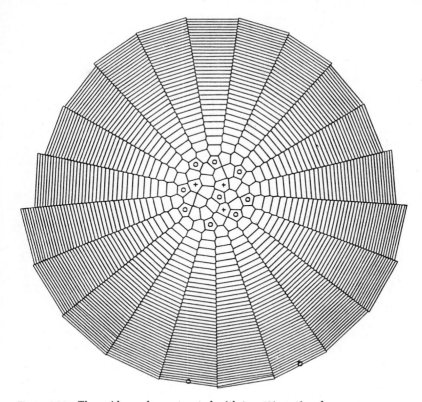

Figure 6.36   The spider web constructed with λ = ¹³/₂₁ rational.

lieved that solar granulation and other patterns in nature are the result of this convective behavior.

6.10   Quasicrystals and Penrose Tiles

According to Nelson [1987],

> In 1984 investigators working at the National Bureau of Standards found that a rapidly cooled sample of an aluminum-manganese alloy, named Schechtmanite after one of its discoverers, seemed to violate one of the oldest and most fundamental theorems of crystallography. Although the material appeared to have the same kind of order that is inherent in a crystal, it also appeared to be symmetrical in ways that are physically impossible for any crystalline substance [Schechtman et al., 1984].

Beams of x-rays directed at the material scattered as if the substance were a crystal with fivefold symmetry, whereas the conventional wisdom of crystallography says that only two-, three-, four-, and

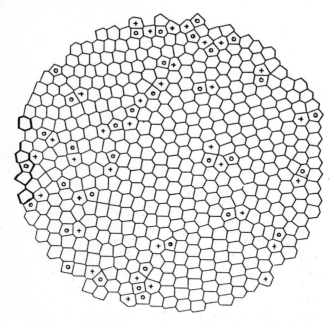

Figure 6.37 Computer-generated reconstruction of the cellular structure in a standard Benard-Marangoni experiment. o = pentagon, + = heptagon, x = octagon, o + = dislocation.

sixfold symmetry can occur in crystals. (The reasons for this will be made clear in Section 12.13.)

Further investigations into the microstructure of this material have shown that it embodies a new kind of order, neither crystalline nor completely amorphous. Materials structured around this new kind of order seem to forge a link between conventional crystals and the materials called *metallic glasses*, which are solids formed when molten metals are frozen so rapidly that their constituent atoms have no time to form a crystalline lattice. The new materials have therefore been called *quasicrystals*.

The nonperiodic Penrose tilings with kites and darts introduced in Section 5.11 provide an excellent two-dimensional model of how pentagonal symmetry can arise in x-ray patterns. Actually, quasicrystal structure is illustrated more clearly by an alternative to the kites-and-darts tiling. This new tiling employs the two rhombic shapes shown in Figure 6.38 [Penrose, 1979], and the tiles are combined by matching the arrows on their edges. Although Penrose tilings are not crystalline in a conventional sense, they do have many crystalline properties. For example, in a Penrose tiling it is possible to pick out many regular 10-sided polygons (decagons), several of which are evi-

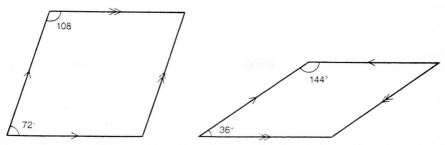

Figure 6.38 Two Penrose rhombuses. They are fitted together so that the arrows superposed on their edges match.

dent in Figure 6.39. Like hexagons, which are the unit cells of a two-dimensional lattice (see Section 6.7), all the decagons have precisely the same orientation. Like Schechtmanite, the Penrose tiling has the long-range orientational order that is usually associated with conventional crystal lattices.

In a subtler way Penrose tilings also have a kind of translational order as well. One way to see this is to shade all the rhombuses that have sides parallel to a given direction. The shaded rhombuses form a series of jagged irregular lines each of which, on the average, approximates a straight line as shown in Figure 6.40. All the lines are parallel and, approximately, evenly spaced. Therefore, in a statistical

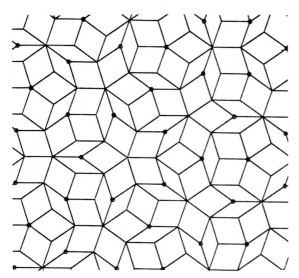

Figure 6.39 Decagons are found throughout the pattern; all have the same orientation demonstrating long-range orientational order.

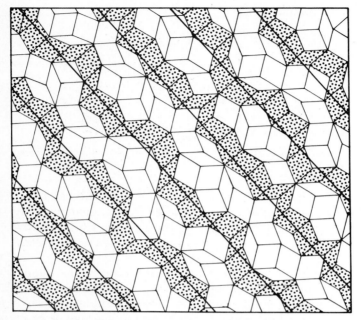

Figure 6.40   One of the five families of parallel and evenly spaced lattice planes formed by jagged lines demonstrating the long-range translational order of the lattice.

sense, a Penrose tiling has long-range translational order as well as orientational order. We also note that the line in Figure 6.40 is tilted at the same angle, 72 degrees, as the ladder in Dürer's *Melancholia* (see Figure 3.17).

Penrose tilings also have a kind of fivefold symmetry. In a Penrose tiling, the shaded rhombuses fall into five families of parallel lines, one of which is shown in Figure 6.40. The lines run in directions that are parallel to the edges of a regular pentagon. They intersect at angles that are multiples of 72 degrees, or one-fifth of a full circle. It can be shown that the lines, like the lattice planes of an ordinary crystal, will scatter beams of x-ray radiation. Beams reflected from a Penrose tiling would have fivefold rotational symmetry no matter where in the pattern they were aimed. The disorderly appearance of the lattice planes is similar to that found in a conventional crystal at temperatures above absolute zero, when the atoms are disordered because of thermal vibrations. In Penrose tilings, of course, the disorder would be present even at a temperature of absolute zero.

H. Lalvani has shown how the two Penrose rhombuses that result in either exact or approximate pentagonal symmetry can be generalized to a wider class of nonperiodic tiles (although nonperiodicity has not yet been proven) [1990]. The rhombuses in Figure 6.38 are derived from the central angle A of a 10-gon, {10}, or $A = \pi/5$ radian. The two rhombuses are the only ones possible with angles that are integral multiples of A; one has angles 1 and 4 times A while the other has angles 2 and 3 times A. Lalvani has discovered that patterns with approximate or exact sevenfold symmetry can be derived from the three rhombuses shown in Figure 6.41. These are the only rhombuses whose angles are integral multiples of the central angle of a 14-gon, {14}, i.e., 1 and 6 times A, 2 and 5 times A, and 3 and 4 times A, where $A = \pi/7$ radian. Notice that the numbers 1 and 6, 2 and 5, and 3 and 4 are the only distinct pairs of integers whose sum is 7, just as the pairs 1 and 4, and 2 and 3 are the only ones whose sum is 5, the condition for Penrose tilings. The patterns derived from these three rhombuses look like standard Penrose tilings (see Figure 6.39) except that {14}-gons, instead of decagons, appear throughout, and their edges are oriented in the seven different directions of the edges of a {14}-gon instead of a pentagon. Figure 6.42 illustrates a distorted image of a standard Penrose tiling that is derived from the tiles of a {14}-gon. Notice that the edges are oriented in the direction of only six of the seven possible directions. Of course, Penrose tilings with eightfold, ninefold, and

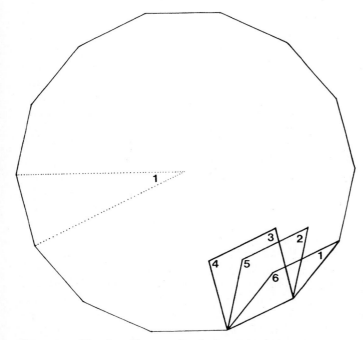

Figure 6.41   The three Penrose rhombuses of the {14}-gon.

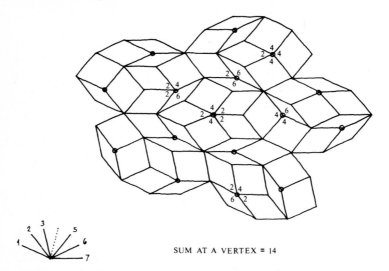

SUM AT A VERTEX = 14

Figure 6.42   Detail from a distorted image of a standard Penrose tiling using tiles derived from a {14}-gon.

higher-fold symmetries can also be created in a similar fashion. In Section 10.14, we will describe a pair of three-dimensional rhombohedrons that are a three-dimensional analogue of the Penrose tiles and serve as an even better model for Shechtmanite.

Lalvani has also noticed that the Penrose tiling with fivefold symmetry shown in Figure 6.43. can be viewed as an increasing sequence of whirling golden triangles (see Section 3.5) where the line segments of the whirling triangles serve as local mirror lines [1989]. Notice how each line segment reflects a portion of the Penrose tiling and that each of these mirror lines is oriented along one of the five families of scattering lines described above. Escher, in his print *Reptiles*, shown in Figure 6.44, also chose a golden triangle as a platform upon which his frogs ascend from a hexagonal tiling of the plane to a polyhedral structure (pentagonal dodecahedron) in three-dimensional space. This brings us to our own study of polyhedra in the next four chapters.

Appendix 6.A. Projective Geometry

The Renaissance artists were the best practicing mathematicians of the fifteenth century. Through the system of perspectivity, they not only developed a more realistic way to represent physical space but also provided the basis for a new area of mathematics, projective geometry, which was developed later. Renaissance artists, such as Alberti, considered themselves to be the most learned and theoretical

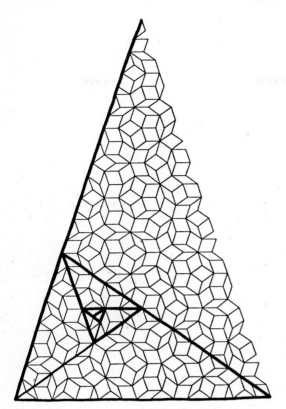

Figure 6.43 A pattern of whirling triangles with local mirror symmetry formed by Penrose rhombuses.

mathematicians of their time. Alberti, in the first written account of the system of perspectivity, which he published in 1435, stated that it was the first requirement of the painter to study geometry [Slawsky, 1977], [Cole, 1976].

The subject of perspective was developed by the fifteenth century artists Alberti, Leonardo da Vinci, and Albrecht Dürer. The system they devised for representing space was fairly simple. The artist imagined that the canvas was a glass screen to be painted as if he were looking through a window at a scene outside. From one eye, which is held fixed, lines of light are imagined to go to each point of the scene. Where each of these lines intersects the glass screen, a point is marked on the screen. The set of lines of light is called a *projection*, and the corresponding set of points is called a *section*. If carried out correctly and when viewed from an appropriate point, the section should create the same impression on the eye as the scene itself does.

Figure 6.44 *Reptiles* by M. C. Escher. (© *M. C. Escher Heirs/Cordon Art-Baarn-Holland.*)

For example, in Figure 6.A.1, a road on the ground plane is transformed to a canvas from a projection point O located at the artist's eye to render a scene as the artist sees it. The road, which recedes in parallel lines l to infinity, converges on the artist's canvas to a single point on the horizon line h. It is also clear from this figure that all paths leading toward the infinite distance on the flat landscape plane will map on the artist's canvas to a "vanishing point" on the horizon line.

What the Greeks were unable to accomplish through rigor, the Renaissance artists accomplished through the imagination. The artists adhered to the following rules in their system of perspective:

1. All horizontal lines in the scene perpendicular to the plane of the canvas must be drawn to meet at the principal vanishing point. This is the way our eyes see parallel lines receding in the distance.

2. Any set of parallel horizontal lines that are not perpendicular to the plane of the canvas but meet it at some angle must be drawn to

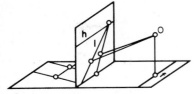

Figure 6.A.1   Projection of a scene onto a viewing plane.

converge at a point that lies somewhere on the horizon, depending on the angle these lines make with the plane of the canvas. Parallel lines that rise or fall as they recede from the viewer must also meet at one point. This point would be the one at which a line from the viewer's eye parallel to the lines described above intersects the canvas.

3. Parallel horizontal lines of the scene are drawn as horizontal and parallel lines. Vertical lines are drawn as vertical and parallel lines.

Figure 6.A.2 is an example of the work of Vredeman de Vries, who very clearly used this technique of having all the vanishing points be collinear on the horizon line. It also shows the effect of exaggerated perspective.

6.A.1   An example of a projected three-dimensional framework

We shall not attempt to give a thorough discussion of projective geometry in this brief space, particularly since there are several books and

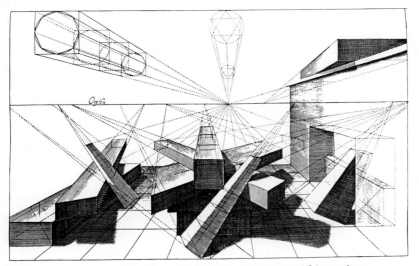

Figure 6.A.2   A work of Vredeman deVries illustrating vanishing points.

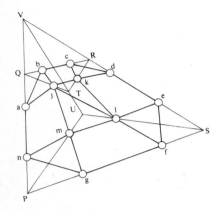

Figure 6.A.3 A framework formed by truncating a tetradedron.

references that can provide such a background [Edwards, 1985], [Kappraff, 1990], [Young, 1930]. Instead we leave the reader with the intuitive notion of a perspective transformation presented above and present an example of the projection of a three-dimensional framework onto a plane.

Consider the framework that is embedded in a tetrahedron shown in Figure 6.A.3. This is a diagram of a tetrahedron *PVSU* truncated by seven planes (*QRT, abj, cdk, efl, hmg, jkl,* and *mjl*). After a series of projections (not shown), the framework is projected in Figure 6.A.4

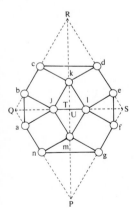

Figure 6.A.4 A nonrigid framework obtained as the end result of a series of projections of the framework in Figure 6.A.3.

onto the base plane of the tetrahedron in such a way that T and U coincide and V is mapped to infinity. This projection is uniquely determined. In Section 7.8 it will be shown that as a consequence of this projection, the projected framework shown in Figure 6.A.4 must not be rigid [Baracs, 1989].

Polyhedra:
Platonic Solids

*On the Platonic solids: We must assume that
the God duly adjusted proportions between
their numbers, their movements, and other
qualities and brought them to the exactest
perfection.* PLATO
Timaeus

7.1 Introduction

Beneath the outer covering of a three-dimensional structure lies a
skeletal frame that absorbs or transmits the external forces that act
upon it. The inner structure of a bridge is evident for all to behold.
Strip away the brick and mortar from a building and what remains
are posts and beams. The metal shell around an airplane masks the
delicate arches which are designed to distribute the dynamic loads experienced in flight just as our own skin is a membrane that surrounds
a structure made up of muscles, tendons, and bones.

On a microscopic level, chemical and biological structures of all
sorts are made up of chains of atoms and molecules linked together in
complex spatially oriented frameworks. In this chapter and the next
two we will study structures that have their basis in complexes of
points linked together by line segments. In this chapter and the next,
the emphasis is on closed structures known as *polyhedra*. Chapter 10
studies open structures consisting of lattices and more general complexes of points.

By defining a polyhedron, we necessarily limit the discourse about
them. Although what we mean by a polyhedron has changed through
the years [Senechal and Fleck, 1988], most authors use a definition
that would have been familiar to Plato. A *polyhedron* is considered to
be a surface made up of a set of plane polygons, called its faces, that
bounds a region of space. The cube and pyramid are the most familiar

examples of polyhedra. The Császár and Szilassi polyhedra described in Section 4.16 are two other examples. Since, in this definition, a polyhedron is thought of as a surface, its polygonal faces are spanned by membranes.

The region of space enclosed by a polyhedron can be either convex or nonconvex, in which case the polyhedron is called *convex* or *nonconvex*. Convex surfaces are natural generalizations of the convex curves introduced in Section 5.2.2. A closed convex surface is defined to be one such that any two points placed within the region bounded by it can be connected by a straight line also lying within that region. If part of the connecting line lies outside of the region bounded by the surface for some pair of internal points, the surface is nonconvex. By this definition, the sphere is convex but the torus (or doughnut) is nonconvex.

This definition of a polyhedron, however, is problematic in that it excludes some important structures that are generally thought of as polyhedra. In fact, experience in constructing models of polyhedra leads us to conclude that the membranes spanning the faces are not only superfluous but actually hide the rich set of inner relationships between the framework of edges and vertices that surround the faces. Nevertheless, this approach applies well to most of the polyhedra discussed in this chapter.

A more modern approach to polyhedra dispenses with the need to consider polyhedra as surfaces and focuses instead on their skeletal structures [Grunbaum, 1977]. A polyhedron is defined as a three-dimensional map consisting of edges, faces, and vertices. The faces are cycles of edges and vertices called polygons, and they no longer have to be planar. Each edge links together exactly two faces in a connected way so that any two edges can be joined by a sequence of faces. For example, two pyramids joined together only at their apexes are not polyhedra by this definition. By contrast with the graphs in Chapter 4, edges are considered to be straight lines of definite length, and two edges meet at vertices with prescribed angles. This definition is more general than the first one, and some of the things that we say about polyhedra will refer to this definition rather than the first.

Let's first become acquainted with polyhedra by building some out of miniature marshmallows, which serve as the vertices, and toothpicks, which play the role of edges of equal length. This activity brings up some sticky problems. We have found Kraft brand marshmallows to be of superior quality at least for constructing polyhedra. Try the following exercises.

Exercise 7.1 Create two equilateral triangles from six marshmallows and six toothpicks. Now rearrange the six toothpicks to form four triangles each the same size as the original.

Exercise 7.2   Construct a square and test it to see that it is not rigid, i.e., the vertices can be moved relative to each other. By adding additional toothpicks and marshmallows, surround the square by the least number of toothpicks and marshmallows to make it rigid.

Exercise 7.3   Construct a cube and notice how it droops, being unable to hold up even its own weight. Now brace each of its faces with a toothpick, letting the squares deform to rhombuses, and notice how the resulting distorted box is quite rigid.

Exercise 7.4   Surround a vertex with six equilateral triangles and notice how the triangles lie in a plane. Now surround a vertex with five equilateral triangles and notice how the central vertex is forced out of the plane to form a cap. Next form triangles on each of the outer edges of the original triangles and connect the unattached vertices of these triangles to form a belt of triangles as shown in Figure 7.1. Finally, complete this figure to a dome with another cap identical to the original.

Exercise 7.5   Construct as many polyhedra as you can that satisfy the following constaints:

1. All faces are identical ordinary planar polygons, e.g., equilateral triangles, rhombuses, hexagons, etc.
2. Each vertex has the same number of incident edges as any other.
3. If the edges were flexible, they could be deformed to a map on a sphere.

For each polyhedron, record in a table the number of faces F, vertices V, edges E, edges incident to each vertex q, edges incident to each face p, the Euler number $F + V - E$, and whether or not the structure is rigid (stands tall or droops after you build it).

7.2   The Platonic Solids

There are five kinds of polyhedra that satisfy the conditions of Exercise 7.5 (including the constraint imposed by the use of toothpicks that

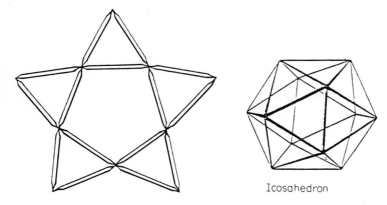

Icosahedron

Figure 7.1 Pattern for constructing an icosahedron dome with marshmallows and toothpicks.

all edges have the same length): the *tetrahedron* (constructed in Exercise 7.1) with 4 triangular faces and 4 vertices, the *hexahedron* (or cube if the faces are squares) with 6 parallelogram faces and 8 vertices, the *octahedron* (the solution to Exercise 7.2) with 8 triangular faces and 6 vertices, the *pentagonal dodecahedron* with 12 pentagonal faces and 20 vertices, and the *icosahedron* (constructed in Exercise 7.4) with 20 triangular faces and 12 vertices. Some of the properties of these polyhedra are listed in Table 7.1.

The three polyhedra with equilateral triangle faces {3}, the tetrahedron, the octahedron, and the icosahedron, are *rigid*. Once they are constructed, their vertices cannot move relative to each other; therefore, they can assume only one form. The other two kinds of polyhedra, the hexahedron and the dodecahedron, are not rigid; once they are constructed they collapse into a continuum of deformed shapes. Some of these shapes do not have planar faces. However, if the faces of the hexahedron are squares {4} and the faces of the dodecahedron are regular pentagons {5} (see Figure 7.2), we get a unique family of five polyhedra known as the platonic solids in honor of Plato who commemorated them in *Timaeus* [1977]. The platonic polyhedra are shown in Figure 7.2 along with their *net diagrams* which show how to fold them up from the plane.

Platonic polyhedra have been studied since the age of ancient Greece [Malkevitch, 1988]. They have sparked the imaginations of creative individuals from Euclid to Kepler to Buckminster Fuller. These polyhedra are rich in connections to the worlds of art, architecture, chemistry, biology, and mathematics. In *Timaeus* four of the solids were related to the four elements: earth, air, fire, and water. The fifth solid, the dodecahedron, represented the cosmos [see Figure 7.3(a)]. In the natural world, the platonic solids present themselves in the form of microscopic organisms known as *radiolaria* [see Figure 7.3(b)]. In this chapter and the next two we will study some of the connections between platonic solids and the natural world along with ex-

TABLE 7.1

Polyhedron	No. of faces F	No. of vertices V	No. of edges E	q edges per vertice	p edges per face	Rigid
Tetrahedron	4	4	6	3	3	Yes
Cube	6	8	12	3	4	No
Octahedron	8	6	12	4	3	Yes
Dodecahedron	12	20	30	3	5	No
Icosahedron	20	12	30	5	3	Yes

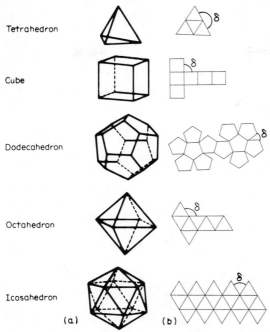

Tetrahedron

Cube

Dodecahedron

Octahedron

Icosahedron

(a)   (b)

Figure 7.2 (a) The five platonic solids; (b) net diagrams to fold them up from the plane along with indications of the angular deficit, δ.

amples of how the platonic polyhedra lead to interesting three-dimensional designs.

It is difficult to understand the nature of three-dimensional objects through verbal descriptions of them or even by looking at two-dimensional images of them. What is needed is an actual model which can be manipulated and viewed from different angles. We strongly recommend that you construct a set of platonic polyhedra as an aid to understanding the material of this chapter. They can be constructed out of sticks and connectors or stiff paper, and methods of construction can be found in *Shapes, Space, and Symmetry* by Alan Holden [1971], *Polyhedra: A Visual Approach* by Anthony Pugh [1976], *Mathematical Models* by H. M. Cundy and A. P. Rollett [1961], and *Polyhedron Models* by M. J. Wenninger [1971].

7.3 The Platonic Solids as Regular Polyhedra

The platonic solids can be considered to be polyhedra at the *limit of perfection*. Throughout this chapter we will see a good deal of evi-

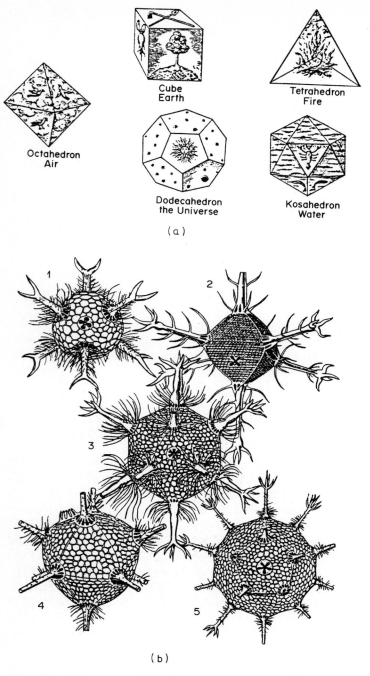

Octahedron Air

Cube Earth

Tetrahedron Fire

Dodecahedron the Universe

Kosahedron Water

(a)

(b)

Figure 7.3 (a) The platonic solids depicted by Johannes Kepler in *Harmonices Mundi*, Book II (1619); (b) the platonic solids in the form of radiolaria.

dence of this perfection. By their very construction, they possess the same kind of perfect symmetry exhibited in Section 4.7 by the regular maps. In fact, the first two conditions of Exercise 7.5 are precisely the defining properties of the regular maps on the sphere (or plane): all faces and vertices are surrounded by an identical number of edges. Any convex polyhedron that can be constructed from regular polygons with these two properties is called a *regular polyhedron* [Coxeter, 1973].

The five regular maps on a sphere listed in Table 4.1 are isomorphic to the five kinds of regular polyhedra constructed in Exercise 7.5 and listed in Table 7.1. For example, the cube has three squares surrounding each vertex, denoted by the Schläfli symbol {4,3}, in which the first number refers to the face valence p (number of edges per face) while the second number refers to the vertex valence q (number of edges per vertex). Likewise, the tetrahedron surrounds each vertex by three triangles {3,3}, the octahedron surrounds each vertex by four triangles, {3,4}, and the dodecahedron and icosahedron are {5,3} and {3,5}, respectively.

There can be no more than five kinds of regular polyhedra that satisfy condition 3 of Exercise 7.5 that the edges be deformable to a map on the sphere. If there were another one, there would be another regular map, in violation of Theorem 4.3.

7.4 Maps of Regular Polyhedra on a Circumscribed Sphere

Besides possessing perfect symmetry in a graphical sense, the platonic solids also have a kind of *perfect geometric symmetry* in the sense that the vertices of each solid are equidistant from a common center and evenly distributed around this center. Thus, they lie upon an imaginary sphere called the *circumscribed sphere,* or *circumsphere.* This prompts us to ask the following question: Can you slice an orange into four congruent (identical) pieces in a way other than the breakfast way? This can be done by *circumscribing* a sphere about the vertices of a tetrahedron as shown in Figure 7.4(a). A source of light is placed at the center of the sphere and the edges are projected onto the sphere where they form a tiling of the sphere by a set of congruent tiles. The congruent slices are represented by the four solid angles obtained by cutting the sphere with planes that include an edge of the tetrahedron and the center of the sphere. Of course, the circumscribing spheres of the other platonic polyhedra divide the sphere into 6, 8, 12, and 20 congruent segments as shown in Figure 7.4 for three of the platonic solids. Spherical stone sculptures of the platonic solids constructed a

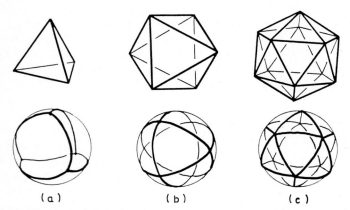

Figure 7.4 Three platonic polyhedra projected onto a sphere as arcs of geodesic circles.

millennium before Plato have been found in the British Isles [Critchlow, 1982].

In this projection, the edges of the platonic polyhedra project onto arcs of great circles (circles on the sphere whose plane includes the center of the sphere, e.g., longitude lines). An arc of a great circle is the path of shortest distance that a bug crawling on the surface would take to get from one point on the sphere to another. It is also the path of airline pilots going from point to point on the globe by the great circle route. In general, curves of minimum distance on any surface are called the *geodesics* of that surface, i.e., the geodesics of a sphere are great circles.

Problem 7.1 A rectangular box of given dimensions is shown in Figure 7.5. A bug is to crawl on the surface of the box from a point *A*, 1 inch below the center of the top edge, to point *B*, 1 inch above the center of the bottom edge on the opposite side of the box. Find the shortest distance from *A* to *B*. (Hint: Cut open the box up in a suitable way and draw the shortest straight line.) [Blake, 1985]

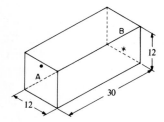

Figure 7.5 Rectangular box of Problem 7.1.

7.5   Maps of the Regular Polyhedra on the Plane—Schlegel Diagrams

In the last section, the platonic polyhedra were projected onto regular maps on the sphere with congruent faces. The platonic solids can also be projected onto the plane. In fact, the five regular maps on the plane, shown in Figure 7.6(*b*), are just such projections. These *Schlegel diagrams* [Loeb, 1976], as they are called, are obtained by projecting the edges of a platonic polyhedron onto the plane from a point directly above the center of one of its faces as shown in Figure 7.6(*a*) for a cube. Visually, this amounts to holding one face of a polyhedron quite close to one's eyes, looking at the structure through that face, and drawing the projection of the structure as seen in this exaggerated perspective. Notice that one of the faces of the polyhedron frames all the others, and this face must be included when counting faces.

Although the projected map has lost its congruent faces, the Schlegel diagram enables us to see a realistic two-dimensional repre-

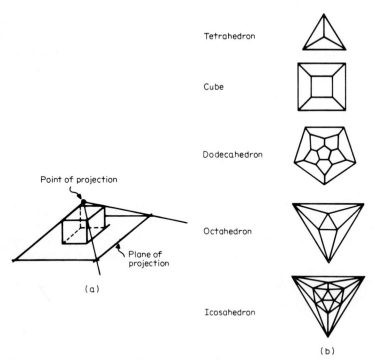

Figure 7.6   (*a*) A cube projected onto the plane of one of its faces as a Schlegel diagram; (*b*) Schlegel diagrams of the platonic polyhedra.

sentation of a three-dimensional object that preserves such important characteristics of the original as its connectivity of edges and vertices and some of its symmetry. In Section 9.9 we will see that the Schlegel diagram provides an excellent tool to help visualize the result of transforming the platonic solids by truncating vertices or edges.

7.6 Duality

7.6.1 The inscribed sphere

At first glance, the platonic polyhedra appear quite different from each other. However, they are related in many ways and form a tightly woven family. In this section we look at perhaps the most basic of relationships between these polyhedra, namely *duality*.

An important observation about the platonic polyhedra can be made by inspecting Table 7.1. There is a natural pairing of the cube with the octahedron, the dodecahedron with the icosahedron, and the tetrahedron with itself. For each face of one of these pairs, there corresponds a vertex of the other reminiscent of the duality of maps described in Section 4.9. In fact, by placing pairs of platonic polyhedra one within the other (you can use marshmallow and toothpick models) you can see that if a vertex of one polyhedron of the pair is placed at the centroid of a face of the other and vertices are connected if their corresponding faces share an edge, the other member of the pair results.

For dual pairs of platonic polyhedra, certain statements that can be made about a polyhedron can also be made about its dual if the following replacements are made:

$$\text{face} \leftrightarrow \text{vertex}$$

$$\text{edge} \leftrightarrow \text{edge}$$

$$p \leftrightarrow q$$

For example, if the Schläfli symbol of a polyhedron is $\{p,q\}$, its dual has the symbol $\{q,p\}$.

Since the face centroids of each of the platonic polyhedra are also vertex points for their duals, they must lie equidistant from a common center. Thus another sphere, the *inscribed sphere* or *insphere,* can be placed within a platonic polyhedron tangent to each of its faces. The inscribed sphere of a platonic solid is then the circumscribed sphere of its dual scaled appropriately in size. Duality for the platonic polyhedra depends on their symmetry. Appendix 7.A is devoted to showing how the concept of duality can be defined for convex polyhedra in general.

At first Kepler believed that the physical structure of the universe was closely connected with geometry. He alternately inscribed and circumscribed spheres about dual pairs of the platonic solids and hypothesized that these spheres represented the orbits of the five planets known in his time, Mercury, Venus, Mars, Jupiter, and Saturn as shown in Figure 7.7 but could not fit his observations precisely to this scheme.

Escher also used inscribed and circumscribed spheres to create the set of nested models of the platonic solids shown in Figure 7.8. He was so fascinated by his creation that when he moved from his home, he gave away most of his belongings, but he took his beloved model of the five solids to his new studio [1971].

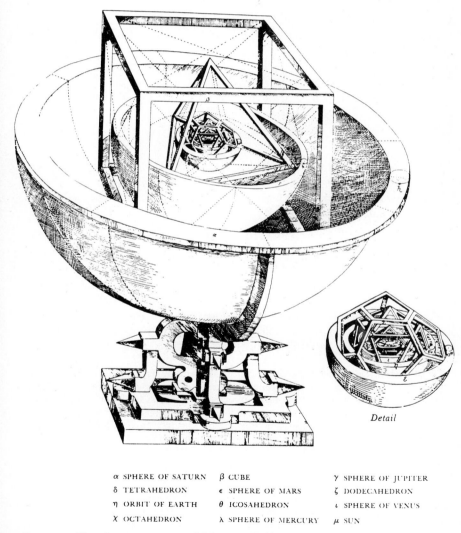

Detail

α SPHERE OF SATURN	β CUBE	γ SPHERE OF JUPITER
δ TETRAHEDRON	ε SPHERE OF MARS	ζ DODECAHEDRON
η ORBIT OF EARTH	θ ICOSAHEDRON	ι SPHERE OF VENUS
χ OCTAHEDRON	λ SPHERE OF MERCURY	μ SUN

Figure 7.7   The planetary system of Johannes Kepler.

Figure 7.8   M. C. Escher contemplating a nested set of platonic polyhedra.

Wenninger has shown how beautiful models of the duals may be constructed, embedded in each other by paper folding [1983]. Lalvani has created many beautiful *transpolyhedra* which demonstrate a continuous transformation from a polyhedron to its dual [1989], [Crapo, 1978].

7.6.2   Interpenetrating duals and the intersphere

The dual pairs of platonic polyhedra can be visualized as interpenetrating each other so that the set of edges of one perpendicularly bisect the corresponding edges of the other. They will be discussed further in the next chapter and if you look ahead to Figure 8.9, you can see pictures of them. We recommend that you construct a set of them.

Construction 7.1   Construct a set of the three interpenetrating pairs of dual platonic polyhedra. This construction can be carried out by placing appropriately sized pyramids on each face of one polyhedron of the pair as shown in Figure 7.9 for the case of interpenetrating tetrahedra. The vertices at the base of the pyramids lie at the midpoints of the sides of the platonic polyhedron. The lateral faces of the pyramids are equilateral triangles and each pyramid may be constructed by folding the triangles up from the plane.

These interpenetrating duals help to define a third sphere that is related to the platonic solids, the *intermediate sphere* or *intersphere*. This sphere intercepts the midpoint of each edge of a platonic polyhedron. An example is shown in Figure 7.10 of the intersphere of the cube and the octahedron combination.

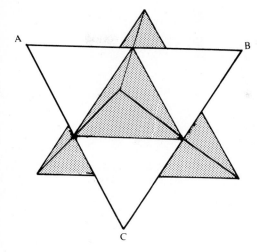

Figure 7.9 Two interpenetrating tetrahedra.

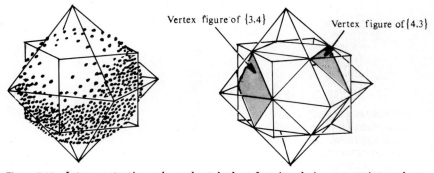

Figure 7.10 Interpenetrating cube and octahedron framing their common intersphere.

Escher was intrigued by the dramatic possibilities that interpenetrating duals offered to design, and he created many designs based on them, one of which is shown in Figure 7.11.

7.6.3 Duals on a Schlegel diagram

A polyhedron and its dual can be represented on the same Schlegel diagram, although this is a little tricky [Loeb, 1976]. We place a vertex inside each face and connect vertices by an edge if two faces of the original Schlegel diagram share an edge. The problem arises in placing a vertex within the framing face. One way around this problem is to imagine that the framing face is the exterior of the Schlegel diagram and that the vertex of the dual on this face is located at infinity.

Figure 7.11   *Stars*, woodcut, 1948. Escher's fantasy based on the platonic solids and their duals. (© *M. C. Escher Heirs/Cordon Art-Baarn-Holland.*)

Then all vertices corresponding to faces bordering on the outside face are connected to the infinite vertex by drawing an edge crossing the boundary of the Schlegel diagram at right angles. This is illustrated in Figure 7.12 for the tetrahedron.

This approach can be justified by imagining the Schlegel diagram drawn on a sphere. The framing face is the remainder of the sphere. The vertices of the dual are placed on each face and connected by the appropriate edges. The vertex in the outer face is punctured, reminiscent of Section 4.4, and stretched to infinity.

7.7   Combinatorial Properties

Since the platonic polyhedra can be thought of as maps on a sphere, the combinatorial results discussed in Sections 4.6 and 4.7 continue to

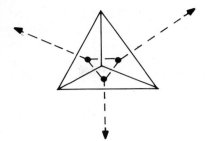

Figure 7.12   Schlegel diagram of a tetrahedron dual.

be valid. In fact, the vertices and edges of every *3-polytope* (convex three-dimensional polyhedron) *P* determines a graph *G(P)*, the graph of *P*. It is relatively easy to prove that *G(P)* is planar and 3-connected (see Section 4.5) for each 3-polytope *P*. The converse statement was first proved by Ernst Stenitz, the most important early twentieth century contributor to the theory of polyhedra, and it constitutes the nontrivial part of the result.

Theorem 7.1 (Steinitz's theorem)   A graph *C* is isomorphic to the graph *G(P)* of some 3-polytope *P* if and only if *C* is planar and 3-connected. (*C* is also referred to as a *polyhedral graph*.)

The proof is not presented here, but we refer the interested reader to [Barnette and Grünbaum, 1969].

The combinatoric formulas of graphs continue to hold for polyhedra, i.e.,

$$\sum_V q = 2E \tag{7.1}$$

$$\sum_F p = 2E \tag{7.2}$$

and

$$F + V - E = 2 \tag{7.3}$$

where summation is taken over all vertices in the first formula and over all faces in the second, and *p* and *q* are the face and vertex valences, respectively. Table 7.1 shows that Euler's formula for the sphere (or plane) holds for the platonic solids. For regular polyhedra the first two formulas can be written as

$$qV = 2E \tag{7.4}$$

$$pF = 2E \tag{7.5}$$

These simple formulas place severe restrictions on the plastic forms that are possible for two- and three-dimensional maps even before the additional constraints of straight edges and face angles of polyhedra are taken into account. These formulas weave a kind of analytic thread through the subject of two- and three-dimensional design, and they lead to some interesting consequences that are worth exploring. For example, Equation (4.13) can be carried over from regular graphs to platonic polyhedra, i.e.,

$$F = \frac{t}{p} \qquad E = \frac{t}{2} \qquad V = \frac{t}{q} \qquad (7.6)$$

where $t = 4pq/(2p + 2q - pq)$

That is, the icosahedron $\{3,5\}$, has $q = 5$, $p = 3$, $t = 60$ Thus,

$$F = \frac{60}{3} = 20 \qquad E = \frac{60}{2} = 30 \qquad V = \frac{60}{5} = 12$$

and the number of faces, vertices, and edges have been determined from knowledge of p and q only. No metric properties (length and angle) are needed. In the next three sections, we will see the effect of introducing length and angle. In the next section we shall see how to interconnect a set of nodes or vertices of a structure by straight rods or edges in order to make it rigid.

Problem 7.2   Apply Equation (7.6) to computing F, V, and E for the other platonic solids.

7.8   Rigidity

When you built the platonic solids out of marshmallows and toothpicks, you noticed that the tetrahedron, octahedron, and icosahedron stood up firmly and rigidly and were even able to support additional weight after their construction. On the other hand, the cube and the dodecahedron drooped over, unable to support even their own weight. How can one explain this behavior? What factors determine the rigidity of a three-dimensional structure? How can a polyhedron that is not rigid be stabilized?

By a *rigid framework*, we mean a structure of vertices and edges whose vertices are not capable of moving relative to each other when its edges are connected to the vertices by swivel joints permitting rotation about the vertex in any direction. We have considered the rigidity of two-dimensional frameworks in Sections 4.18 and 6.6. Now we wish to determine the least number of edges required to make the structure rigid. A. L. Loeb explains that V disconnected vertices need $3V$ coordinates to fix them in three-dimensional space, i.e., three coordinates x, y, z for each vertex [1976]. As a result, we say that V vertices have $3V$ *degrees of freedom*. Each time we add an edge, we constrain the structure so that there is one less degree of freedom. Since a rigid object has no degrees of freedom, we might think that $3V$ edges would be needed to make V vertices rigid. Actually, only $3V - 6$ edges are required, although certain qualifications are mentioned below. For example, the tetrahedron with four vertices is rigid with $E = 3(4) - 6 = 6$ edges.

Why does a rigid body have six fewer degrees of freedom than we would expect? Since a rigid body can be translated to a new location or rotated as a whole in space and still be rigid, all its vertices need not be fixed. Consider three vertices on the rigid structure not all on the same line, as in Figure 7.13. The three coordinates of point 1 account for the freedom of translation. Point 2 is free to rotate on a sphere about point 1, in which case two coordinates are needed to specify its location on this sphere. Finally, point 3 is free to rotate in a circle about the axis through points 1 and 2, and its location on this circle is specified by one more coordinate. Thus, six coordinates need not be fixed on the configuration for it to remain rigid, three for translation and three more for rotation. In general, the formula

$$E \geq 3V - 6 \qquad (7.7)$$

is a good predictor of rigidity, although this formula has its limitations as we shall see.

While Equation (7.7) puts a lower bound on the number of edges needed to make a structure with V vertices rigid, no information is given about where these edges should be placed. If there are fewer edges, the structure is not rigid according to this argument. Once the $3V - 6$ degrees of freedom are removed, additional edges are redundant since they do not further constrain the structure. For example, if $V = 12$, $3V - 6$, or 30, edges are needed for rigidity. This is certainly true for the icosahedron. However, a cube with 8 vertices requires 18 edges to be rigid, six more than it has. A little experimentation shows

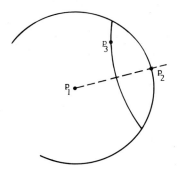

Figure 7.13 Three noncollinear points on a rigid body.

that if an additional edge is added along one of the diagonals to each of the cube's six faces, the new structure is rigid, as Exercise 7.3 shows.

Exercise 7.6 Equation (7.7) is a reliable predictor of rigidity of three-dimensional structures. However, it does not guarantee rigidity, and a structure may be rigid even when it is violated. Construct several rigid and nonrigid polyhedra out of marshmallows and toothpicks to test this formula. Try to find nonrigid structures that, nevertheless, satisfy the formula and rigid structures that violate it.

The problem with Equation (7.7) as a predictor of rigidity is that it can be violated, and yet the structure may still be rigid as shown in Figure 7.14 for an octahedron with a vertex and two edges connecting opposite vertices. Surely this structure is rigid since the octahedron is and segments that make up the diagonal cannot move without changing length. A calculation shows that $3V - 6 = 15$, but this rigid structure has only 14 bars. Nevertheless, the violation of the rigidity condition does underscore a dangerous condition. Although this structure is rigid, it is *infinitesimally nonrigid*, which means that under stress certain vertices slightly alter their positions, and it is of cardinal importance for structural engineers to avoid such circumstances.

In two dimensions, $E \geq 2V - 3$ is a predictor of rigidity provided the edges are properly placed. Figure 7.15(*a*) shows a portion of the $3^2.4.3.4$ tiling (see Section 5.5) built out of marshmallows and toothpicks that is not rigid even though $E = 2V - 3$. If certain edges are removed and reinserted at the positions represented by the dotted lines, the same configuration of vertices is rigid. Figure 7.15(*b*) shows how such a stable configuration can be constructed from marshmallows and toothpicks [Loeb, 1988].

In Section 6.6 we presented another approach to predicting the rigidity of structures based on the projective properties of polyhedra. Henry Crapo and others have delved deeper into the subject of rigidity and have come up with the following general condition for the rigidity of a two-dimensional framework:

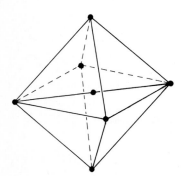

Figure 7.14 An octahedron with two edges along a diagonal. It forms a rigid but noninfinitesmally rigid structure.

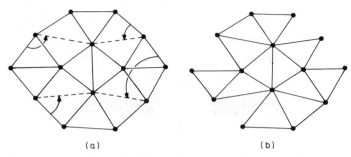

(a)   (b)

Figure 7.15 (a) A nonrigid two-dimensional framework that, never-theless, satisfies $E = 2V - 3$; (b) it can be made rigid by changing the locations of the indicated edges to the those of the dotted lines while maintaining the positions of the vertices.

Theorem 7.2 A two-dimensional framework with the configuration of a polyhedral graph (3-connected and planar) is not rigid if and only if it is the projected image of a polyhedron and satisfies $E = 2V - 3$.

For example, a slightly modified version (but still nonrigid) form of the nonrigid configuration in Figure 7.15(a) is shown in Figure 6.A.3. The polyhedron from which it was projected is shown in Figure 6.A.4. Theorem 7.2 implies that if the lengths of the toothpicks were slightly altered in Figure 7.15(a), the structure would be rigid since it is unlikely to be the projection of a polyhedron. Figures 6.A.3 and 6.A.4 illustrate that the end result of a series of projections of a polyhedral framework embedded in a truncated tetrahedron yields a nonrigid two-dimensional framework according to Theorem 7.2. If this two-dimensional framework is not rigid, it is easy to see that the closely related framework in Figure 7.15(a) is also not rigid. It is a little more difficult to predict the nonrigidity of three-dimensional structures, but Janos Baracs and Crapo have extended Theorem 7.2 to partially cover this case [1989].

Theorem 7.3 A three-dimensional framework with the configuration of a polytopal graph is not rigid if it is the projection of a 4-polytope (four-dimensional convex polyhedron; see Section 4.20) and satisfies $E = 3V - 6$.

Theorems 7.2 and 7.3 place the study of the rigidity of structures squarely within projective geometry, a place that this study enjoyed a century ago. Further discussion in great depth can be found in *Structural Topology*, a journal edited by Henry Crapo [1978].

Finally, it can be proven that any convex polyhedron is rigid if and only if each of its faces is a triangle.

7.9 The Angular Deficit

A total angle of 360 degrees surrounds a point in the plane. This is also true around a point on a sphere since the locality of any point on a sphere can be approximated as closely as you wish by a tangent plane. How does the sum of angles around a vertex of a polyhedron compare with the 360-degree angle around a point on the plane or sphere? The *angular deficit*, or *spherical deviation*, at a vertex of a polyhedron is defined as the difference between the sum of the angles of the polygons surrounding the vertex and 360 degrees and given the symbol δ, i.e.,

$$\delta = 360 - \text{sum of angles around a vertex}$$

In other words, it is the gap which results if the vertex is opened out flat as in the net diagrams of Figure 7.2. The angular deficits of typical vertices of the platonic polyhedra are shown on these net diagrams. The smaller the angular deficit, the more sphere-like the polyhedron. Of course, if the polyhedron degenerates to a plane or a sphere, the angular deficit at each vertex is zero. The summation of the angular deficits over all the vertices of a polyhedron is the total angular deficit.

René Descartes made some important contributions to geometry in a treatise entitled *De Solidorum Elementis* [Frederico, 1982]. An important formula is stated in this manuscript known as *Descartes' formula*. It states that the total angular deficit, or the sum of all the angular deficits, taken over each vertex of a convex polyhedron equals 720 degrees, i.e.,

$$\sum_V \delta = 720 \qquad \text{or} \qquad \delta V = 720 \text{ for regular polyhedra} \qquad (7.8)$$

where summation is over all the vertices of the polyhedron.

Descartes' formula is a remarkable constraint on space. Only when it is satisfied can a set of vertex patterns close up to form a convex polyhedron. This formula deserves a proof, which is given in Appendix 7.B where we show that it is equivalent to Euler's formula.

The fact that the sum of the face angles around any vertex of a polyhedron is less than 360 degrees leads to another proof that there are only five platonic polyhedra.

proof For any regular polyhedron {*p,q*}, its faces are regular polygons of *p* sides. From Equation (5.1), the internal angle of a regular polygon is

$$\theta = 180 \left(\frac{p-2}{p} \right) \text{ degrees} \qquad (7.9)$$

Since *q* such regular polygons surround each vertex, and the sum of the face angles meeting at a vertex is less than 360 degrees,

$$\frac{180 \, q \, (p-2)}{p} < 360$$

Using algebra, it follows that

$$(p - 2)(q - 2) < 4$$

The only values of p and q (which must be integers) satisfying this inequality are

$$\{3,3\}, \{3,4\}, \{4,3\}, \{3,5\}, \text{ and } \{5,3\}$$

where the bracketed numbers represent the Schläfli symbols for the platonic polyhedra.

Descartes' formula can also be used to determine the number of faces, edges, and vertices of a polyhedron if the pattern of polygons surrounding each vertex is the same, as it is for regular polyhedra. Thus, for an icosahedron, it follows from Equations (7.3), (7.4), and (7.8) that:

$$\delta = 360 - (5)(60) = 60 \text{ degrees}$$

$$V = \frac{720}{\delta} = 12$$

$$E = \frac{qV}{2} = 30 \text{ since } q = 5$$

$$F = 2 + E - V = 20$$

which agrees with our expectations.

7.10   From Maps to Polyhedra—The Dihedral Angle

Most of what we have said about polyhedra in the last two sections takes into account only the connections of vertices by edges, i.e., its qualitative, not quantitative, properties such as length and angle. To determine whether or not a structure is rigid we must connect vertices by just the right number of rods of the appropriate length, but nothing has been said about what those lengths are. Even Descartes' formula holds equally well for all the drooping shapes of a nonrigid polyhedron and says more about the ability of the polyhedron to form a closed figure than about the angles between its edges and faces.

In an actual polyhedron with planar faces, each face is oriented in a particular direction. The direction of a line in space that is perpendicular to every line in a given plane is called a *normal vector* to the plane (see Figure 7.16). All planes perpendicular to the direction specified by a normal vector are said to have the same orientation.

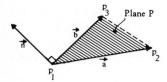

Figure 7.16 The normal vector to a plane.

The orientation in space of the plane faces of a polyhedron can be determined from information about the angle between pairs of its faces, the *dihedral angle*, θ_D. The angle between the pages of the open book shown in Figure 7.17(*a*) shows what we mean by the dihedral angle. The dihedral angle between two planes is the apparent angle between them when they are viewed in such a way that their line of intersection appears as a point (edge view or plane view) as shown in Figure 7.17(*b*). In other words, the dihedral angle is the angle between the traces of the planes on a cutting plane normal to their line of intersection. It is also the angle between the normal vectors to the planes if the vectors are threaded in the direction of the angle as Figure 7.17(*c*) shows.

The symmetry of the platonic solids makes the dihedral angle between any pair of bordering faces the same. We would like to compute the dihedral angles for the platonic solids. But first we will show that there is a relation between the dihedral angle, the angular deficit, and the vertex figure. The *vertex figure* of a given vertex is the polygon formed by all the vertices that are connected by an edge to the given vertex (see Figure 7.10). All the vertex figures of a platonic polyhedron are identical. In fact, the vertex figures of a platonic polyhedron are the faces of its dual.

A pattern of six equilateral triangles surrounds a vertex and lies flat in the plane. However, if we connect five equilateral triangles around a vertex, the pattern bulges out of the plane to form a three-dimensional cap as shown in Exercise 7.4. In the first case, the angular deficit of the vertex and the dihedral angle between the triangular faces are both zero. In the second case, the angular deficit equals 60 degrees, but the dihedral angle is not fixed since the cap is not rigid

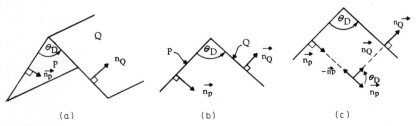

Figure 7.17 (*a*) The dihedral angle between two planes; (*b*), (*c*) edge view of the planes showing the dihedral angle.

(the vertices can be moved relative to each other). However, if we insist that the vertex figure be a planar regular pentagon, the cap is uniquely defined and is a typical segment of an icosahedron.

More generally, if any number of isosceles triangles surround a vertex, and if the corresponding vertex figure is a regular polygon, the dihedral angle between any two adjacent faces is

$$\cos \theta_D = \frac{2 \cos \beta + \cos \alpha - 1}{1 + \cos \alpha} \tag{7.10}$$

where α = the angle of the isosceles triangle that is incident to the vertex

β = the internal angle of the vertex figure (a regular planar polygon)

This formula can be proven from the definition of the dihedral angle and a little vector analysis. (Try it!)

The icosahedral cap has five equilateral triangles around a vertex and a regular pentagon vertex figure so that α is 60 degrees and β is 108 degrees. Replacing these values in Equation (7.10) gives a dihedral angle, θ_D = 138.18 degrees. Although the cube has three squares surrounding a vertex, these squares contain three isosceles triangles surrounding the vertex so that α = 90 degrees and β = 60 degrees, in which case Equation (7.10) yields θ_D = 90 degrees. Similarly, the three pentagons surrounding a vertex of a dodecahedron can be divided into three isosceles triangles. The dihedral angles for the platonic polyhedra are computed using Equation (7.10), and they are listed in Table 7.2.

7.11 Space-Filling Properties

A polyhedron is more useful for creating models of biological, chemical, or architectural forms if it can be combined with others to form a larger aggregate. For example, cubes stack to completely fill space. This explains why cubes or their close relatives, parallelopipeds, are

TABLE 7.2

Polyhedron	δ	α	β	θ_D
Tetrahedron	180	60	60	70.53
Octahedron	120	60	90	109.47
Cube	90	90	60	90.00
Dodecahedron	36	108	60	116.57
Icosahedron	60	60	108	138.18

ideal structures with which to subdivide the space inside or outside of a building.

Can other platonic polyhedra stack to fill space? Try as we may, only the cube among the platonic polyhedra can fill space by itself. In a *space-filling* array of polyhedra no gaps can remain and the edges and vertices of any polyhedron from the array must coincide with the edges and vertices of adjacent polyhedra, i.e., no vertex lies within an edge or face. Whether or not a set of polyhedra can fill space is determined by the following self-evident necessary condition:

The sum of the dihedral angles between all faces meeting at a common edge is 360 degrees, i.e.,

$$\sum_E \theta_D = 360 \tag{7.11}$$

where summation is over the set of edges.

Problem 7.3 Check the dihedral angles in Table 7.2 to see that, of all the platonic polyhedra, only an integral number of cubes are able to surround an edge to satisfy the criteria for space filling.

Although tetrahedra cannot fill space by themselves, they can combine with octahedra in a ratio of two tetrahedra for each octahedron to fill space. In fact, as we saw in Exercise 7.3, bracing each face of a cube with an edge deforms the cube into a parallelopiped shown in Figure 7.18 made up of two tetrahedra and one octahedron, which can stack as well as cubes. From Table 7.2 we see that the dihedral angles for the tetrahedron and octahedron are 70.54 and 109.46 degrees, respectively. Thus, Equation (7.11) requires two tetrahedra and two octahedra to surround each edge in a space-filling array.

In Chapter 10 we will see how tetrahedral and octahedral arrangements form the underlying structure of metallic crystals. Buckminster Fuller used the stackability of tetrahedra and octahedra to create a structural module called the *octet truss*, which is the basis of very rigid structures known as *spaceframes* shown in Figure 7.19 and discussed further in Chapter 10 [1975], [Edmondson, 1987].

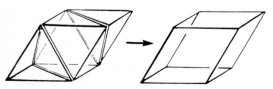

Figure 7.18 Two tetrahedra and one octahedron form a parallelopiped that fills space.

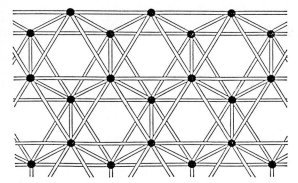

Figure 7.19 An octet truss.

7.12 Juxtapositions

There are three ways polyhedra can be joined one to another: *vertex to vertex, edge to edge,* and *face to face* (see Figure 7.20). If one polyhedron is held fixed in space, in the vertex-to-vertex arrangement the other is free to move in a sphere about the first; in the edge-to-edge juxtaposition the second polyhedron can move in a circle about the edge as axis; while in the face-to-face configuration no relative movement is possible. Fuller likened the freedom of the vertex-to-vertex configuration to the vapor state of molecules, edge to edge to the liquid state, and the most constrained face-to-face arrangement to the solid state. Although this analogy is metaphorical, many of the properties of actual molecules can be explained by these different kinds of bonding of molecules [Wells, 1956], [Pauling and Hayward, 1964].

Figure 7.21 shows a model of a compound with chemical formula of the type ABX_3, known as a *perovskite*, a compound capable of storing

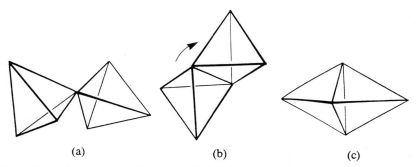

(a)   (b)   (c)

Figure 7.20 Three juxtapositions of polyhedra. (*a*) Vertex to vertex; (*b*) edge to edge; (*c*) face to face.

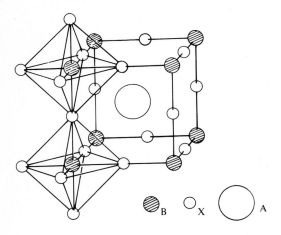

Figure 7.21 The basic structural unit of perovskites is a cube. One metallic atom A lies at the center, 8 smaller metallic atoms B occupy the corners, and 12 nonmetallic atoms X are the midpoints of the edges; the 6 X anions that surround each B cation form an octahedron; the basic structural model becomes a group of 8 corner-linked octahedra around an A cation.

electric energy and possessing other remarkable electrical properties such as high-temperature superconductivity. The three species of perovskite ions are arranged in one cell of a cubic lattice in Figure 7.21 with a large A ion at the cube centers, a smaller B ion at the vertices, and X ions in the center of the edges. It would appear at first that the formula for this compound should be AB_8X_{12}. However, since each vertex is shared by eight adjacent cubes and each edge is shared by four adjacent cubes, the actual formula must be ABX_3. Each B ion at the corners of the cube is surrounded by an octahedral configuration of six X ions at the midpoints of the edges incident to each of these vertices. This collection of octahedra can be connected vertex to vertex and is responsible for many of the remarkable electrical properties of perovskites [Hazen, 1988]. For example, if the A ion is slightly undersized, the octahedra respond to mechanical pressure by displacing themselves from their equilibrium positions, thus setting up an electric field capable of storing electric energy. Robert M. Hazen and his group at the Geophysical Laboratory of the Carnegie Institution of Washington has shown that a class of high-temperature superconducting materials is made up of structurally flawed perovskites. The structure of perovskite crystals will be discussed further in Section 10.7.3.

Construction 7.2 [Pugh, 1976]   The opposite edges of a regular tetrahedron are at right angles to each other. A *ring of tetrahedra* can be formed by joining the opposite edges of eight tetrahedra as shown in Figure 7.22. If the joints between the figures are flexible enough to allow each tetrahedron to rotate about its neighbors, the whole ring of tetrahedra can rotate as a smoke ring rotates in the air. A similar ring can be constructed from 16 octahedra joined edge to edge.

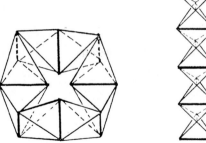

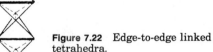

Figure 7.22   Edge-to-edge linked tetrahedra.

Construction 7.3 [Pugh, 1976] Tetrahedra can also be joined face to face to create a form which can be likened to a twisted column with triangular faces as shown in Figure 7.23(a). The edges of this arrangement follow helical lines, so the figure is referred to as a *tetrahelix*.

Besides the tetrahelix, elongated structures called *masts* can be built out of octahedra or icosahedra as shown in Figure 7.23(b) and (c). A seven-frequency octetmast is shown in Figure 7.23(d). The *frequency* is the number of units that combine to form the length. If each rod of the octet truss in Figure 7.19 is replaced by an octetmast, a truss of lower

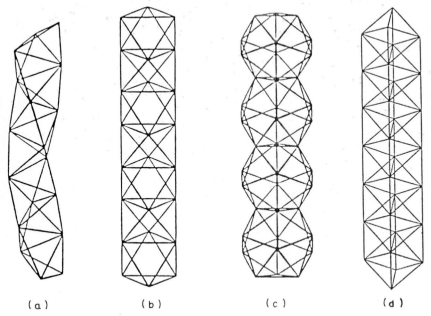

(a)   (b)   (c)   (d)

Figure 7.23   (a) Tetrahelix; (b) octamast; (c) icosamast; (d) octet-mast.

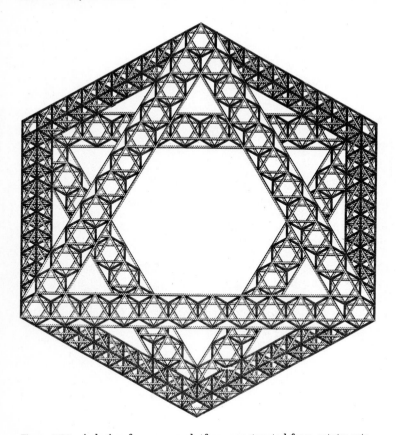

Figure 7.24 A design for a space platform constructed from octetmasts.

mass-to-volume ratio is produced. If each rod of the new truss is again replaced by an octetmast, the density can be reduced even further. Russell Chu [1986] has proposed that this octet truss expansion system be used as the structure of a large lightweight space station in the form of an octahedron (see Figure 7.24). Although each rod of the structure measures only 5 feet in length and 3 inches in diameter, a 14-frequency expansion followed by a 12-frequency expansion produces an expanded mast that is 840 feet in length by 70 feet in diameter.

7.13 Symmetry

Up to now we have vaguely referred to the "perfect symmetry" of the platonic solids. Now we become a little more precise. When we look at a model of a polyhedron, what we are most likely to notice at first are physical aspects of the model such as the positions of its vertices, the connectivity of its edges, and the shapes and orientations of its faces. But if we continue to examine the model for a while, turning it every which way, we note other more subtle attributes of the polyhedron known as its symmetries. When the model is oriented in special directions, the jumble of edges and vertices coalesces into a highly ordered pattern.

7.13.1 Rotational symmetry

Most people have an intuitive notion of symmetry; generally we recognize when a geometric pattern is or is not symmetric. However, we can formalize this concept to make it a little more precise. An object is said to have *rotational symmetry* if a rotation of the object about some axis results in precisely the same overall configuration of points, although these points may be in new positions. The polyhedron is said to be invariant under this rotation.

The cube is invariant under rotation about an axis through the center of two opposite faces under four rotations of 90, 180, 270, and 360 degrees. This is called a fourfold rotational symmetry of the cube and the axis is known as a *fourfold axis* of rotational symmetry and corresponds to a *face-on* view of the cube. Figure 7.25(*a*) shows the three fourfold axes of the cube which are responsible for a total of nine rotations in addition to the identity transformation which leaves the cube unchanged (or rotates it 360 degrees). The cube also has four *threefold axes* (responsible for eight rotations) through opposite vertices which result in a *vertex-on* view of the cube [see Figure 7.25(*b*)], and six *twofold axes* (responsible for six rotations) through the centers of opposite edges, as shown in Figure 7.25(*c*), resulting in an *edge-on* view. When all 13 rotational axes are placed in cube [see Figure 7.25 (*d*)], the cube has a total of 24 rotations (including the identity transformation) that leave its configuration invariant.

Because the cube possesses four-, three-, and twofold axes, it is said to have 4.3.2 symmetry. It turns out that all the platonic solids fall

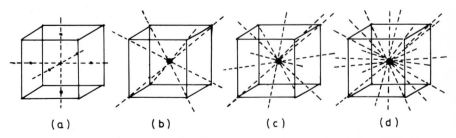

    (a)   (b)   (c)   (d)

Figure 7.25 (*a*) The three fourfold axes of rotational symmetry for a cube; (*b*) the four threefold axes; (*c*) the six twofold axes; (*d*) the combined 13 axes of rotational symmetry.

into three symmetry families 4.3.2, 5.3.2, and 3.3.2. Dual platonic polyhedra possess identical symmetries: the cube and octahedron are 4.3.2, the icosahedron and the dodecahedron are 5.3.2, and the tetrahedron shares the twofold and threefold axes of the other two systems. Table 7.3 lists all the rotational symmetries of the platonic polyhedra.

If a polyhedron has rotational symmetry with respect to some axis, its plane projection in the direction of this axis must have the same symmetry. Thus, by turning the polyhedron about and observing its planar profiles, rotational symmetries can be detected. For example, Figure 7.26 shows the profiles of a cube projected in the directions of its four-, three-, and twofold axes, i.e., the face, vertex, and edge views.

Problem 7.4   Locate the axes of symmetry of all the platonic polyhedra in this way and draw their planar profiles. You will see that this amounts to drawing face-on, vertex-on, and edge-on views of the platonic solids.

7.13.2   The principal directions of the cube and 4.3.2 symmetry

The system of 4.3.2 rotational symmetry is characterized by the three *principal directions* of the cube, namely, the edge direction E, face diagonal direction FD, and the body diagonal direction BD, shown in

TABLE 7.3

Polyhedron	Rotational axes				Planes of reflection
	Twofold	Threefold	Fourfold	Fivefold	
Tetrahedron	3	4	6		
Cube	6	4	3		9
Octahedron	6	4	3		9
Dodecahedron	15	10		6	15
Icosahedron	15	10		6	15

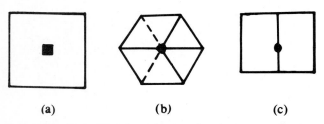

(a)   (b)   (c)

Figure 7.26   Projection of a cube onto the plane perpendicular to its axes of rotation. (*a*) Fourfold axis; (*b*) threefold axis; (*c*) twofold axis.

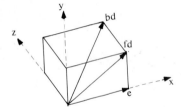

Figure 7.27   The principal directions of a cube.

Figure 7.27. Thus, the fourfold axes of a cube are in the E direction, the threefold axes are in the BD direction, and the twofold axes are in the FD directions. If the cube is placed into a cartesian coordinate system, the principal directions can be abbreviated by the points that their tips intercept when they are anchored to the origin. Thus, according to Figure 7.27, $E \leftrightarrow [1,0,0]$, $FD \leftrightarrow [1,1,0]$, and $BD \leftrightarrow [1,1,1]$.

Peter Pearce [1978] has designed a set of sticks and connectors, which he calls the universal node system, to exploit the relationship of 4.3.2 symmetry to the principal directions of a cube. In Pearce's system, edges are color- and shape-coded to match the 26 protrusions on the connectors in the directions of the 13 axes of rotational symmetry. With the universal node system any polyhedron from the 4.3.2 system can be built quickly yielding a graphic demonstration of its structure. Thus, the cube uses only E directions, whereas the tetrahedron and octahedron are constructed from FD directions. The system is of particular importance to crystallographers and architects since all space-filling polyhedra with equal edge lengths, as we shall see in Chapters 8 and 9, are members of the 4.3.2 symmetry class.

Steven Baer has designed a set of sticks and connectors to build polyhedra from the 5.3.2 system [1970]. The connectors in Baer's system are spheres punctured by the 31 axes of rotational symmetry of the 5.3.2 system. This system is turning out to be useful for studying quasicrystals (see Sections 6.10, 10.13, and 10.14).

Problem 7.5   Use the pythagorean theorem and trigonometry to find the angle between the following directions incident to a vertex of the cube: E and BD, E and FD, BD and FD, FD and FD, and BD and BD meeting at the center of the cube and called the *Miraldi angle* (see Section 8.9).

7.13.3   Reflection symmetry

An object is said to have reflection, or mirror, symmetry if half of the object reflects to the other half in a mirror which lies on the *plane of reflection*. Likewise, if a two-way mirror is placed on a plane of reflection symmetry, the points of the object on one side of the mirror reflect to the points on the other side, leaving the original configuration invariant, although points are interchanged across the mirror.

The cube has three planes of reflection symmetry each parallel to a pair of faces. In addition, six planes of reflection symmetry slice through opposite edges and are perpendicular to the two faces that do not contain these two edges. Find and determine the number of planes of reflection for the other platonic polyhedra (see Table 7.3).

We saw that axes of rotational symmetry of an object can be found by physically manipulating the object. On the other hand, it is difficult to detect planes of reflection symmetry physically since it is not easy to insert mirrors into an object and physically observe the reflections; therefore reflection symmetry is sometimes called a *nonperformable symmetry*. However, the existence of at least one plane of reflection symmetry can be detected by placing the object before a mirror so that its suspected mirror plane is perpendicular to the mirror. If the object could be physically moved behind the mirror and imagined to match up point for point with its image, the suspected plane is indeed a plane of reflection symmetry. For example, humans and other land animals have an approximate plane of mirror symmetry, on the exterior of their bodies, separating left from right (but not up from down).

Problem 7.6   Place the following objects before a mirror and detect mirror symmetry if it exists: a cube, cone, tetrahedron, spiral, and glove.

Conversely, if an object does not possess reflection symmetry, its mirror image is distinctly different from the object and the two cannot be matched up through a movement of the object in three-dimensional space. We generally distinguish such objects as being left or right handed. For example, we talk of a left and right hand or a left- and right-handed spiral or molecular arrangement. The assignment of left and right, while arbitrary since it depends on the viewer's perspective, is generally established to everyone's agreement according to some convention. However, the arbitrariness of the convention means that there is no way to convey our meaning of left and right to someone located in a remote corner of the universe. More will be said about this in Section 11.9.

7.13.4   Orthoschemes and the dihedral kaleidoscope

Although rotation and reflection symmetry of platonic polyhedra appear to be entirely different, there is a profound connection between them. This is demonstrated for a cube.

Circumscribe a sphere around a cube and cut the sphere and cube by

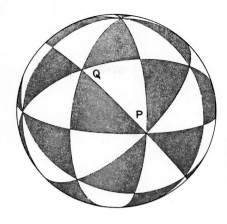

Figure 7.28 Great circles related to the symmetry group 4.3.2. Drawn by Patrick Du Val for his book *Homographies, Quaternions* and *Rotations*.

the nine planes of reflection symmetry of the cube [Coxeter, 1988]. This divides the surface of the sphere into 24 spherical triangles and their mirror images. Half of these triangles are colored grey in Figure 7.28 to distinguish them from their mirror images which are colored white. Each of these triangles has one right angle. All the vertices at point P are 45 degrees, and P along with its antipode comprise a fourfold axis of rotation. The angles at Q are 60 degrees, and Q is one of the axes of threefold rotation. The twofold axes are at the position of the right angles and their antipodes. The six points equivalent to P comprise the vertices of an octahedron while the eight points equivalent to Q are the vertices of a cube. Each of the 24 rotational symmetries of the cube transforms a grey triangle into one of the other 23 grey triangles or to itself in the case of the identity. The same goes for the white triangles. A grey triangle can be transformed to a white triangle by either a single reflection or a rotation followed by a reflection and there are 24 of these transformations. We will return to this sphere in Section 9.9 to see how it can be used to generate other polyhedra with cubic symmetry.

A similar construction can be carried out for the other platonic polyhedra $\{p,q\}$. The planes of reflection decompose these polyhedra into oppositely congruent tetrahedra called *orthoschemes*, first conceived of by Ludwig Schläfli [Williams, 1972]. In this decomposition, the four faces of these tetrahedra are right triangles, and the lengths of the three edges meeting at the polyhedron center are radii of the circumsphere, insphere, and intersphere as illustrated for the cube in Figure 7.29.

Schläfli showed that the angles and radii of this orthoscheme shown in Figure 7.29(*b*) can be expressed in terms of p and q as follows:

$$\cos \phi = \frac{O_1 O_3}{O_0 O_3} = \frac{1^R}{O^R} = \cos \frac{\pi}{p} \csc \frac{\pi}{q} \qquad O^R = e \sin \frac{\pi}{q} \csc \frac{\pi}{h}$$

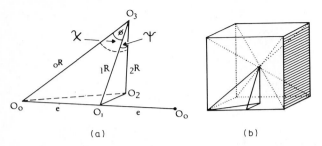

Figure 7.29   The orthoscheme for the cube.

$$\cos \psi = \frac{O_2 O_3}{O_1 O_3} = \frac{2^R}{1^R} = \csc \frac{\pi}{p} \cos \frac{\pi}{q} \qquad 1^R = e \cos \frac{\pi}{p} \csc \frac{\pi}{h}$$

$$\cos \chi = \frac{O_2 O_3}{O_0 O_3} = \frac{2^R}{0^R} = \cot \frac{\pi}{p} \cot \frac{\pi}{q} \qquad 2^R = e \cot \frac{\pi}{p} \cos \frac{\pi}{q} \csc \frac{\pi}{h} \qquad (7.12)$$

where O^R represents the radius of the circumsphere, 2^R is the radius of the insphere, and 1^R is the radius of the intersphere, e is the semiedge length of the platonic polyhedron, and h is the number of lengths into which a great circle is divided by an edge.

In this decomposition, an orthoscheme and its mirror image border each edge of the polyhedron. Thus, the cube is decomposed into 24 right- and 24 left-handed orthoschemes.

Construction 7.4   Construct a large tetrahedron of the same shape as the orthoscheme of a cube out of reflecting surfaces. Only the three faces of the orthoscheme that meet at the center of the platonic polyhedron are needed; one side is open. In Figure 7.30 a small cardboard model of an orthoscheme has been placed into this tetrahedral chamber of mirrors called a *dihedral kaleidoscope*. Observe that the missing 47 orthoschemes appear in the mirrors and reassemble the cube.

The orthoscheme can also be used as a unit of structure. For example, Kapusta's K-dron (see Section 5.10.2) can be constructed from 12 left- and right-handed pairs of orthoschemes of a cube. This may account for the striking optical properties that the K-dron possesses and explain why K-dron structures exhibit such a strong relationship between form and function.

Construction 7.5   Construct a set of orthoschemes and their mirror images and use them as building blocks to construct interesting designs, or buy one of the Rhombics, Inc. kits of prefabricated orthoschemes.

7.14   Star Polyhedra

In Section 5.2.2, the edges of a regular polygon were extended, and for a polygon with five or more sides, this extension enclosed additional

Figure 7.30 Dihedral kaleidoscope based on the symmetry of the cube.

regions of the plane forming a star polygon. If we now try extending the face planes of a platonic solid, no new regions are defined for the tetrahedron and cube; however, the face planes of the octahedron enclose eight additional tetrahedra as we can see by looking again at Figure 7.9. The faces of this *star polyhedron* are the large equilateral triangles of the interpenetrating tetrahedra, one of which is labeled *ABC*. The vertices of this star polyhedron are the eight apexes of the tetrahedra. The points at which the eight large equilateral triangle faces self-intersect are not considered to be vertices of the star polyhedron.

Extending the face planes of the dodecahedron leads to three distinct types of cells inside the intersecting planes and three stellated forms, two of which were discovered by Kepler (1619) and the other by Poinsot (1809). These three star polyhedra and one additional one derived from the icosahedron and also discovered by Poinsot have all the properties of the platonic solids, namely, each face is a regular polygon (or star polygon) and each vertex is surrounded identically. They differ from the platonic solids in that their graphical structures cannot be deformed to a graph on a sphere (i.e., they are not convex).

Wenninger gives details on how to construct beautiful models of star polyhedra [1971]. Models of two of the platonic star polyhedra, the small stellated dodecahedron {5/2,5} and the great dodecahedron {5,5/2}, are shown in Figures 7.31 and 7.32. The first number in the Schläfli symbol stands for the kind of polygon face ({5/2} is a star pentagon face as we saw in Section 5.2.2) while the second number stands

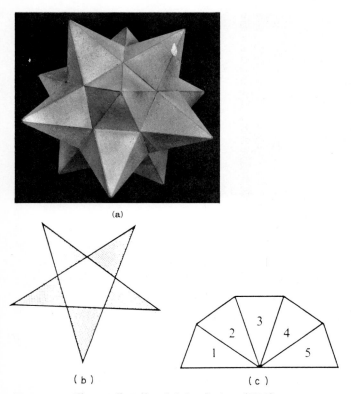

(a)

(b)   (c)

Figure 7.31   The small stellated dodecahedron {5/2,5}.

for the kind of polygon that makes up the vertex figure. From their Schläfli symbols, it is not surprising that these two polyhedra are duals in the sense described in Section 7.6.1.

Notice in Figure 7.31 that {5/2,5} can be assembled by gluing together identical golden triangles of type 1 (see Section 3.5) derived from the shaded portion of the accompanying star pentagram. Five isosceles triangles are glued together to form a pentagonal pyramid, and 12 pyramids are then glued to the faces of a pentagonal dodecahedron to form the star polyhedron. In Figure 7.32, 20 trihedral dimples, constructed from golden triangles of type 2 shaded in the accompanying star pentagram, are cemented together to create {5,5/2} whose faces are 12 interpenetrating pentagons with pentagonal stars embossed on them.

Construction 7.6   It is a good exercise in visual thinking to construct models of the *Kepler-Poinsot* polyhedra and observe the variety of their appearances as they are viewed from different angles.

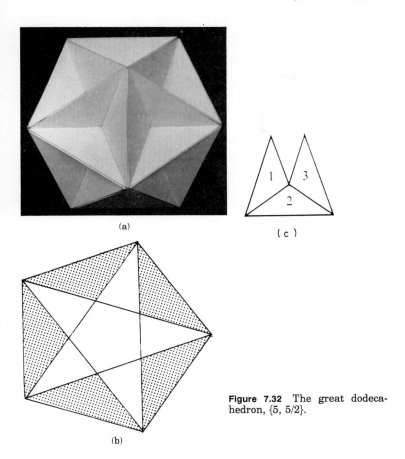

(a)

(c)

(b)

Figure 7.32 The great dodeca-hedron, {5, 5/2}.

Appendix 7.A. Duals

It is a bit surprising that the concept of duality, which is so well-defined for maps and so natural for the platonic solids, proves to be elusive for polyhedra in general [Grünbaum and Shepherd, 1988]. However, if we restrict ourselves to convex polyhedra, there is a natural way to define a dual. But first we must define what is meant by the pole and polar to a circle and a sphere.

In Appendix 2.B we defined two points P and Q to be inverse with respect to a circle with center at O and radius r if they satisfy the relation

$$OP \cdot OQ = r^2$$

However, if we rewrite this equation as

$$\frac{OP}{r} = \frac{r}{OQ}$$

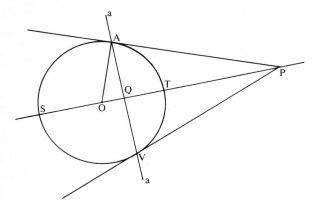

Figure 7.A.1   Pole and polar of a circle.

then by referring to Figure 7.A.1, we see that triangle OAQ ~ triangle OAP where AP is a tangent line to the circle from P. But, since ∢ OAP = 90 degrees (why?), it follows by similarity that ∢ OQA = 90 degrees and consequently AQ is perpendicular to OP. Line AQ is called the *polar* of *pole P* with respect to the circle. The polar to a sphere is defined analogously. It is the plane through Q perpendicular to the line connecting point P with the center of the sphere.

We now define a polyhedron dual to a given convex polyhedron. Take any point within the polyhedron as the center of a sphere of arbitrary radius. Let the vertices of the polyhedron be poles with respect to this sphere. The dual is defined to be the envelope formed by the planes polar to the vertices. Note that the dual polyhedron is not uniquely defined in the euclidean sense of length and angle. If we restrict ourselves to symmetric polyhedra that have circumscribing spheres, such as the platonic solids, the poles can be taken to be the vertices intercepted by the circumsphere while the polars are tangent planes to this sphere (the inscribed sphere of the dual). The envelope of these tangent planes constitutes the dual polyhedron. In Section 9.7 duals are described with respect to yet another sphere, the intersphere.

Appendix 7.B   A Proof of Descartes Formula

The following proof of Descartes' formula is by Alan Stewart [1986]. For the polygons around a vertex of a two- or three-dimensional map, each polygon with face valence p shares p vertices. We may take the contribution of each polygon at a vertex as $1/p$. That is, since when we count over all vertices we count the same face p times, we take $1/p$ of it per vertex in order to get one face when all are added up. Summing over all faces per vertex and all vertices,

$$F = \sum_V \left(\sum_F \frac{1}{p} \right) \text{ faces} \tag{7.B.1}$$

where Σ_F is the sum over each polygon face incident to vertex V and Σ_V is the sum over all vertices.

Using similar arguments, or swapping $F \leftrightarrow V, p \leftrightarrow q$ (see Section 7.6.1), gives

$$V = \sum_F \left(\sum_V \frac{1}{q} \right) \text{ vertices} \tag{7.B.2}$$

Now assume that all polygons are regular, i.e., all internal angles are equal. (Descartes' formula can be proved without this restriction, but the proof is more technical.)

For any polygon, the sum of the external angles equals 360 degrees. Therefore, if the polygon is regular and has p angles (and edges), the internal angle is, according to Equation (7.9),

$$\theta = 180 - \frac{360}{p} \tag{7.B.3}$$

If there are q polygons surrounding a vertex V,

$$\delta = 360 - \sum_F \theta = 360 - \sum_F \left(180 - \frac{360}{p} \right)$$

$$\delta = 360 - 180q + 360 \sum_F \frac{1}{p}$$

where, as before, $\Sigma_F \, 1/p$ is the sum of $1/p$ for each of the faces incident to vertex V, or

$$\delta = 360 \left(1 + \sum_F \frac{1}{p} - \frac{q}{2} \right) \tag{7.B.4}$$

The total angular deficit over all vertices is

$$\sum_V \delta = 360 \sum_V \left(1 + \sum_F \frac{1}{p} - \frac{q}{2} \right)$$

or

$$\sum_V \delta = 360 \left[V + \sum_V \left(\sum_F \frac{1}{p} \right) \right] - \frac{1}{2} \sum_V q \tag{7.B.5}$$

But using Equations (7.B.1) and (7.1), the last equation becomes

$$\sum_V \delta = 360(V + F - E)$$

or

$$\sum_V \delta = 360\chi$$

where χ is Euler's number. On a sphere, $\chi = 2$; therefore $\Sigma_V\delta = 720$ (sphere). On a torus, $\chi = 0$; therefore $\Sigma_V\delta = 0$ (torus).

8

Transformations of the Platonic Solids I

On the platonic solids: Their combinations with themselves and with each other give rise to endless complexities, which anyone who is to give a likely account of reality must survey.   PLATO
Timaeus

8.1   Introduction

The platonic solids are fascinating structures in their own right. We can think of them as a group of primitive structures capable of generating an unlimited variety of other shapes, much as the primary colors form the base of other colors. In this chapter we explore some of the ways that the platonic polyhedra relate to each other. The next chapter is devoted to the ways in which platonic solids can be transformed to other classes of polyhedra. To get a better sense of the transformability of the platonic solids try the following exercises [Laycock, 1989]:

Exercise 8.1   From six 20-centimeter straws assemble a tetrahedron with hairpins. Then connect the midpoints with twelve 10-centimeter straws (preferably of a different color) as shown in Figure 8.1. What did you create?

Exercise 8.2   From twelve 20-centimeter straws assemble an octahedron. Connect the midpoints with twenty-four 10-centimeter straws (of a different color) as shown in Figure 8.2. Describe the new polyhedron that results.

Exercise 8.3   From thirty 20-centimeter straws assemble an icosahedron. Connect the midpoints with sixty 10-centimeter straws as shown in Figure 8.3. Describe the new polyhedron that results.

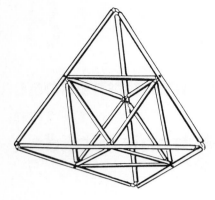

Figure 8.1 The edge centers of a tetrahedron are joined to form an octahedron.

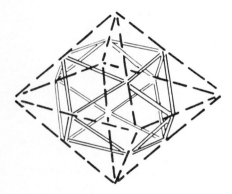

Figure 8.2 The edge centers of an octahedron are joined to form a cuboctahedron.

Figure 8.3 The edge centers of an icosahedron are joined to form an icosidodecahedron.

8.2 Intermediate Polyhedra

In the above exercises, the three rigid platonic solids—the tetrahedron, octahedron, and icosahedron—were used as frames to create new polyhedra. In the first exercise you discovered something of fundamental importance about the tetrahedron: It can be subdivided into an octahedron surrounded by four tetrahedra whose edges are half the length of the parent.

The second exercise resulted in a polyhedron with six square faces parallel (but with different orientations) to the six square faces of the cube and eight triangular faces parallel to the triangular faces of the octahedron. This amalgam of cube and octahedron, shown in Figure 8.4, is called a *cuboctahedron*. It shares a property with the platonic solids, namely, each vertex is surrounded by an identical collection of regular polygons; however, the polygons are not all the same. For the cuboctahedron, each vertex is surrounded by the sequence triangle, square, triangle, square, symbolized by the Schläfli notation, 3.4.3.4. Such polyhedra are called *semiregular*, or *archimedean*, in analogy to the semiregular tilings of the plane. (The next chapter will study the wider class of archimedean polyhedra.) The cuboctahedron is also referred to as an *intermediate polyhedron* since it is midway between a cube and an octahedron as we shall soon see. Although cuboctahedra do not stack to fill space by themselves, since they are formed by removing ⅛ octahedra from the vertices of a cube, they leave octahedral gaps and therefore fill space in a 1:1 ratio along with octahedra as shown in the polyhedral sculpture of Figure 8.5 [Loeb, 1985]. The cuboctahedron was fundamental to the world system of Buckminster Fuller [1975], [Edmondson, 1987].

The result of the third exercise is another semiregular polyhedron called the *icosidodecahedron*, shown in Figure 8.6. This polyhedron has 12 pentagon faces parallel to the faces of the dodecahedron and 20 triangle faces parallel to the faces of the icosahedron surrounding each vertex in the sequence, 3.5.3.5. It is another intermediate polyhedron.

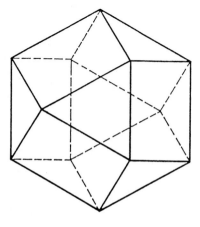

Figure 8.4 A cuboctahedron.

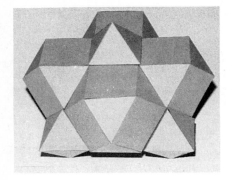

Figure 8.5 A polyhedral sculpture with space filling octahedra and cuboctohedra.

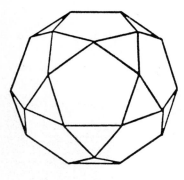

Figure 8.6 (*a*) An icosidodecahedron.

Construction 8.1 William Varney has extended his movable tilings of the plane (see Section 5.10.1) to the creation of polyhedra with movable faces. For example, Figure 8.7(*a*) shows an icosidodecahedron with its triangle and pentagon faces hinged together in such a way that the faces can move apart. In their most extreme position an open space in the form of a square appears at each vertex to form a polyhedron called the *small rhombicosidodecahedron*, one of the 13 pos-

Figure 8.7 (*a*) An icosidodecahedron with movable hinged faces opened to its extreme position of the small rhombicosidodecahedron (*courtesy of William Varney*).

sible semiregular polyhedra that will be discussed in Section 9.2. The trick to creating such a movable polyhedron lies in Dennis Dreher's hinge design. The hinge has the dihedral angle of the icosidodecahedron built into it. A similar movable polyhedron can be constructed for the cuboctahedron which produces a small rhombicuboctahedron (see Section 9.2) in its most extreme position. A construction kit is available from Tensegrity Systems [1990].

Exercise 8.4 To get a first-hand look at the internal structure of the cuboctahedron, construct one from marshmallows and toothpicks. Just keep surrounding vertices by the 3.4.3.4 pattern until the polyhedron closes up. Place one additional marshmallow at the center and connect it to each of the 12 vertices by additional toothpicks [see Figure 8.8].

Notice how the 12 radial toothpicks form equilateral triangles with the edges of the cuboctahedron and divide the cuboctahedron into eight tetrahedra and six half-octahedra (corresponding to the triangle and square faces).

Each radial toothpick is aligned with another in the opposite direction; the radial toothpicks form four groups of six with each group lying in one of four regular hexagons also defined by the edges shown in Figure 8.8. These four planes are parallel to the faces of a tetrahedron and the six pairs of radial toothpicks are parallel to the edges of a tetrahedron. For this reason Fuller said that space is intrinsically related to the tetrahedron. We shall say more about this in Section 10.3 and in the next section.

8.3 Interpenetrating Duals Revisited

Let's examine the interpenetrating duals of Section 7.6.2 more closely. Understanding this section will be easier if you have constructed a set of these structures to illustrate the points we shall discuss. Besides being visually attractive, they are wonderfully subtle.

The two interpenetrating tetrahedra, shown in Figure 8.9, form a stellated polyhedron called the *stella octangula*. Its surface is made up of eight pyramids in the form of regular tetrahedra. If these pyramids

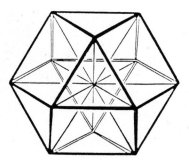

Figure 8.8 The cuboctahedron with vertices joined to its center and made up of four intersecting hexagons.

are removed, the octahedron of Exercise 8.1 remains. Since the eight vertices of the stella octangula lie above the faces of an octahedron, they must be vertices of its dual, a cube. In fact the six perpendicularly intersecting pairs of edges from these duals are diagonals of the faces of this cube. Finally, three square belts along the base of the pyramids envelop this polyhedron. These belts correspond to the three geodesics of the sphere upon which the octahedron projects [see Figure 8.10(a)]. It should be noted that, as for the star polyhedra of Section 7.14, the faces of the stella octangula self-intersect and carry us beyond our strict definition of a polyhedron.

In a similar fashion, the interpenetrating octahedron-cube pair, shown in Figure 8.9(b), forms another stellated polyhedron. Remove its eight triangular pyramids and six square pyramids and the cuboctahedron of Exercise 8.2 remains. From a model of the interpentrating pair it is evident that the cuboctahedron is made up of four belts of hexagons, seen in Exercise 8.4, that project to the four great circle geodesics on its circumscribing sphere [see Figure 8.10(b)]. Perpendicularly bisecting pairs of edges on the interpenetrating duals are the diagonals of a set of rhombic-shaped faces. The diagonals of these rhombuses are proportioned as $\sqrt{2}$:1 and are the faces of the polyhedron dual to the cuboctahedron (in a sense described in Section 9.6) called the *rhombic dodecahedron* (RD), which can be seen by looking ahead to Figures 8.12 and 8.13. The rhombic dodechedron is symbolized by $V(3.4.3.4)$ since its faces are related to the vertex figures of 3.4.3.4 as we will show in Section 9.7.

The third interpenetrating dual, shown in Figure 8.9(c), is formed by the icosahedron-dodecahedron pair. Remove its 20 triangular pyramids and 12 pentagonal pyramids and what remains is the

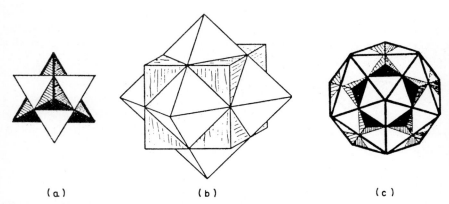

(a)   (b)   (c)

Figure 8.9 Interpenetrating duals. (a) Tetrahedra; (b) cube and octahedron; (c) icosahedron and dodecahedron.

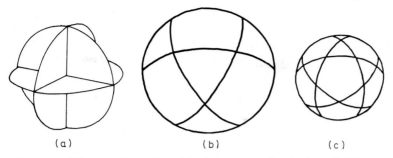

Figure 8.10   Edges on the surface of the interpenetrating duals project to geodesics on a sphere. (*a*) Octahedron; (*b*) cuboctahedron; (*c*) icosidodecahedron.

icosidodecahedron of Exercise 8.3. The icosidodecahedron is made up of six belts of decagons that project to six great circle geodesics on its circumscribing sphere, shown in Figure 8.10(*c*). Perpendicularly bisecting edges are in the ratio of φ:1 and are the diagonals of the set of rhombic-shaped faces that make up the *rhombic triacontahedron* V (3.5.3.5) dual to the icosidodecahedron which can be seen by looking ahead to Figures 10.29 and 10.33(*b*). H. F. Verheyen has shown that this golden rhombus is the very rhombus formed by two triangular faces of the pyramid of Cheops (see Section 3.2). This is an interesting result since it relates the Great Pyramid directly to the icosohedral system [Verheyen, 1989].

Construction 8.2 [Edmondson, 1987]   The four geodesics of the cuboctahedron can be constructed by an ingenious method of Fuller's. Rather than gluing four circles together directly (very difficult), Fuller constructs four bow ties from each of four circles as follows: sharply fold a 6-inch-diameter circle in half and then in thirds [see Figure 8.11(*a*)]. If the folds have been made properly, when the circle is unfolded [see Figure 8.11(*b*)] there will be one mountain fold toward you and two valley folds away from you. By bringing point *a* to point *b* you create the bow ties in Figure 8.11(*c*) which can then be connected with a hairpin to form the geodesics, shown in Figure 8.11(*d*). A surprising spatial cooperation must take place between length and angle to enable this construction to work.

Six circles folded in half with each half folded in fifths create the geodesics of the icosidodecahedron. The details of this construction are left to the curious reader by way of the above reference. This construction is basic to the "flow of energy" in Fuller's "world system" (see Section 8.11.3).

8.4   The Rhombic Dodecahedron

The rhombic dodecahedron is the dual of the cuboctahedron, and it fills space by itself (see Section 10.7). It is shown in Figure 8.12(*b*) to be the augmentation of an octahedron by eight triangular pyramids, each of which is one of the four congruent sectors of a tetrahedron as shown in Figure 8.11(a). The RD can also be obtained by stellating a cube with each of its six congruent sectors as shown in Figure 8.12. This stellation also shows why the RD is able to stack to fill space. As you can see, the RD reallocates the space within a space-filling stack of cubes [Loeb, 1976].

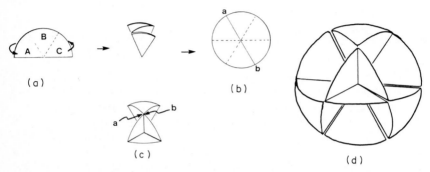

Figure 8.11   (a, b, c) Folding up a bowtie from a circle; (d) four bowties create the four great circles of the cuboctahedron.

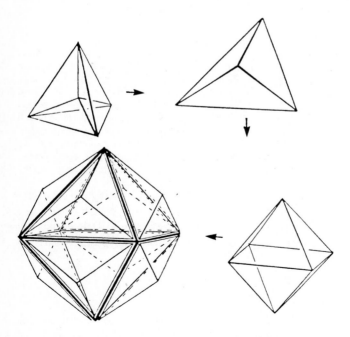

Figure 8.12   Eight quartants (1/4 sectors) of a tetrahedron added to the faces of an octahedron form a rhombic dodecahedron.

Construction 8.3   Generate the RD by stellating an octahedron and a cube as follows:

1. Construct six square pyramids hinged together in the manner shown in Figure 8.14(a). When they are folded inward, they form a cube; when they are placed around a cube with square faces joined to square faces, an RD is formed.

2. Construct eight triangular pyramids of the kind shown in Figure 8.12 and hinge four of them together as shown in Figure 8.14(b). When they are folded inward, they form a tetrahedron. When they are combined with four identical hinged triangular pyramids and placed around the sides of an octahedron, another RD is formed.

Detailed instructions for sizing and hinging the pyramids can be found in an article by Arthur Loeb and Jack Gray that appears in *Shaping Space* [1988b].

8.5   Embeddings Based on Symmetry

The many family relationships between the platonic polyhedra come from their shared symmetries (see Section 7.13). Alan Holden, in

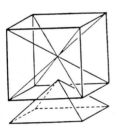
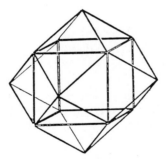

Figure 8.13   Six sextants (1/6 sectors) of a cube added to the faces of a cube form a rhombic dodecahedron.

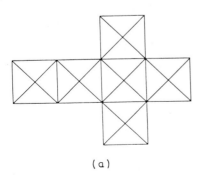
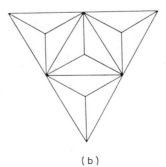

(a)   (b)

Figure 8.14   Patterns for adding pyramids to a (a) cube and (b) octahedron to construct a rhombic dodecahedron.

Shapes, Space, and Symmetry [1971], shows lovely designs that result from taking advantage of these shared symmetries. We present two examples from his book:

1. A tetrahedron can be placed in a cube so that the four threefold axes of the cube and tetrahedron coincide. The tetrahedron can be expanded until its corners fall on four of the cube's eight corners and its edges are embedded in the faces of the cube [see Figure 8.15(*a*)]. The duality of cube and octahedron suggests how to inscribe the tetrahedron correspondingly in an octahedron. The corners of the tetrahedron now fall on the faces of the octahedron [see Figure 8.15(*b*)].

2. A tetrahedron can be inscribed in a dodecahedron by following the same principle: the four threefold axes of the tetrahedron can be aligned with 4 of the 10 threefold axes of the dodecahedron. The vertices of the tetrahedron are then expanded out until they fall on four vertices of the dodecahedron (see Figure 8.16). The three twofold axes of the tetrahedron are then automatically aligned with three of the twofold axes of the dodecahedron. In this case the edges of the tetrahedron do not lie on the faces of the dodecahedron.

There is another important difference between a tetrahedron in a cube and a tetrahedron in a dodecahedron. The six reflection planes of an inscribed tetrahedron coincide with six of the cube's reflection planes, but they do not coincide with any of the reflection planes of the inscribing dodecahedron. Hence the combination of tetrahedron and dodecahedron has no surviving reflection planes. Since the combination lacks nonperformable symmetries, it can appear in two essentially different forms, right handed and left handed, each the mirror

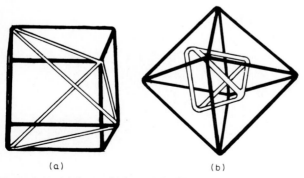

(a)   (b)

Figure 8.15   (*a*) A tetrahedron embedded in a cube; (*b*) a tetrahedron embedded in an octahedron.

Figure 8.16 A tetrahedron embedded in a dodecahedron.

image of the other. (Analogously, the principle of duality can be used to inscribe a tetrahedron in an icosahedron.)

Construction 8.4 [Holden, 1971]. The inscription of a tetrahedron in a dodecahedron suggests a way of compounding five tetrahedra in a symmetrical fashion. After a single tetrahedron has been inscribed, four more appear when the assembly is turned about a fivefold rotation axis of the dodecahedron. The 20 (5 times 4) corners of the tetrahedra therefore occupy the 20 corners of the dodecahedron. Since there are two different ways to inscribe five tetrahedra in a dodecahedron, providing a left-handed and right-handed inscription, turning the assembly generates either a left-handed or right-handed compound [see Figure 8.17(a)], each a mirror image of the other. Each of the compounds has all the

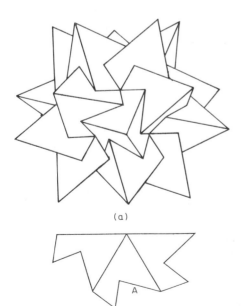

(a)

(b)

Figure 8.17 (a) Five tetrahedrons embedded in a dodecahedron lead to this figure. Since it has no plane of mirror symmetry, it has an enantiomorphic copy. (b) Paper-folding pattern.

rotational axes of a dodecahedron but none of its reflection planes. In Figure 8.17(b) we give the fundamental pattern from which a model of this complex figure can be constructed.

8.6   Designs Based on Symmetry Breaking

The constructions of the previous section were based on the symmetries in common to pairs of platonic polyhedra. Designs can also be created by destroying symmetry. As Holden describes,

> Truncating only one corner of a cube leaves one of its threefold axes of rotation unmolested but destroys the other threefold axes. All fourfold and all twofold axes disappear. Only three of the nine planes of reflection remain. Truncating the opposite corner as well restores the three twofold axes that are perpendicular to the surviving threefold axis [see Figure 8.18].
>
> When four nonadjacent corners of a cube are truncated, the remaining symmetry is that of a regular tetrahedron, and it is interesting to see how the truncation has degraded the cubic into the tetrahedral symmetry. The cube's three fourfold axes have degenerated into the tetrahedron's three twofold axes and the cube's twofold axes have disappeared. The four threefold axes are still untouched. The six reflection planes

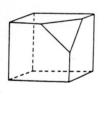

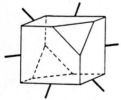

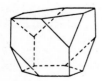

Figure 8.18   Symmetry breaking for a tetrahedron.

'through opposite edges of the cube remain, but the three planes parallel to the cube's faces have been destroyed.

Problem 8.1 [Holden, 1971]   Each of the solids pictured in Figure 8.19(a) can be made by trimming some of the edges of a cube. Can you visualize which edges have been truncated to produce each solid? The eight solids are symbolized in

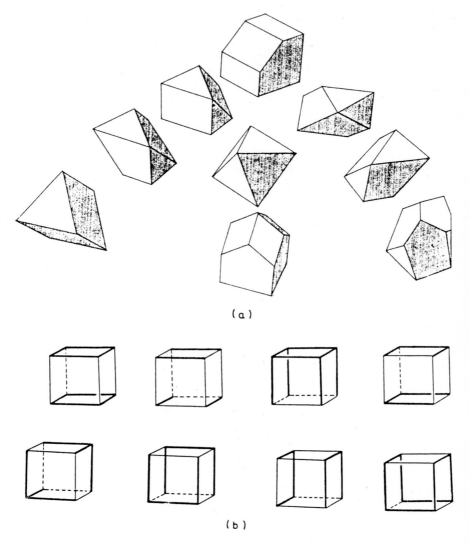

(a)

(b)

Figure 8.19   (a) Nine edges of a cube (two are mirror images); (b) indications of the truncated edges.

Figure 8.19(*b*) by outlined cubes with heavy lines along the trimmed edges. When you have paired off the illustrated solids with these cubes, you can go on to specify the symmetry of each solid. Of course, if you were to build the solids, you would find that handling them is the best way to understand them. Notice that in two of these solids, which look much alike, all planes of reflection have been destroyed: they are mirror images of each other.

8.7   Relation to the Golden Mean

At this point in our study of polyhedra we should begin to wonder what lies at the basis of the platonic solids that enables them to unfold in such a variety of ways. We have already seen that symmetry plays a role. The golden mean has also made its entry at several key points. Perhaps the most striking example was given in Section 3.4 where three mutually perpendicular golden rectangles were shown to span the vertices of an icosahedron. The crucial role of the golden mean to the platonic solids was clear to Euclid, who devoted a good part of Book XIII of the *Elements* to describing it.

The golden ratio of diagonal to side of a pentagon is shown in Figure 8.20(*a*) where a cube is embedded in a dodecahedron. Another example is shown in Figure 8.20(*b*) where the edges of an octahedron are divided in the golden section. When division points are joined, an icosahedron results. Since these two embeddings are dual to each other, it is not surprising that while the icosahedron and the octahedron share common facial planes, the dodecahedron and the cube share common vertices.

In Section 7.14 two of the Kepler-Poinsot star polyhedra were shown to be constructed from identical golden mean triangles (see Section

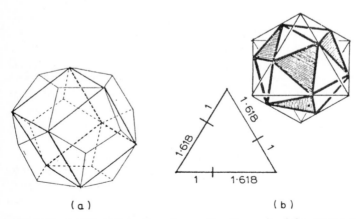

(a)   (b)

Figure 8.20   (*a*) The edges of a cube are the diagonals of the pentagon faces of the dodechedron; (*b*) the vertices of an icosahedron are the golden section of the edges of an octahedron.

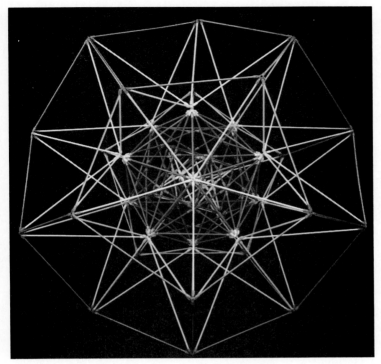

Figure 8.21 Dan Winter's star crystal with platonic solids embedded within each other.

3.5). In fact if we extend the edges of a dodecahedron, they meet in the 12 vertices of an icosahedron above the center of each face. It turns out that, in this way, the dodecahedron is stellated to the *small stellated dodecahedron* {5/2,5} (see Figure 7.31) by 12 pentagonal pyramids. Each pyramid is made up of five golden triangles. Now take the resulting icosahedron and extend its edges until they meet at the 20 vertices of another dodecahedron. Each of the resulting pyramids is again made up of three golden triangles and forms the *great stellated dodecahedron* {5/2,3}. The process can of course be repeated indefinitely.

Figure 8.21 illustrates a construction called a *star crystal* by its creator, Dan Winter [1990]. The golden mean stellation of dodecahedron to icosahedron to dodecahedron is evident in the figure. The central region of the "crystal" is a nested sequence of platonic polyhedra beginning with an octahedron "seed" surrounded in order by a tetrahedron, cube, and dodecahedron. A kit for building the star crystal is available from Winter [1990].

Figure 8.22   *Easy Landing*, a tensegrity structure by Kenneth Snelson.

8.8 Tensegrities

Polyhedral forms have inspired those who have studied them to create an incredible variety of structures that capture various aspects of their forms in unusual ways. A good example of this are tensegrities, or structures of tensional integrity [Pugh, 1976], [Kenner, 1976], [Minke, 1971], which were first conceived of by the sculptor Kenneth Snelson and then popularized by Fuller, his teacher. They are structures composed of a combination of struts under compression and ties under tension as Snelson's sculpture *Easy Landing*, which stands in Baltimore Harbor, shows (see Figure 8.22). They can be thought of as discrete analogues of balloons in which air, under pressure within the balloon, is balanced by the skin of the balloon under tension. The balance between tension and compression results in light, airy structures like Snelson's *Needle Tower*, which adorns the garden of the Hirschorn Museum in Washington, D.C., and is shown in Figure 8.23, in contrast to the bulky structures that result when structural elements are primarily under compression as they are in brick buildings.

Perhaps the first tensegrity structures were Egyptian seagoing vessels dating from 2500 B.C., on which stout rope was passed over the top of a series of vertical struts, its two ends being looped under the ends of the ship so as to prevent them from drooping [Gordon, 1978]. To some extent, our own bodies can also be thought of as tensegrities in which the vertebrae are under compression balanced by the tendons [Thompson, 1966].

8.8.1 Tensegrity models

A tensegrity model of an icosahedron is constructed by making use of the golden mean property of the icosahedron (see Section 3.4).

Figure 8.23 *Needle Tower*, a tensegrity structure by Kenneth Snelson.

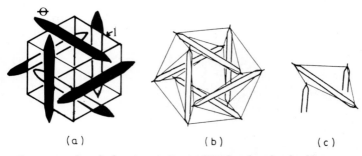

(a)   (b)   (c)

Figure 8.24 Icosahedron tensegrity. (a) Wooden dowels of golden mean length are placed on a unit cube; (b, c) the vertices are connected by 24 strings to form diamond patterns.

Construction 8.5 Fasten a set of wooden dowels φ units long to the sides of a unit cube as shown in Figure 8.24(a). Connect the 12 ends of the dowels with string (fishing line is good) to form the 20 faces of the icosahedron. Tighten the strings so that they are all under tension and then destroy the supporting cube as shown in Figure 8.24(b). You will find that some of the strings are redundant and can be removed without affecting the rigidity of the tensegrity. In fact only 24 tendons are needed instead of the 30 edges that form the icosahedron. These 24 tendons can be arranged to form nonplanar diamonds around each strut, with four tensions for each of the six struts as shown in Figure 8.24(b) and (c).

Figure 8.25 A tensegrity model of a cuboctahedron with the inner structure of four interpenetrating equilateral triangles.

Tensegrities can also be made by connecting the compression members in the plane with strings and folding them into three-dimensional space. Patterns for assembling the rods and tendons are given in *Tensegrity* by Anthony Pugh [1976].

Construction 8.6   A fascinating tensegrity is shown in Figure 8.25, in which the vertices of four mutually interlocking but nontouching equilateral triangles coincide with the vertices of a cuboctahedron. The triangles serve as the struts while the tendons lie along the edges of the cuboctahedron. Since the vertices of

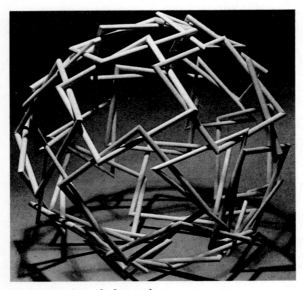

Figure 8.26   An orderly tangle.

the cuboctahedron lie at the midpoints of a cube, this tensegrity can be constructed by first erecting a cube as scaffolding. This tensegrity brings out the underlying structure of the cuboctahedron in which its 12 vertices lie at the corners of four regular hexagons like the one shown in Figure 8.8.

8.8.2   Orderly tangles

Holden has created a class of polyhedra, which he calls *orderly tangles*, that are woven out of wooden dowels [1983]. Edges of the platonic or archimedean polyhedra interlock in an under-over pattern as shown in Figure 8.26 for the small rhombic cuboctahedron 3.4^3 (see Section 9.2). When the ratio of length to diameter of the dowels is just right, the orderly tangle is rigid and the cycles of edges hold each other up by leaning on each other. Removing even a single edge causes the whole structure to collapse.

8.9   The Tetrahedron—Methane Molecule and Soap Bubble

In a certain sense, the tetrahedron is the most fundamental of forms. Through the stella octangula [see Figure 8.9(*a*)], it demarcates the domain of a cube and contains the octahedron as a substrate, it can be truncated to form an icosahedron (see Section 9.5), and along with four other tetrahedra, it shares the vertices of a dodecahedron (see Figure 8.16). Exercise 8.4 also shows that the internal structure of the cuboctahedron is closely related to the tetrahedron. In fact, in Chapter 10 we shall see how, through the mediation of the cuboctahedron, the tetrahedron plays a fundamental role in describing the packing of atoms in a metallic crystal. Finally, in Section 9.2 we shall see how the tetrahedron is used to define the class of archimedean polyhedra.

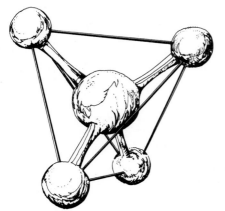

Figure 8.27   The methane molecule. Redrawn from Pauling and Hayward [1964].

From a physical point of view, the tetrahedron serves as the basis for visualizing the methane molecule and primary soap bubble configurations. For the methane molecule shown in Figure 8.27, the four hydrogen atoms are situated at the vertices of a tetrahedron and share one electron each, with a carbon atom located at the centroid of the tetrahedron. Since the direction from vertex to center lies in the direction of the body diagonals of a cube, the angle between two carbon-hydrogen bonds is 109.48 degrees, i.e., the angle between two body diagonals (see Section 7.13.2). This angle, which is characteristic of organic molecules and therefore of life itself, is called the *Miraldi angle.*

Construction 8.7 Build a tetrahedron out of wire and dip the frame into a soap solution [Stevens, 1974]. What do you observe? When you withdraw the tetrahedron frame, six films extend from the wire frame inward to the center of the tetrahedron as shown in Figure 8.28(a). Each of the six films is a triangle enclosed by an edge of the tetrahedron and two edges running from the center to a vertex of the tetrahedron. If you look closely, you will notice that each edge of the film is the junction of three films that meet each other at 120 degrees and that each edge joins three other edges to make a corner that unites six films. Four edges meet at each vertex, and the angle between any pair of edges is the Miraldi angle. The angles between edges and faces in the tetrahedral frame are also generic to all freestanding three-dimensional bubble configurations (see Section 10.11). The angles are the result of the need to create an equilibrium between the tensile forces that the soap films exert on the meeting point of its faces and edges.

If an additional small bubble is blown at the center of the tetrahedron, the bubble must assume the shape of a curvilinear tetrahedron with spherical faces, as shown in Figure 8.28(b), in order to conform to the generic soap bubble form. D'Arcy Thompson noted that this form represents one species of microscopic organism known as *radiolaria,* shown in Figure 8.28(c) [1966]. Again, if a wire frame in the form of a cube is dipped into soap solution and a small bubble is blown at its center, another species of radiolaria results. The Soviet mathema-

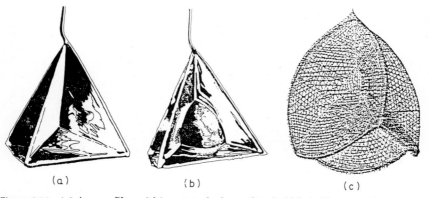

(a)   (b)   (c)

Figure 8.28 (a) A soap film within a tetrahedron; (b) a bubble is blown at the center of the film; (c) a radiolarian with a similar configuration.

tician and artist A. T. Fomenko describes all the possible geometrical forms that soap bubbles can assume [1986].

8.10   Tetrahedron as the Atom of Structure

We live in a world dominated by the right angle. Our streets meet along orthogonal grids. The shapes of our buildings are usually rectangular parallelopipeds. The fact that we choose to measure area and volume in square and cubic units reflects our feelings that it is the square and cube which are the most natural units of measure. Yet, Fuller felt that it is actually the tetrahedron that is the fundamental measure of volume. He felt that the very nature of space requires the tetrahedron to supplant the cube as the unit of space [Edmondson, 1987], [Loeb, 1975; 1965]. Let's see why.

Volume has to be measured relative to something. Why not measure it relative to a tetrahedron? Table 8.1 compares the volumes of a tetrahedron, octahedron, cube, and cuboctahedron when the cube and tetrahedron are taken as units of measure. For comparison, two different cubes are used, one with unit edge and the other with unit diagonal. By looking at Table 8.1 you can see that the tetrahedron distinguishes itself as a natural building block of form by having a volume that divides evenly into the volumes of other polyhedra. Let's see why these polyhedra are integral multiples of the tetrahedron's volume.

Look at an octahedron and a tetrahedron with the same edge lengths. It certainly does not appear as though the volume of the octahedron is 4 times that of the tetrahedron. This follows from decomposing the tetrahedron into four tetrahedra of half the edge length of the parent and one octahedron as shown in Figure 8.29 and as we did in Exercise 8.1. Thus,

$$T = 4t + O \tag{8.1}$$

where T stands for the volume of the large tetrahedron, t for the volume of the smaller tetrahedron, and O for the volume of the octahedron.

TABLE 8.1*

Polyhedron measured	Unit edge	Unit face diagonal	Tetrahedron
Tetrahedron	0.11785	0.33333	1
Octahedron	0.47140	1.33333	4
Cube (unit diagonal)	0.35356	1.00000	3
RD	0.70710	2.00000	6
Cuboctahedron	2.35700	6.66666	20

*From [Edmondson, 1987].

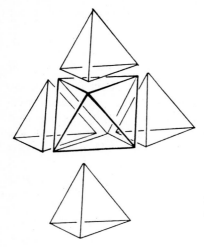

Figure 8.29   Four tetrahedra added to one octahedron produce a larger tetrahedron.

As we saw in Section 2.2, the volumes of two similar figures are related by Equation (2.2b), which is restated as

$$\frac{V_2}{V_1} = \left(\frac{L_2}{L_1}\right)^3 \qquad \text{where } \frac{L_2}{L_1} = 2$$

Thus, $T = 8t$ (8.2)

Substitution from Equation (8.2) into Equation (8.1) leads to the result

$$\frac{O}{t} = 4$$

That the volume of a cube is 3 times the volume of a tetrahedron with edges equal to the cube's diagonal also appears to confound our perceptions. Yet this can be shown by pulling apart a cube into four octants—an octant is the ⅛ sector of an octahedron—and one tetrahedron, as shown in Figure 8.30. Thus,

$$C = 4\frac{O}{8} + T$$

where as before,

$$O = 4T$$

Substituting for O and rearranging the terms yields the result

$$\frac{C}{T} = 3$$

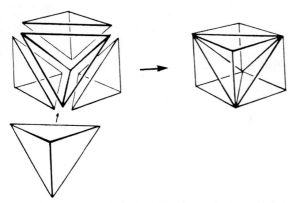

Figure 8.30   Four octants added to a tetrahedron produce a cube.

Problem 8.2   From the decomposition of the cuboctahedron into half-octahedra and tetrahedra given by Exercise 8.4 show that the cuboctahedron has the volume of 20 tetrahedra with the same edge.

Construction 8.8   Dissect a cube into a tetrahedron and four octants of an octahedron as shown in Figure 8.30. It makes a good puzzle to reassemble the cube from its parts.

8.11   Packing of Spheres

Perhaps the most fundamental context in which to study the relationships between the platonic polyhedra is the *packing of spheres*. Keith Critchlow, in *Order in Space* [1987], considers the hierarchy of geometrical concepts: point, line, plane, and three-dimensional space. This unfolding of dimensions can be made tangible, as illustrated in Figure 8.31, by representing a *point* by a sphere surrounding it, a line by the *line segment* joining the centers of two identical tangent

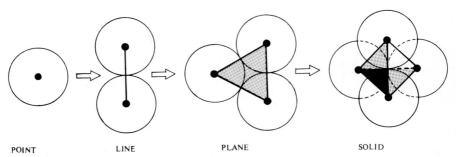

POINT   LINE   PLANE   SOLID

Figure 8.31   The unfolding of the dimensions of space. (*a*) Point; (*b*) line; (*c*) plane; (*d*) solid.

spheres, a *plane* by the region spanned by the triangle joining the centers of three mutually touching identical spheres, and *three-dimensional space* by the solid space spanning the tetrahedron formed by the joins of four identical spheres in mutual contact.

8.11.1 Evolution of platonic polyhedra from sphere configurations

A tetrahedron is formed by placing a sphere atop an equilateral triangle arrangement of three spheres. The four spheres making up the tetrahedral configuration of Figure 8.32(a) are the greatest number of spheres that can be in mutual contact. If a second set of spheres is placed in the interstices of the first set, a spherical model of a dual tetrahedron is formed by the sphere centers of this second set as shown in Figure 8.32(b), and (c).

An octahedron is formed by placing a sphere above and below a square arrangement of four spheres. The six spheres are then arranged so that each sphere touches four others as shown in Figure 8.32(d). The joins of these spheres result in an octahedron. Once again, if a second set of nontouching spheres is introduced into the interstices, the dual figure shown in Figure 8.32(e) and (f), a cube results from their joins.

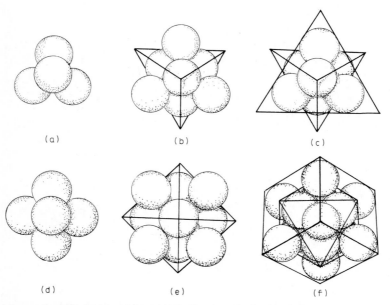

Figure 8.32 Evolution of the basic sphere point configurations. (a) Tetrahedron; (b, c) dual tetrahedron; (d) octahedron; (e, f) the cube dual to the octahedron.

The *closest packing* of equal spheres about a central nucleus involves 12 spheres in contact with the central sphere so that each sphere touches four neighbors in addition to the nucleus. In this arrangement, six spheres surround the nucleus [see Figure 8.33(*a*)], with three spheres lying in the interstices above and three spheres below [see Figure 8.33(*b*)]. Thus the spheres group themselves into three layers. If the three spheres in the top and bottom layers are oppositely oriented, the result of connecting the twelve surrounding spheres is a cuboctahedron, and the spheres are said to be *cubically close-packed*. The model of a cuboctahedron constructed in Exercise 8.4 from marshmallows and toothpicks illustrates this packing quite well if you imagine the marshmallows to be spheres of radius half the length of the toothpicks. Notice how the surface spheres group into triangular and square arrangements.

If the three spheres in the top and bottom layers are oriented similarly so that a sphere from the bottom layer lies directly beneath a sphere from the top layer, a polyhedron called an *orthobicupola* is formed and the spheres are said to be *hexagonally close-packed*. The relation between these close-packing arrangements and the structure of metallic crystals will be explored in Chapter 10.

If the central sphere is removed, the close-packed arrangement becomes unstable and the cuboctahedron collapses into an icosahedron with each sphere touching five others as shown in Figure 8.34. If a sphere is reintroduced into each of the interstices of the icosahedral arrangement, a dual figure, the dodecahedron, is formed upon connecting the 20 outer nontouching spheres (not shown).

Exercise 8.5   Arrange 16 spheres into a square pattern. Notice that the gaps between the spheres are curvy squares (see Figure 8.37). Now add successive layers of 9, 4, and 1 spheres. Notice that the gaps along the triangular faces of the resulting pyramid are curvy triangles.

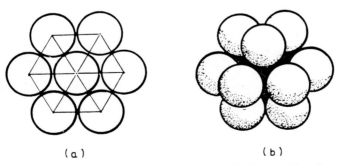

(a)   (b)

Figure 8.33   (*a*) Six spheres surround a central sphere in the plane;
(*b*) twelve spheres surround a central sphere in the close-packing of
spheres in three-dimensional space.

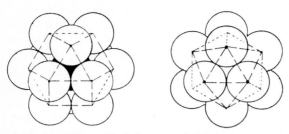

Figure 8.34 Removal of the central sphere from a cuboctahedron produces an icosahedron.

8.11.2 A hierarchy of platonic polyhedra

The evolution of platonic polyhedra from the packing of spheres reveals the natural hierarchy

tetrahedron → octahedron → icosahedron

Some years ago this hierarchy was demonstrated to me by someone who had a rubberized model of a cuboctahedron. When he twisted it to the left or to the right, an icosahedron resulted (imagine that the diagonals of the rhombic faces are added to complete the necessary triangles). Then he collapsed the icosahedron, first into an octahedron, then to an equilateral triangle and finally to a tetrahedron, as shown in Figure 8.35. At the time, I thought this demonstration was the epit-

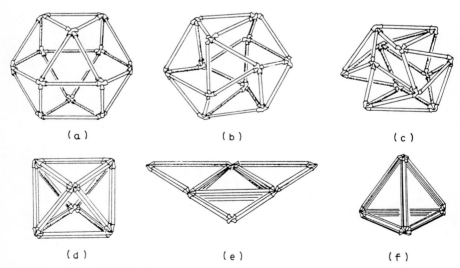

Figure 8.35 Stages in Fuller's jitterbug. (*a*) Cuboctahedron; (*b*) icosahedron; (*c, d*) transition to an octahedron; (*e*) equilateral triangle; (*f*) tetrahedron.

ome of magic. Years later I learned that it was discovered by Fuller, who called it the *jitterbug* [Edmondson, 1987].

Construction 8.9 [Frank, 1987].   To construct a jitterbug build a cuboctahedron out of ⅜-inch wooden dowels about 6 inches long. Connect the dowels with about 2 inches of surgical tubing to ensure flexibility. Since four dowels meet at each vertex, you will have to connect one pair of dowels with the other. This can be done by punching a hole through one of the rubber connectors with an awl (or leather punch) and pulling the other pair through the hole (wetting the connector with water makes this operation a little easier).

8.11.3   Frequency

The six squares and eight triangles that outline the cuboctahedron in the cubic close-packing arrangement become more and more distinct if the cuboctahedron is surrounded by additional layers as Figure 8.36 shows for 1-, 2- and 3-frequency cuboctahedra. Here the *frequency f* is defined as the number of tangency points between spheres lying along an edge of the figure. The high-frequency cuboctahedra also show how cubic close-packing figures are defined by planes of spheres arranged in both square and triangular patterns. There is no limit to the number of layers that may be added to the cuboctahedron. In fact, the layers may grow to fill all of the space, as we shall show in Chapter 10. Fuller determined that the number of spheres N on the surface of each frequency cuboctahedron could be determined by the formula

$$N = 10f^2 + 2 \qquad (8.3)$$

which has been derived by Loeb [1975].

Let's compare this layering property of the cubic close packing of spheres with the sphere packing that results in an icosahedron. Con-

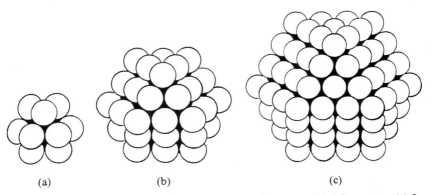

(a)   (b)   (c)

Figure 8.36   Higher-frequency cuboctahedra. (*a*) 1-frequency; (*b*) 2-frequency; (*c*) 3-frequency.

secutive higher-frequency icosahedral shells cannot surround a previous layer as they did for the cuboctahedron. Higher-frequency or
larger icosahedra must be built one by one with one more sphere per
edge and always of a single thickness. Nevertheless, the number of
spheres on the surface of the single-thickness icosahedron is also derived from Equation (8.3). This may seem surprising since this equation pertains to a cuboctahedron, but remember that in the jitterbug
transformation each square of the cuboctahedron twists to form two
triangular faces of the icosahedron as shown in Figure 8.37(a). Since
spheres are more closely packed in a triangular arrangement than in
a square configuration, as shown in Figure 8.37(b), the surface of the
icosahedron encompasses greater volume for the same surface area
than the cuboctahedron.

The endless spacefilling layering of cuboctahedra make this structure open in comparison with the shell-like closed structure of the
icosahedron. In fact, Fuller felt that the vector-equilibrium (cuboctahedron) models a universal structure in which energy propagates
along "cosmic railroad tracks" represented by the encircling or "trans-
universal" geodesics of the cuboctahedron [Edmondson, 1987]. However, the geodesics of the icosahedron are not encircling [see Figure
7.4(c)]. Thus when a single sphere is removed from the center of a
cuboctahedron, the energy flow is disconnected from the closest packing "railroad tracks" and directed into local orbits. The free flow of energy in the cuboctahedron might be likened to a kind of harmony of
the spheres, in contrast to the dissonance resulting from the blockage
of energy in the icosahedron.

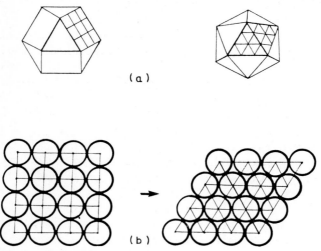

Figure 8.37 Transformation from square packing to triangular
packing of spheres.

Fuller explains that the closed system of the icosahedron is immersed within the harmony of the spheres and serves as the origin of life itself. Although this claim may appear to be exaggerated, there is also some truth to it since the icosahedron does appear as one of the geometric forms of viruses, as we shall see in the next section.

When thinking about connections between polyhedra and the natural world, it is difficult not to engage, as Fuller did, in mystical speculation. In dealing with these objects, far-flung associations are continually suggesting themselves. For example, in Sections 1.7.3 and 1.8.3 we introduced different notions of open and closed systems of proportion based on the proportions $\phi:1$ and $\sqrt{2}:1$. It is curious that $\sqrt{2}:1$ is the ratio of the sides of the cuboctahedron's vertex figure, while $\phi:1$ is the ratio of diagonal to side of the icosahedron's vertex figure, a regular pentagon. The duality of Béla Bartók's closed system based on the golden mean, in contrast to his open system based on the acoustic scale (see Section 3.8), also reminds one of the organic shell of an icosahedron emerging from the crystalline matrix of the close-packed spheres.

A tantalizing analogy to Fuller's world system involves a modern view of the collective nature of bacteria described by Sorin Sonea [1988]:

> In contrast to plants and animals, which are multicellular and exhibit a tremendous variety of configurations, most bacteria are one-celled and possess little morphological diversity.... Their range of metabolisms enables bacteria to colonize every environmental niche on the planet.... Their genetic material is not bound by a nuclear membrane, a characteristic that has earned them the name prokaryote, meaning before the nucleus.... They behave as if they were not discrete organisms; they are able to shuffle genetic information (among themselves) virtually overnight (resistance to an antibiotic in Tokyo will manifest in New York in a matter of days or weeks without direct transmission). In this respect the bacterial world resembles a vast computerized communications network—a superorganism whose myriad parts shift and share genetic information to accommodate any and all circumstances.... As the bacteria multiplied and colonized more of the earth's surface the superorganism created the environmental conditions that favored an entirely new form of life: the eukaryotes.... The eukaryotic cells adapted by gathering into one bound cell as much DNA as possible and [disconnecting itself from the universal "communications system"] to live in near solitude—in the restricted gene pools of their respective species.

8.12 Geodesic Domes and Viruses

Fuller is best known for his *geodesic dome* [1973]. In its most rudimentary form, this dome is an icosahedron with its faces subdivided into equilateral triangles as shown in Figure 8.38 for a 3-frequency dome.

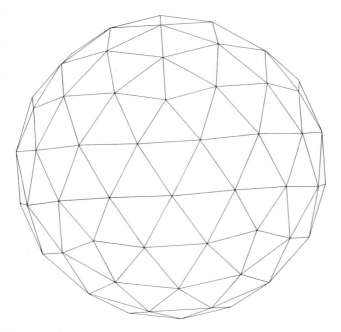

Figure 8.38 A 3-frequency icosahedron geodesic dome.

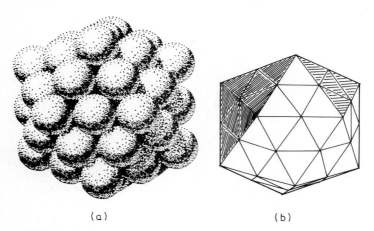

(a)   (b)

Figure 8.39 A cluster of 42 spheres with certain sphere centers interconnected to define a 2-frequency icosahedron.

The vertices are then projected onto the surface of the circumscribing sphere. The projected triangles are no longer congruent but are of two different kinds. Finally, the dome is truncated to give the desired height. Since a sphere has the least surface area for a given volume, the geodesic dome both creates a great amount of internal space and

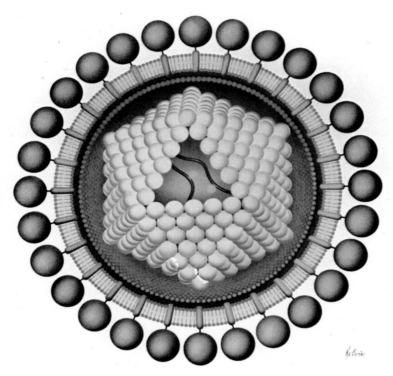

Figure 8.40   A portrait of the HTLV-1 virus. (*From R. C. Gallo, "The First Human Retrovirus." Photo by George V. Kelvin, Science Graphics. © 1986 by Scientific American, Inc. All rights reserved.*)

minimizes heat loss because of its decreased outer skin surface. The number of vertices V on the surface of an icosahedron of frequency f is given by Equation (8.3). Thus, for the dome in Figure 8.38 $f = 3$ and $V = 92$ (before it is truncated).

One class of virus with icosahedral symmetry is reminiscent of geodesic domes. A virus particle is composed of a basic infective agent, a nucleic acid core of either DNA or RNA, and a protective shell called a *capsid* composed of protein units called *capsomers*. In some virus particles the shell is encased in an outer membrane or envelope [Williams, 1972].

The virus class with icosahedral symmetry has the structure of a geodesic dome in which each capsomer attaches to a vertex of the dome. For example, the Simian virus (SV39), the K-virus, and the Polyoma virus with one capsomer at each vertex of a 2-frequency icosahedron, 42 in all, look very much like the spheres of Figure 8.39. A portrait of the AIDS virus (HTLV-1) is shown in Figure 8.40.

Transformations of the Platonic Solids II

*There is an old formula for beauty in nature
and art: Unity and variety.* JOHN DEWEY

9.1 Introduction

The platonic solids can also be transformed by cutting off their edges or vertices or by placing pyramids or other structures on their faces. These operations of truncation and stellation generally result in polyhedra that are no longer members of the platonic family. In this chapter we show how the truncation operation leads to a family of semiregular polyhedra known as the *archimedean solids* while the stellation operation leads to the *archimedean duals*. We then introduce two other families of semiregular polyhedra, prisms and antiprisms.

9.2 Archimedean Solids

The platonic solids satisfy a severe constraint: The same number of identical regular polygons must meet at each vertex. If this condition is relaxed to allow similar arrangements of more than one polygon at each vertex, new possibilities—referred to as *semiregular polyhedra*—result, of which the cuboctahedron and the icosidodecahedron are two examples. Varney's movable polyhedron with a triangle, pentagon, and square surrounding each vertex, the small rhombicosadodecahedron, shown in Figure 8.7 is another example. Heron in the first century B.C. said that Archimedes discovered 13 polyhedra that meet this requirement. They are illustrated in Figure 9.1. These polyhedra

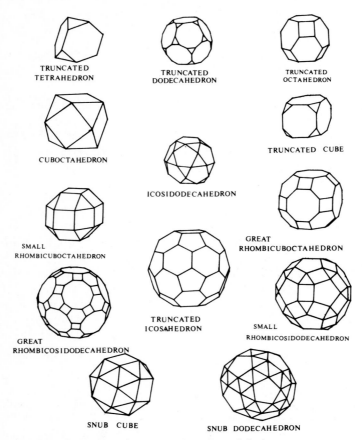

Figure 9.1   The archimedean or semiregular polyhedra.

were rediscovered during the Renaissance; the first written descriptions appeared in Kepler's *Harmonices Mundi* in 1619 [Malkevitch, 1988], [Grünbaum, 1977a].

Archimedes' original 13 polyhedra can be inscribed in a regular tetrahedron so that four appropriate faces share the faces of a regular tetrahedron as shown in Figure 9.2 for the cuboctahedron. This distinguishes them from prisms and antiprisms (see Section 9.10) and from one additional polyhedron called the *pseudo-rhombicuboctahedron*, 3.4.4.4, which is also semiregular. These 13 polyhedra are the archimedean solids.

Although the faces of the archimedean solids are of more than one kind, they are distributed in such a way that each vertex is equidistant from the geometrical center of the solid. Thus, a circumscibing sphere can be placed around each of the archimedean polyhedra. How-

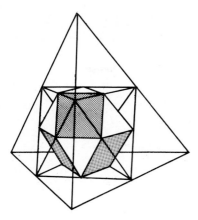

Figure 9.2 An archimedean solid framed by a tetrahedron.

ever, since the faces are not all alike, these polyhedra do not have inscribed spheres.

What if we relax the constraint placed on the archimedean solids to include all convex polyhedra with regular polygon faces? Does this widen the membership beyond bound? Only 92 polyhedra can be constructed in addition to the 13 archimedean solids, five platonic solids, and two infinite families of prisms and antiprisms discussed in Section 9.10. This result was proven by V. A. Zalgaller.

Certain space-filling combinations of archimedean solids in conjunction with the platonic solids have been used by architects such as Zvi Hecker [1970], Keith Critchlow [1971], and Safdie [1969] as a source of building forms. A number of space-filling combinations are catalogued by R. W. Williams [1982]. One such combination is shown in Figure 9.3.

9.3 Truncation

The archimedean polyhedra illustrate the chameleon-like characteristics of the platonic solids. Every one of them can be obtained by slicing off either the vertices or edges of a platonic polyhedron with a cutting plane [Pugh, 1976], [Williams, 1972], [Loeb, 1976]. Such an operation is called *truncation*. Six of the archimedean solids are "children" of the cube-octahedron pairs since they can be obtained by truncating either a cube or octahedron. Six others are related to the icosahedron-dodecahedron pair, and a single archimedean solid is obtained by truncating the tetrahedron [Critchlow, 1987]. H. Lalvani has discovered additional ways in which the polyhedra within each of these three families are related in a unified manner such that these polyhedra can transform to one another continuously [1981].

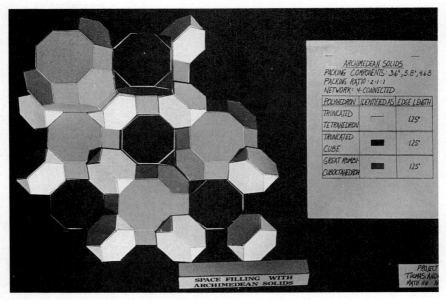

Figure 9.3 An archimedean space filling.

Let's look at one sequence of truncations. In a sense, the faces of the octahedron on the cube-octahedron interpenetrating pair [see Figure 8.9(*b*)] are cutting planes which truncate the vertices of the cube at the midpoints of the edges to obtain the cuboctahedron. If these eight cutting planes are moved parallel to each other toward the vertex so that the vertices are now truncated at appropriate points, another archimedean solid called the *truncated cube* is created. Figure 9.4 shows the traces on the cube of a sequence of cutting planes situated at the 0, ⅛, ½, ¾, and 1 positions along an edge. These result in the

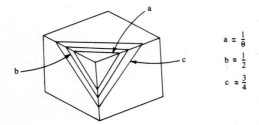

$$a = \frac{1}{\theta}$$

$$b = \frac{1}{2}$$

$$c = \frac{3}{4}$$

Figure 9.4 A cube showing three truncation planes at the ⅛, ½, and ¾, points of the edges ($\theta = 1 + \sqrt{2}$).

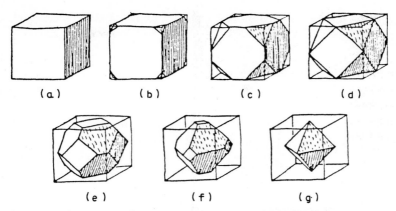

Figure 9.5 A sequence of truncations of a cube. (*a*) cube; (*b,c*) transition to a truncated cube; (*d*) cuboctahedron; (*e*) truncated octahedron; (*f, g*) transition to an octahedron.

sequence of truncations from cube → truncated cube → cuboctahedron → truncated octahedron → octahedron as shown in Figure 9.5.

Notice that up to the ½ position of the cuboctahedron (intermediate polyhedron) the truncation planes do not interfere with each other. After this point the planes intersect within the cube. For example in Figure 9.5(*e*), the eight cutting planes intersect to form the eight hexagonal faces of the truncated octahedron, leaving its six square faces embedded in the faces of the cube. Finally the square faces disappear as further truncation produces an octahedron in Figure 9.5(*g*).

Schlegel diagrams (see Section 7.6.3) can be useful in visualizing the result of transforming a polyhedron. For example, the cuboctahedron 3.4.3.4 in Figure 9.6(*c*) is gotten by truncating the vertices of the cube in Figure 9.6(*a*) at the center of its edges while the

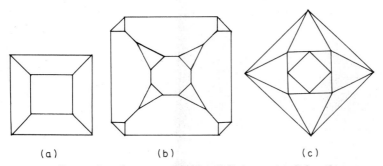

Figure 9.6 Truncation shown on a Schlegel diagram. (*a*) Cube; (*b*) truncated cube; (*c*) cuboctahedron.

truncated cube 3.8^2 in Figure 9.6(b) is gotten by truncating the edges of the cube as in Figure 9.5(c).

9.4   The Truncated Octahedron

A similar sequence of truncations could also have been obtained by considering the six faces of a cube to be the truncating planes of the octahedron in the interpenetrating pair. In this way, as Figure 9.7 shows, the truncated octahedron results from truncating a 3-frequency octahedron at the one-third points of its edges. The truncated octahedron is the only space filler among the archimedean solids, although this is not evident by looking at its shape. Its space-filling capability is better seen by reconsidering the truncation of the cube in Figure 9.5(e) that produced it. Such a slice divides the cube into two congruent halves each with an hexagonal profile. Figure 9.8 shows eight of these half-cubes joined together to form the truncated octahedron. Its space filling is thus the result of reallocating the space within a stack of cubes.

Construction 9.1 [Loeb, 1986]   Loeb has used the creation of the truncated octahedron from the truncation of eight cubes shown in Figure 9.8 as the basis of a remarkable construction. The eight half-cubes are hinged together as shown in Figure 9.9 to form a cube that envelops the other eight half-cubes that form the truncated octahedron. The hinges are so arranged on the cube that its eight segments can fold inward to form another truncated octahedron while the truncated octahedron is able to fold outward to form another enveloping cube.

The ability to fill space by itself places stringent conditions on the form of a polyhedron. Construction 9.1 shows the intricate geometry required of a truncated octahedron that enables it to pass the test. Figure 9.10 shows in another way that the truncated octahedron is a natural candidate to fill space. Here, bricks are laid down in a regular space-filling pattern. The edges are colored with paint. Do you see

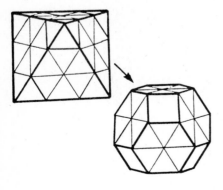

Figure 9.7   Truncation of a 3-frequency octahedron to a truncated octahedron.

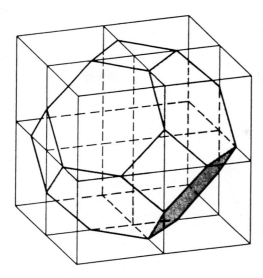

Figure 9.8 The hexagonal faces of a truncated octahedron bisect eight cubes.

Figure 9.9   A hinged cube transforms to a truncated octahedron.

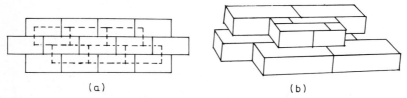

Figure 9.10   Two layers of bricks with painted edges imprint a polyhedron topologically equivalent to a truncated octahedron upon each brick.

that each brick borders 14 adjacent bricks? If all these surrounding bricks are removed, the imprints of their edges on the central brick is a pattern with the six rectangular and eight hexagonal faces of the truncated octahedron [Steinhaus, 1969]. The brick can now be topologically deformed (see Section 4.11). In this deformation, opposite faces remain parallel (see Section 10.13).

9.5   The Snub Figures

Perhaps the most unusual of the archimedean solids are the two *snub figures*. These are the only archimedean solids that do not have a plane of reflection, and therefore they occur in enantiomorphic (see Section 2.2) pairs. But even these polyhedra can be obtained by truncation of the platonic solids. Jean-François Rotgé [1984] has shown that the icosahedron, snub cube, and snub dodecahedron can be obtained by truncating a tetrahedron, octahedron, and icosahedron, respectively, in a special way. Each triangular face of the parent platonic solid is subdivided as shown in Figure 9.11. When the crucial ratio r is assigned the value 1.618 (the golden mean), the tetrahedron is truncated to an icosahedron [see Figure 9.12(a)]. A value of $r = 1.839$ results in transforming the octahedron to a snub cube [see Figure 9.12(b)], and the value $r = 1.943$ produces a snub

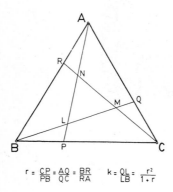

Figure 9.11   Construction of the fundamental triangle for snub figures.

$$r = \frac{CP}{PB} = \frac{AQ}{QC} = \frac{BR}{RA} \qquad k = \frac{QL}{LB} = \frac{r^2}{1+r}$$

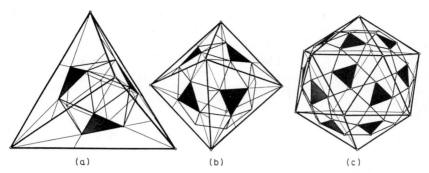

Figure 9.12   (*a*) An icosahedron is produced from *r* = 1.618 (golden mean); (*b*) a snub cube is produced from *r* = 1.839; (*c*) a snub dodecahedron is produced from *r* = 1.943.

dodecahedron after truncation of an icosahedron [see Figure 9.12(*c*)]. These snub figures also have many interesting relatives, as Rotgé shows.

9.6   Archimedean Duals

Although the archimedean solids are ancient, all their duals with the exception of the RD were only discovered in the nineteenth century [Malkevitch, 1988]. The duals are known as the *Catalan solids*. Just as the archimedean solids have identical vertices, their duals have congruent faces. One way to define the duals is to place tangent planes to the circumscribing sphere at the vertices of the corresponding archimedean solid. The archimedean dual will then be the envelope formed by the intersection of these tangent planes, and the circumscribing sphere of the archimedean solid will be the inscribed sphere of the dual. In this way the archimedean duals are related to the more general definition of dual polyhedra in terms of the pole and polar to a sphere given in Appendix 7.A.

Just as every archimedean solid is obtained by truncating a platonic solid, each of the archimedean duals is gotten by adding the appropriate platonic solid by placing identical pyramids on its faces as Figures 8.12 and 8.13 illustrate for the RD [Critchlow, 1987]. Another method of constructing the faces of any one of the duals is presented in the next section. M. Wenninger has shown how to construct fascinating designs of archimedean polyhedra and their duals embedded one within the other [1983].

9.7   Maps on a Sphere

We have seen that a sphere can be circumscribed about each of the archimedean solids and inscribed within each of their duals. If the edges are projected onto the surface of these spheres from the polyhedral centers, the edges of the polyhedra once again map to arcs of geodesics of the sphere. In this way patterns of curvilinear polygons are created on a sphere somewhat reminiscent of the ancient Ukrainian tradition of Pysanki by which elaborate tilings on the surface of an egg are created and which inspired Ron Resch to build a 25-foot-long egg-shaped dome in a Ukrainian farming community in Vegreville, Alberta (see Figure 9.13). Figure 8.10 shows how the edges of the octahedron, cuboctahedron, and icosidodecahedron project to three, four, and six great circles, respectively.

Since the edges of any archimedean polyhedron are equal chords on

Figure 9.13 Ron Resch's egg-shaped dome.

its circumscribing sphere, the centers of these edges lie on another sphere (why?) called the *intersphere*. In fact, archimedean polyhedra and their duals can be constructed to share a common intersphere as we did with the interpenetrating platonic duals. This results in the *Dorman Luke method* for constructing the faces of the duals [Pugh, 1976]. Draw the vertex figure of an archimedean solid which is obtained by connecting the midpoints of the edges incident to a vertex as shown in Figure 9.14(a) for a truncated octahedron. Circumscribe the vertex figure by a circle. This circle must be a small circle on the surface of the intersphere. Since the dual polyhedron shares this intersphere, its faces are defined by edges that are tangent to this circumscribed circle at the points where the archimedean solid touches it as shown in Figure 9.14(b) and (c).

9.8 Combinatorial Properties

Because an archimedean polyhedron has identical environments surrounding each vertex, its global properties such as total number of edges, faces, and vertices can be determined from its local properties characterized by its Schläfli symbol.

As we showed in Section 7.7, V, E, and F can be determined from Descartes' formula, $qV = 2E$, and from Euler's formula. For example, for 3.4.3.4

$$\delta = 360 - (60 + 90 + 60 + 90) = 60$$

$$V = \frac{720}{\delta} = 12$$

$$E = \frac{qV}{2} = 4 \times \frac{12}{2} = 24$$

$$F = 2 + E - V = 14$$

The number of triangles and squares can also be obtained in the same way in which the module for the semiregular tilings of the plane

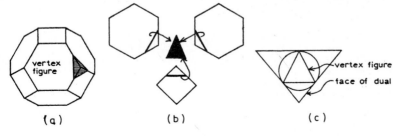

(a)   (b)   (c)

Figure 9.14 Dorman Luke construction of faces of the archimedean duals.

was obtained in Section 5.8. The dual tiling is superimposed on the projection of an archimedean polyhedron onto a circumscribing sphere. Since the dual tiles are congruent and cover the sphere, we can determine the proportions of each species of face within a typical tile of the dual. Thus, for the cuboctahedron, 3.4.3.4, there are

$\frac{2}{3}$ triangle:$\frac{2}{4}$ square   or   4 triangles:3 squares

Therefore,

$$\text{No. of triangles} = \frac{4}{7} \times 14 = 8$$

$$\text{No. of squares} = \frac{3}{7} \times 14 = 6$$

As usual, the dual polyhedra have their edges paired but their faces and vertices interchanged. Thus for the RD, $V(3.4.3.4)$,

$$E = 24 \qquad F = 12 \qquad V = 14$$

Although all faces are congruent, there are two species of vertices in the same ratio as the two species of faces of the cuboctahedron: eight vertices with vertex valence 3 and six vertices with valence 4.

One of the prominent themes of this book is that the nature of space places severe constraints on the possibilities of the forms that can be created in that space. One such constraint is described by A. L. Loeb [1976]. Consider polyhedra constrained to have the same number of edges incident to each vertex (as for the archimedean solids) and two kinds of faces (two different face valences). From Equations (7.1) and (7.2),

$$p_1 F_1 + p_2 F_2 = qV \tag{9.1}$$

From Equations (7.1) and (7.3),

$$F_1 + F_2 = 2 + \left(\frac{q}{2} - 1\right)V \tag{9.2}$$

Solving these two equations simultaneously, we get:

$$F_1 = \frac{2p_2 + \left[\left(\frac{q}{2} - 1\right)p_2 - q\right]F_2}{p_2 - p_1} \tag{9.3}$$

This equation reveals a number of interesting constraints. If each vertex has three incident edges, i.e., $q = 3$ and we wish to build a polyhedron with only pentagons and hexagons, i.e., $p_1 = 5$ and $p_2 = 6$, we see that Equation (9.3) yields

$$F_1 = 12 + (0)F_2 = 12$$

Thus Equation (9.3) says that any polyhedron with vertex valence $q = 3$ and with pentagon and hexagon faces can have an arbitrary number of hexagons but only 12 pentagons. Of course, the dodecahedron satisfies this constraint with 12 pentagons and no hexagons. The soccer ball, i.e., the truncated icosahedron 5.6^2, also satisfies it.

Problem 9.1   Use Equation (9.3) to show that for $q = 3$, four triangles can combine with any number of hexagons to form polyhedra (e.g., the tetrahedron and the truncated tetrahedron). Also, six squares can combine with any number of hexagons (e.g., cubes and truncated octahedra). If $q = 4$, show that any number of squares can combine with exactly eight triangles (e.g., the cuboctahedron).

If you replace $F \leftrightarrow V$, $q \leftrightarrow p$ in Equation (9.3), an equally correct but dual relation is obtained for the case of polyhedra with one kind of polygon but two kinds of vertices.

9.9   Symmetry Revisited

The operations of truncation and stellation do not alter the symmetry of the parent platonic solid. Thus, all archimedean polyhedra and their duals have either the symmetry of the octahedron-cube 4.3.2, icosahedron-dodecahedron 5.3.2, or tetrahedron 3.3.2 (see Section 7.13). This fact was made clear to me one day when my (empty) coffee cup fell into the dihedral kaleidoscope of a cube (see Section 7.13.4) and was visually multiplied by the 48 elements of the symmetry group of the cube to the 48 vertices of the great rhombicuboctahedron grouped in the pattern 4.6.8 about each vertex (coffee cup).

This kind of relationship between symmetry and three-dimensional pattern generation is shown in Figure 9.15(b) where a single point has been placed at point W within one of the 48 spherical triangles of Fig-

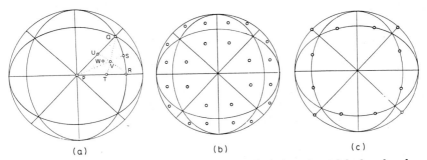

(a)   (b)   (c)

Figure 9.15   (a) Typical vertices of seven regular and semiregular polyhedra placed in a fundamental domain of the cube; (b) vertices V of the great rhombicuboctahedron; (c) vertices S of the truncated cube.

ure 9.15(a). V is the meeting point of the angle bisectors of a typical spherical triangle. The boundaries of these triangles are the great circles where the nine planes of mirror symmetry of the cube intersect its circumscribing sphere (see Section 7.13.4). By reflecting the point successively in each of these mirror lines, the point is replicated once in each of the triangles to create the projection onto a sphere of 4.6.8. If point S in Figure 9.15(a), located on one of the edges of the spherical triangle, is multiply reflected in all the mirror lines of the sphere, the truncated cube shown in Figure 9.15(c) is the result. In this way all archimedean polyhedra excluding the snub figures can be generated by the appropriate set of reflections [Coxeter, 1988], [Burt, 1982], [Lalvani, 1987].

Vedder Wright led a group of elementary school students from Cambridge, MA, in the exploration of the symmetry of polyhedra by having them play with a set of kaleidoscopes related to the platonic solids. Each kaleidoscope is a pyramid whose base is a face of one of the platonic solids and whose sides are triangles that connect an edge of the base with the center of that platonic solid. The inner faces of the pyramid are lined with a reflective mylar tape. For example, an icosahedron kaleidoscope is constructed from three isosceles triangles with central angle arccos $\phi/(1 + \phi^2) = 63.43$ degrees as shown in Figure 9.16. (These edge lengths and angles can be determined by using the three golden rectangles that lie within an icosahedron as a coordinate system. Try it! See Section 3.4.) This isosceles triangle is identical to the triangles on the face of the Pyramid of Cheops (see Sections 3.2 and 8.3) [Verhayen, 1989]. This angle is also the supplement of the dihedral angle of the dodecahedron (see Section 7.10). The central angles for the octahedron and tetrahedron kaleidoscopes are 90 and 109.5 degrees (the Miraldi angle), respectively.

Construction 9.2 [Wright, 1989]   Construct kaleidoscopes for the tetrahedron, octahedron, and icosahedron, and place polygons of various shapes inside the kaleidoscope. Images of all of the archimedean polyhedra except the snubs can be created in this way. Appropriately sized pyramids will result in the archimedean duals and star polyhedra. For example, the star polyhedra {5/2,5}

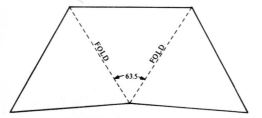

Figure 9.16   Construction of a kaleidoscope with icosahedral symmetry.

and {5,5/2} of Section 7.14 are created by placing pyramids constructed from three golden triangles of type 1 and type 2, respectively (see Section 3.5).

9.10   Prisms and Antiprisms

Besides the archimedean polyhedra, there are two additional families of semiregular polyhedra: prisms and antiprisms. Many of these polyhedra and their duals are defined by the external shapes of crystals and classical architectural structures. A prism is a polygonal cylinder. An example of an hexagonal prism is shown in Figure 9.17(a). If the

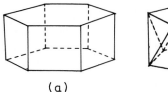
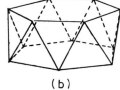

(a)   (b)

Figure 9.17   (a) Hexagonal prism; (b) hexagonal antiprism.

upper face in this figure is rotated through half of the central angle between adjacent vertices, as shown in Figure 9.17(b), an antiprism is formed when adjacent vertices from the top and bottom faces are connected. Among the platonic polyhedra, the cube is a prism and the octahedron is an antiprism. Figure 9.17 shows that while the lateral faces of a prism are all squares, the lateral faces of an antiprism are equilateral triangles. The prism duals are dipyramids and have all triangulated faces as illustrated for the hexagonal dipyramid in Figure 9.18(a). The dual of the hexagonal antiprism, called the hexagonal trapezohedron, is shown in Figure 9.18(b).

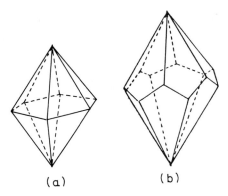

(a)   (b)

Figure 9.18   (a) Hexagonal dipyramid; (b) hexagonal trapezohedron.

Prisms have no triangular faces (with the exception of the triangular prism); thus they are not rigid according to Section 7.8. They will droop like cubes and dodecahedra if they are constructed with marshmallows and toothpicks. However, antiprisms are rigid in response to lateral loads (forces perpendicular to their axes), although they will fold up like accordions when acted upon by axial loads (forces in the direction of their axes).

9.10.1   A paper structure constructed from antiprisms

The barrel vault, shown in Figure 9.19, has been an important structural element of architecture since the time of the Romans because of its great structural efficiency [Ackland, 1972]. It is able to withstand lateral external loads many times its own weight. V. Sedlak has constructed lightweight prefabricated paper shelters made from weatherized paper folded into a series of half-antiprisms [Sedlak, 1973]. Braces can be added between the vertices of successive antiprisms to prevent their tendency to fold up in response to axial loads.

Construction 9.3

1. Construct a vault by folding paper in a plane without cutting, as shown in Figure 9.20(a). In this figure, three families of lines are drawn on a piece of

Figure 9.19 A gothic vault.

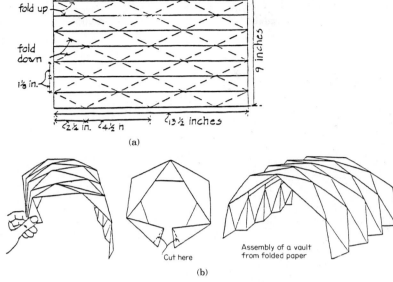

(a)

(b)

Figure 9.20   Construction of a vault made from antiprisms. (*a*) Scoring the paper; (*b*) assembling the vault.

stiff paper. The two dotted families are folded down, while the solid family of lines is folded up. The vault may then be assembled as shown in Figure 9.20(*b*).

2. In the previous construction, the vault was made up entirely of a sequence of identical isosceles triangles arranged in a parallel configuration as illustrated in Figure 9.21(*a*). There are two other arrangements of isosceles triangles: the pyramidal and the radial [see Figure 9.21(*b*) and (*c*)]. These arrangements figure in the construction of two other structures: the semidome and the intersection of vaults. The semidome is used to cap off a vault, while the intersection of vaults is the meeting point of three vaults that surround it radially and enables the structure to be continued in another direction. De-

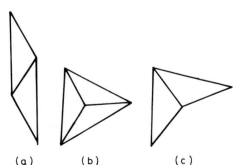

(a)   (b)   (c)

Figure 9.21   Three ways to join isosceles triangles. (*a*) Parallel; (*b*) pyramidal; (*c*) radial.

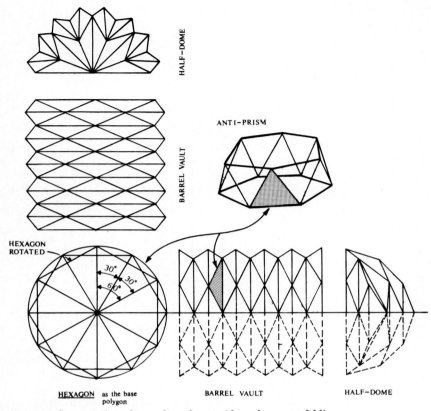

Figure 9.22   Construction of a vault and a semidome by paper folding.

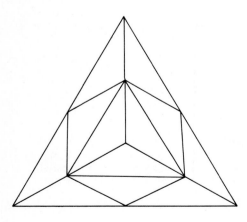

Figure 9.23 Construction of an intersection of three vaults (IV). The edges of the IV must abut with the edges of the three vaults.

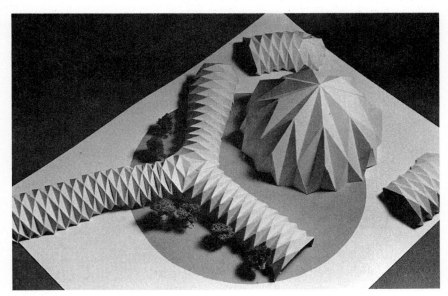

Figure 9.24   A paper folded structure.

tails for constructing the semidome and intersection of vaults are shown in Figures 9.22 and 9.23. All isosceles triangles are identical with the exception of the three in the center of the intersection of vaults. Good models can be constructed with isosceles triangles that have 5-inch bases and 2¾-inch sides. Once you have constructed several vaults, semidomes, and intersections of vaults, you can arrange them in an indefinite number of ways to achieve some interesting tilings of the plane as Figure 9.24 illustrates.

Polyhedra: Space Filling

The rules will be seen to grow out of the
parts and the parts out of the rules.
                              PETER PEARCE

10.1   Introduction

The sphere and the cube are certainly the most familiar of all three-dimensional geometric objects. The sphere is distinguished by being the most symmetric of all of them while the cube is known for its ability to fill space when stacked. The apparent simplicity of spheres and cubes may make them seem uninteresting; however, careful analysis discloses that they are the building blocks of complex geometric structures and lie at the basis of biological forms, soap bubble froths, crystal patterns, and architectural forms. In this chapter, we shall show how forms as simple as a sphere and a cube lead to complexity and diversity. First try the following exercises.

Exercise 10.1   Get a bunch of 3-inch styrofoam spheres and toothpicks. Arrange the spheres on several levels, connecting adjacent spheres by toothpicks, and create some interesting designs.

Exercise 10.2   Get some soap solution and a bubble blower and blow a froth of bubbles on a wetted surface. Observe the polyhedral configurations in the soap froth. Count the number of faces that meet at an edge of the film, the number of edges that are incident to a meeting point of films, and the number of different kinds of polygonal faces that make up various cells of the froth.

10.2   Close Packing of Spheres

As you can see from Exercise 10.1, although the sphere is extremely simple, the patterns it forms in juxtaposition with other spheres can be quite complex. A. L. Loeb shows that each sphere on the first level of a stack of spheres can be arranged so that it is surrounded by six others, forming the familiar triangular grid of Islamic patterns (see Section 5.13), as shown in Figure 10.1(*a*) and schematically in Figure 10.1(*b*) with the spheres occupying position *D* [1966]. Spheres in the second layer are then placed in the interstices of the spheres from the first layer (see Figure 10.2). Here, the spheres may be placed in the curvy triangles of either the *E* meshes or the *F* meshes where each mesh is related to itself and the other mesh by a rotation about *D*. Also note that the *D, E, F* centers each form their own triangular mesh.

When a third layer is placed on the second one, a choice must again be made. If the *E* mesh is chosen for the second layer, the choices for the third layer are either *D* or *F* positions. In the *D* case a sphere in the third layer lies directly above a sphere in the first layer. This DED (or DFD) pattern is the hexagonally close-packed system (hcp) referred to in Section 8.11.1 while the DEF pattern is the cubically close-packed system (ccp) that resulted in the cuboctahedron when we limited ourselves to surrounding a single sphere. Now we continue the pattern set by the first three layers, i.e., DEDED...(or DFDFD...) or DEFDEF...(or DFEDFE...), to fill all of space with spheres.

Although the packing was constructed with a triangular grid of spheres [see Figure 10.3(*a*)], diagonal planes through the packing reveal curvy square-shaped interstices [see Figure 10.3(*b*)]. But this is no surprise, since Figure 8.36 shows how both triangular and square interstices arise naturally from the close packing of spheres. The

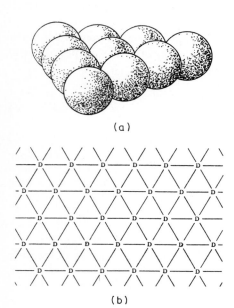

(a)

(b)

Figure 10.1 The centers of a triangular arrangement of spheres lie on a triangular grid.

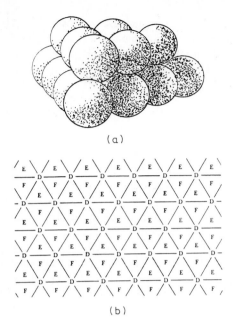

(a)

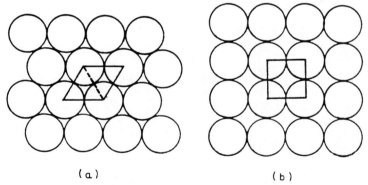

(b)

Figure 10.2 (a) Arrangements of spheres on two levels; (b) schematic of three layers of sphere centers are organized on three triangular grids labeled D, E, and F.

(a)   (b)

Figure 10.3 Comparison of square and triangular packings of equal circles in a given area.

spheres group around the triangular gaps to form tetrahedra and around the square gaps to form octahedra. If we now concentrate on the toothpicks connecting adjacent sphere centers and forget about the spheres themselves, as we showed in Section 7.11, these octahedra and tetrahedra combine to fill all of space in a ratio of two tetrahedra to each octahedron. We will say more about this octet configuration in the following sections.

Both the hexagonal and cubic close-packed systems form the basis for the structure of mineral crystals where one species of atom occupies the sphere positions while other species are located in the gaps between the spheres. Section 10.7 describes this in more detail.

10.3   The Shape of Space

Since the close-packing arrangement of spheres occurs quite naturally by stacking identical spheres and since, like a sphere, space itself is symmetric in all directions, Buckminster Fuller felt that the ccp, or vector equilibrium configuration as he called it, represented the shape of space. Let's see how Fuller used this ccp to give shape to space.

If the centers of touching spheres in a ccp arrangement are connected, a space-filling set of octahedra and tetrahedra remain corresponding to tetrahedral and octahedral interstices between the closely packed spheres. Figure 10.4 shows how each octahedron is completely surrounded by eight tetrahedra sharing its faces while each tetrahedron is completely surrounded by four octahedra sharing its faces. This configuration formed the basis of Fuller's octet truss (see Section

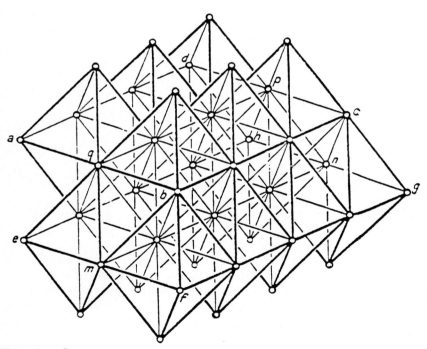

Figure 10.4   Octet truss.

7.11). The space frame can either span a volume, as Figure 10.4 shows, or it can span a planar region as shown in Figure 10.5.

Each vertex of the ccp is surrounded by 12 edges corresponding to the 12 close-packed spheres that touch the sphere represented by it. As we saw in Section 8.2, these are the 12 vertices of the cuboctahedron. Looked at another way, the 12 edges divide this cuboctahedron into 8 tetrahedral and 6 half-octahedral cells, as Figure 8.8 shows.

Furthermore, four planes defined by the cuboctahedron's hexagonal belts are centered at each vertex and correspond to the triangular grid of Section 10.2. Recall that these hexagonal belts appeared as geodesics on the surface of the interpenetrating cube-octahedron pair in Section 8.3 and that they are oriented parallel to the four faces of a tetrahedron.

At each vertex, three additional planes slice through the square cross sections of the six octahedra that are incident to that vertex. These are planes of square sphere packing shown in Figure 10.3(b). Again, these square belts are evident as the geodesics of the interpenetrating tetrahedron duals (stella octangula) of Figure 8.9(a). In fact, the stella octangula represents a local view of the octet configuration with its central octahedron surrounded by eight tetrahedra.

Construction 10.1 If the octahedra and tetrahedra are no longer regular but have edges of varying sizes, the planar space frame develops curvature as shown in Figure 10.6. Architects have used this form to create an impressive variety of architectural structures [Gabriel, 1985].

Now grab a handful of soft spherical pellets and squeeze them together. The 12 points at which these close-packed spheres contact the central sphere widen into plane faces which eventually eliminate the gaps. The result (ideally if not practically) is a space-filling polyhedral version of the close-packed spheres shown in Figure 10.7. Although Fuller called these polyhedra *spherics*, they are actually rhombic dodecahedra—the duals of the cuboctahedra (see Section 8.4).

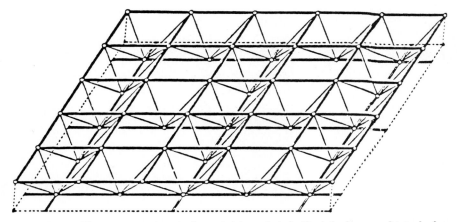

Figure 10.5 Planar arrangement of an octet truss into half octahedrons and tetrahedra.

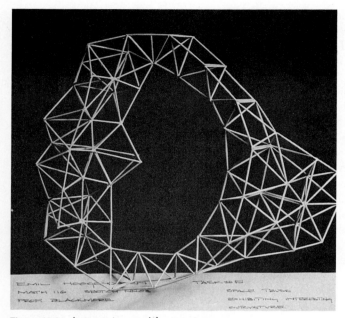

Figure 10.6   A space truss with curvature.

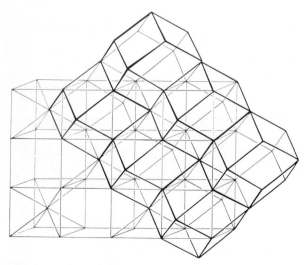

Figure 10.7   Space-filling collection of rhombic dodeca-hedra.

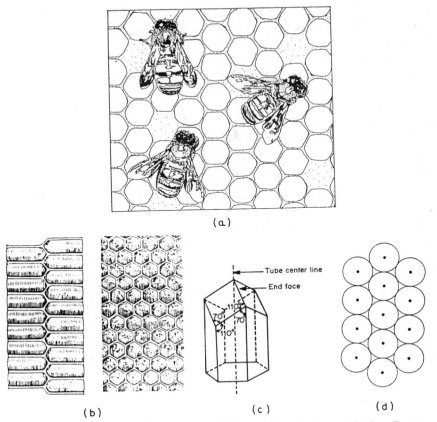

Figure 10.8 The geometry of a beehive. (*a*) Plane section of a close-packed configuration of bees; (*b*) edges of neighboring chambers are flattened to form a hexagonal pattern; (*c*) detail of rhombic dodecahedron end attached to an hexagonal prism; (*d*) each bee working in its own sphere.

Each vertex of the RD occupies the centroid of the cavities between the close-packed spheres. The six vertices of a typical RD in a space-filling array that are surrounded by the acute angles of the rhombic faces fit in the octahedral interstices while the eight vertices at the obtuse angles lie in the tetrahedral interstices. These RD also occur as the shape of garnet crystals.

It is perhaps in this way that bees construct their hives, each working in its own sphere, yet always ending up with beehives in the form of hexagonal prisms capped by half-RDs, as shown in Figure 10.8.

10.4 Packing Ratios

The two arrangements of circles shown in Figure 10.3 are two planar cross sections of the ccp spheres. However, they differ significantly from each other in their utilization of space. For the square pattern, the *packing ratio* is the ratio of circles to squares in a typical unit of the pattern, as shown in Figure 10.3(*b*). If the circles are taken to have unit radius,

$$\text{Packing ratio} = \frac{\text{area of circle}}{\text{area of square}} = \frac{\pi}{4} = 0.785$$

so that about 79 percent of the space in the plane is occupied by the circles. Compare this with the triangular pattern shown in Figure 10.3(*a*) in which packing ratio $= \pi/2\sqrt{3} = 0.905$. Thus, about 91 percent of the plane is covered in this close-packing arrangement of circles.

Now let's compare packing ratios for three different space-filling arrangements of spheres. In *simple cubic packing* (scp) each sphere is positioned at the vertices of an infinite aggregation of cubes. A single cell is shown in Figure 10.9(*a*). Thus each unit cube contains ⅛ of a sphere of radius ½ at each vertex. The density d of the scp arrangement is defined to be the ratio of the volume of the spheres contained in the cube divided by the volume of the cube, i.e.,

$$d = \frac{\frac{4}{3}\pi(\frac{1}{2})^3}{1} = \frac{\pi}{6} = 0.5236$$

A similar computation can be carried out for a *body-centered cubic* packing (bcc) in which each sphere lies at the center of a cube and contacts eight other spheres, each located at the vertices of a typical cube from an infinite aggregation of cubes [see Figure 10.9(*b*)]. Each sphere has a radius one-half the body diagonal of a cube, or $\sqrt{3}/2$. It is easy to verify that the packing density is $d = \pi\sqrt{3}/8 = 0.6801$. (Try it!)

The *face-centered cubic* packing (fcc) places a sphere at the center of each face of a cube so that it touches other spheres located at the ver-

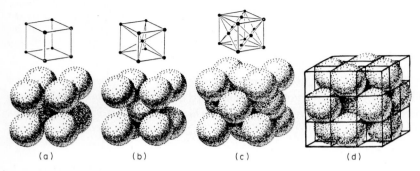

(a)   (b)   (c)   (d)

Figure 10.9   (*a*) Simple cubic packing of spheres; (*b*) body-centered cubic packing of spheres; (*c*) face-centered cubic packing of spheres; another fcc packing.

tices of the cube [see Figure 10.9(c)]. In this packing, each sphere has a diameter of $\sqrt{2}/2$ units if the cube has edge of 1 unit. You can check to see that the packing density is $d = \pi/\sqrt{18} = 0.7408$. Although it is not obvious, this packing of spheres is equivalent to the ccp [see Figure 10.9(d), notice the cuboctahedron]. No other regular packing of spheres has a larger packing ratio, which justifies the reference to fcc as the lattice of *cubic-close packing*.

10.5   Three-Dimensional Lattices

Each of the three arrangements of spheres from the last section forms a three-dimensional lattice structure, obvious generalizations of the two-dimensional lattices of Section 6.7 in that they are invariant under translation in three nonparallel directions. These are three of the fourteen possible lattice types catalogued by Auguste Bravais in 1848 [Bloss, 1971]. However, they are the most prevalent types found in the structure of metallic crystals and salts. The arrangement of spheres in these lattices follows closely the geometry of stacked cubes (see Figure 10.10). For example, in an infinite array of cubes, each corner of a cube is connected to six other corners by edges incident to that corner in the [1,0,0] directions (edge direction). This defines the three nonparallel lattice directions of the scp lattice. Decorating each corner of the cubes in Figure 10.10 with a lattice point [see Figure 10.9(a)] results in a scp lattice.

Each corner of a stack of cubes is also incident to eight cubic cells. The directions from a corner to the centers of these eight cells are along the [1,1,1] direction (body diagonal) of a cube and define the points of the bcc lattice. Placing a lattice point in each corner of the

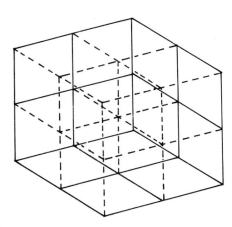

Figure 10.10   Space-filling cubes.

cubes in Figure 10.10 and also at their centers results in a bcc lattice [see Figure 10.9(*b*)].

Each corner of a cube is incident to 12 faces of the adjacent cubes. The directions from the corners to the center of these faces, along the [1,1,0] direction (face diagonal) of a cube, define the fcc lattice. Placing a lattice point at the corners and face centers of each cube of Figure 10.10 results in a fcc lattice. Can you see in Figure 10.9(*c*) that the unit cell of the lattice is the octet parallelopiped formed from two tetrahedra and one octahedron made up of six lattice points at the face centers and two opposite vertices of the cube.?

10.6   Dirichlet Domains

Just as for two-dimensional lattices, the Dirichlet domains (D-domains) of a point from a lattice of points are defined to be the points of space nearer to it than to any of the other lattice points. As for the D domains of two-dimensional lattices (see Section 6.5), to construct the D domain of a lattice point, the lines between that point and all other points of the lattice are perpendicularly bisected by planes, and the innermost envelope of planes is extracted. The resulting polyhedron comprising this innermost envelope is the D domain, and the mirror image points on opposite sides of the faces of the D domain are said to be neighbors of the original point. In fact, these neighbors form another polyhedron called the *coordination polyhedron* [Loeb, 1986].

Since each point of a lattice must belong to some D domain and since all domains are identical for a lattice, the D domains are space fillers. Also, the collection of points that lie at the vertices of the D domains of a lattice are known as *point complexes*. Point complexes are invariant under translations in three nonparallel directions, but they are not, in general, lattices.

The D domains of the scp lattice are cubes, as can be seen by bisecting the six [1,0,0] directions on a cube (see Section 7.13.2) that connect a lattice point to its nearest neighbors. The coordination polyhedron is the octahedron formed by the vertices to which these six edges connect.

Each point of the fcc lattice is directed to its nearest neighbors along the 12 [1,1,0] directions. The envelope of planes that are the perpendicular bisectors of these directions form rhombic dodecahedron D domains, i.e., Fuller's spherics. The coordination polyhedron is the cuboctahedron.

What are the D domains of the bcc lattice? Each lattice point is directed to eight nearest neighbors in the directions of the body diagonals of the cube as we showed in the last section. The eight bisecting planes might form an octahedron. However, octahedra do not fill space by themselves. In constructing the D domains of the bcc lattice, in addition to the eight nearest neighbors, we must consider the six next nearest neighbors located at the centers of the surrounding cubes since these also contribute to the innermost envelope. The resulting polyhedra are truncated octahedra which are space fillers according to Section 9.4. The coordination polyhedron consists of the eight vertices corresponding to the nearest neighbors and the six centers of the surrounding cubes corresponding to the next nearest neighbors. These are the vertices of an RD by the construction of Figure 8.13. So we see that the RD serves as both the D domain of the fcc lattice and the coordination polyhedron of the bcc lattice.

10.7   Crystal Structure

Looking at the configuration of ions that make up a crystal is a little like looking at the stars in the sky. Unless you have a clear ordering principle, the patterns appear chaotic or random. Just as constellations are formed from arbitrary orderings of the stars, there are many possible ordering principles that can describe the positions of ions in a crystal. Now we are in a position to better understand a system that Loeb developed to unify and simplify the interpretation of the crystal structure of common minerals [1975; 1970; 1966].

10.7.1   Cubically close-packed crystals

In many metals (e.g., copper), the atoms arrange themselves as if they were closely packed spheres. Sometimes the packing is hexagonal (DED) and sometimes it is cubic (DEF). In the case of ccp each atom surrounds itself with 12 other identical atoms in the form of a cuboctahedron.

When more than one species of atom is present, a great many of these crystals conform to a model in which one species is close packed while the other species occupies the curvy tetrahedral or octahedral interstices. For example, in sodium chloride (NaCl) crystals, the chlorine ions play the role of the cubically close-packed spheres and the sodium ions fill up all of the octahedral interstices. Since there is one octahedral cavity for each sphere in the packing, this creates just the right setting to house the equal numbers of sodium and chlorine ions present in the crystal. Likewise, potassium oxide (K_2O) crystals are represented by a ccp arrangement of the oxygen ions with the potassium ions occupying the two tetrahedral cavities that are available for each close-packed ion. Loeb has constructed plastic tetrahedron and octahedron chambers called Moduledra with vertices representing the positions of ccp or hcp atoms and interiors modeling the interstices.

He places spheres into the appropriate chamber whenever a species of atom occupies that interstice [1963; 1965].

10.7.2 Cubic close packing with incompletely filled interstices

For the examples of NaCl and K_2O, all of either the octahedral or tetrahedral cavities are filled to capacity by either Na or K ions. But what of situations in which the interstices are only partly filled? To answer this question, first observe that each of the ccp positions, octahedral cavities, and tetrahedral cavities forms a triangular grid oriented parallel to the four faces of a tetrahedron or four geodesics that make up the vector equilibrium, i.e., the four directions or "dimensions" characteristic of space according to Fuller (see Section 10.3). Figure 10.11 shows how such a grid can be subdivided into four equivalent subnets where each subnet occupies the positions of a triangular grid [Loeb, 1958; 1964; 1966] [Morris and Loeb, 1960]. In this network each point of one subnet is surrounded by six equidistant points from the same subnet and two equidistant points from each of the other three subnets. In this way, the maximum number of each species of

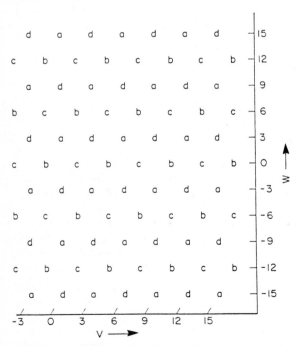

Figure 10.11 Subdivision of hexagonal net into four equivalent subnets.

ion is equidistant from each other. Also, by the geometry of triangular networks, you can see that there is no way to subdivide the nets into two congruent parts.

Now consider the structure of the mineral sphalerite (ZnS). The sulfur ions are cubically close packed while the zinc ions occupy half of the tetrahedral sites. Since the tetrahedral positions cannot be subdivided into congruent halves, only the upright tetrahedra are filled. The emptiness of the downward tetrahedra accounts for sphalerite's electrical polarity. Perovskites (with chemical formula ABX_3; see Section 7.12) offer another example of partially occupied sites. As Figures 7.21 and 10.9(*d*) show, the ccp sites are occupied by both the X and A ions, with A filling the positions in one of the four subnets while X occupies the positions of the other three subnets. The B ions fill one of the four subnets of the octahedral interstices, leaving three subnets open [Loeb, 1970]. Robert Hazen and his group have recently discovered that a superconducting material, a compound of barium, yttrium, copper, and oxygen with the formula $YBa_2Cu_3O_7$ (called 1-2-3 because of the ratio of Y to Ba to Cu ions) has the structure of a perovskite with Y, Ba, and O occupying the ccp sites and Cu occupying the octahedral cavities. This perovskite has an incomplete complement of oxygen atoms since nine atoms are predicted by the formula. Hazen [1988] describes the ingenious way in which the atoms are arranged in 1-2-3 to give rise to its remarkable electronic properties.

10.7.3 The vector equilibrium principle

Unfortunately, the close-packing model does not explain all mineral crystal structures. For example, bcc structures are almost twice as prevalent in metals as the fcc structures described by the close-packing model. Loeb has postulated an organizing principle, the *vector equilibrium principle* (VEP), that agrees with the close-packing model when it is relevant and generalizes it to cases in which it no longer holds [1970]:

> Crystal structures tend to assume configurations in which a maximum number of identical atoms or ions are equidistant from each other. If more than a single type of atom is present, then each atom tends to be equidistant from as many as possible of each type of atom.

Loeb says of this principle that, "the VEP marks tendencies, recognizing the need for compromises in satisfying geometric and stoichiometric constraints." This principle explains the structures of cubically close-packed copper atoms and NaCl crystals on the one hand and crystals such as β-tungsten arranged in a bcc lattice on the other. In both of these cases, one species of atom is arranged in a lat-

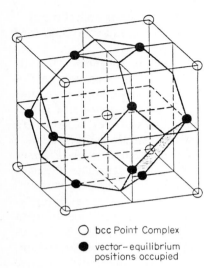

○ bcc Point Complex

● vector–equilibrium
   positions occupied

Figure 10.12 Relation of β-tungsten structure to the D domain of the bcc point complex.

tice (fcc or bcc) while the other species lies at the vertices of the D domain corresponding to the lattice. These vertices of the D domain play the role of the interstices in the close-packing model. For example, the Cl atom lies at the vertices of an fcc lattice while the Na atoms sit at the vertices of the D domain of this lattice, the RD. Since each vertex of a space-filling collection of RDs is surrounded by either four or six cells, each vertex of the RD corresponds to either one of the tetrahedral or octahedral interstices of the close-packed spheres. A picture of β-tungsten is shown in Figure 10.12 with one species of atom lying at the positions of the bcc lattice, and the other species occupying one-half of the sites of its D domain, the truncated octahedron.

This picture conforms to Loeb's general principle in that each atom is a neighbor—in the D domain sense discussed in Section 6.5—of the maximum number of atoms of the same type. For the fcc lattice, the 12 neighbors are equidistant, forming the vertices of the coordination polyhedron of the fcc lattice, the cuboctahedron. The 14 nearest neighbors of the bcc lattice lie at the vertices of the bcc lattice's coordination polyhedron, the RD. These 14 nearest neighbors cannot all be equidistant because the constraints of space permit only 12 equidistant points; rather, 8 neighbors are equidistant while the other 6 are only 15 percent further away. The hexagonal grid of Section 10.7.2 also conforms nicely to the VEP.

10.8   Networks

To characterize metallic crystal structures we focused on the points of the lattice, or point complex, rather than on the network of edges that joins point to point. The lattice model is a good one because the points representing the atoms are held together in their matrix by a field of interatomic forces rather than by bonds from atom to atom, which would be modeled better by connectors.

We would now like to shift attention to the nets joining the points of the lattices and point complexes since these also lead to crystal structures when chemical bonding is important. But of more direct interest to us, they lead to interesting polyhedral forms.

If we connect the lattice points of the bcc lattice to form the 8-connected bcc net, we notice that unlike the fcc network in which connectors define a space-filling set of octahedra and tetrahedra, the bcc net does not define any plane-faced polyhedra. However, if the six edge directions are added to the net to form a 14-connected regular net at each lattice point, the set of space-filling tetragonal disphenoids (not shown) is formed [Pearce, 1978].

However, the bcc network can also define a set of space-filling four-faced polyhedra if we permit these polyhedra to have skew polygonal faces. These generalized tetrahedra are shown in Figure 10.13. Each face is a skew quadrilateral, i.e., a quadrilateral whose edges do not all lie in the same plane. Some authors have chosen to span these skew faces by soap films in order to subdivide space into discrete cells [Pearce, 1978], [Burt, 1973]. So, we see that the bcc lattice can be associated with two distinctly different nets, with each net defining a space-filling set of polyhedra: tetragonal disphenoids in the first case and skew-faced tetrahedra in the second case.

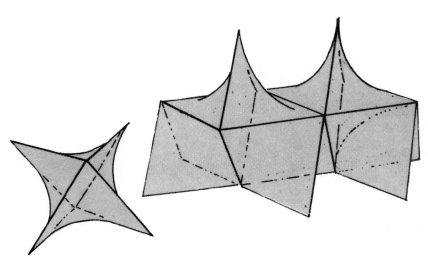

Figure 10.13   Space-filling bcc saddle tetrahedra.

10.9   Infinite Regular Surfaces

M. Burt, M. Kleinmann, and A. Wachman [1974] discovered a large family of infinite regular polyhedra based on nets derived from lattices and point complexes and their duals. Whenever a net defines a space-filling collection of polyhedra with central symmetry, Burt et al. define a dual net as follows: place a vertex at the center of symmetry of the polyhedron defined by the net and pair an edge of the dual net with each face of the polyhedron so that the edges connect the center of symmetry of adjacent polyhedra through the centroid of the face. For example, by this definition the dual net of the fcc connects the centers of each tetrahedron to the four surrounding octahedron centers and the centers of the octahedron to the eight surrounding tetrahedron centers. This dual net is made up of the edges of the space-filling RDs.

The dual net to the bcc net, on the other hand, is the space-filling collection of truncated octahedra when the net is defined in terms of the tetragonal disphenoids. However, when the bcc net is defined in terms of skew tetrahedra, the dual network defines a space-filling array of skew octahedra shown in Figure 10.14, in which each face is a skew hexagon with edges intersecting at right angles. The bcc network and its skew octahedron dual is shown in Figure 10.15.

The infinite regular polyhedral surfaces of Burt et al. are constructed so that all of space is separated by the surface into two tunnels through which a net and its dual net wind. Each tunnel is connected in the sense that any pair of points in one tunnel can be connected by a path lying within the tunnel, but points in different tunnels cannot connect without breaking through the surface. The surface is also regular in the sense that each of its vertices is sur-

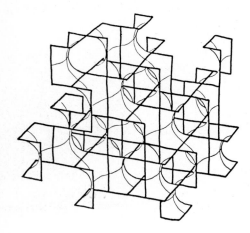

Figure 10.14 Skew octahedral network dual to the bcc network.

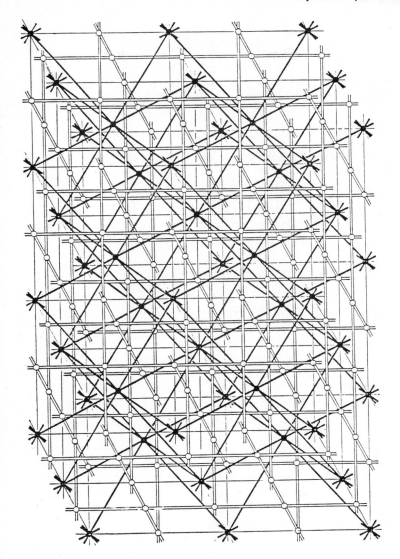

Figure 10.15 Body-centered cubic network and its dual.

rounded identically by regular polygons, i.e., the surface is semiregular. One such surface based on the bcc net and its skew octahedron dual net is shown in Figure 10.16.

Sometimes the infinite regular surface can be constructed by removing selected faces from space-filling polyhedra. For example, Figure 10.17 shows space-filling great rhombicuboctahedra 4.6.8 and octagonal prisms with the octagon faces removed. This is the structure of

Figure 10.16   Infinite regular surface with tunnels along a bcc lattice and its dual.

certain zeolite crystals. A net derived from an scp lattice and its dual net fits through the tunnels. Since the scp net is its own dual, it is not surprising—but nonetheless amazing—that the surface divides all of space into two congruent tunnels.

Figure 10.18(a) shows another surface formed by removing the square faces from a space-filling bunch of truncated octahedra. Again, this surface forms along a scp net and its self-dual and divides space into two congruent connected segments. Four regular hexagons surround each vertex on this surface {6,4}. This infinite polyhedron has all the properties of a platonic polyhedron, namely, identical faces and all vertices surrounded alike. It was discovered by H. S. M. Coxeter in 1937 along with two other infinite platonic polyhedra. Its dual, also an infinite platonic polyhedron with six squares surrounding each vertex {4,6} is shown in Figure 10.19(a). The vertex figures of these polyhedra are also shown in Figures 10.18(b) and 10.19(b) along with the dihedral angles between the faces.

The third Coxeter polyhedron {6,6} is made up of a space-filling combination of truncated tetrahedra and tetrahedra in a 1:1 proportion with the faces of the tetrahedra removed (see Figure 10.20).

Infinite regular polyhedra can be constructed by assembling and then combining the vertex figures or, in certain cases, by constructing

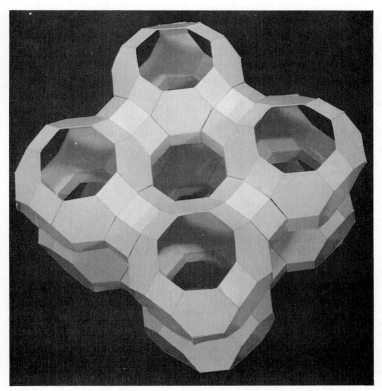

Figure 10.17   Infinite regular surface built from octagonal prisms and great rhombicuboctahedra.

space-filling polyhedra and removing certain faces. [Burt et al., 1974] gives the necessary information to construct many interesting examples.

10.10   The Diamond and Graphite Networks

Carbon appears in nature in two very different forms, graphite and diamond. The first, as we all know, is a soft material whereas the second is the hardest known material. That a single element can be found with properties that are so different can be attributed to the different ways in which the atoms of each bond together to form networks.

Each of the point complexes of diamond and graphite forms a network with four connectors at each point representing the bonds between carbon atoms [Loeb, 1966]. The graphite net has three of its connectors lying in a plane forming a two-dimensional space-filling tiling of regular hexagons while the fourth connector points in the di-

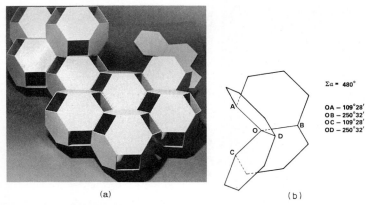

(a)   (b)

Figure 10.18 Infinite regular surfaces of Coxeter. (a) Square faces removed from truncated octahedra to form {6,4}; (b) vertex figure.

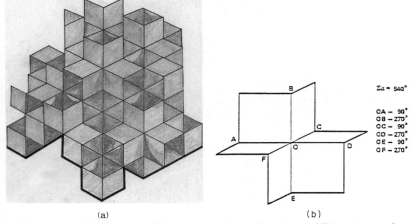

(a)   (b)

Figure 10.19 (a) Faces removed from every other cube of space-filling array to form {4,6}; (b) vertex figure.

rection perpendicular to this plane and has a different length from the others. Each carbon atom lying in the plane is situated both at the vertex of an upright equilateral triangle and at the centroid of a down-pointing equilateral triangle, as shown in Figure 10.21.

On the other hand, in a diamond the carbons atoms occupy half the sites of the bcc lattice. Each carbon has four connectors meeting at the Miraldi angle (see Section 8.9) and pointing in the body-diagonal direction of the cube to the sites of four symmetrically placed nearest neighbors at the vertices of a regular tetrahedron [see Figure 10.22(a). Its dual network is an identical diamond net. The D domains of the

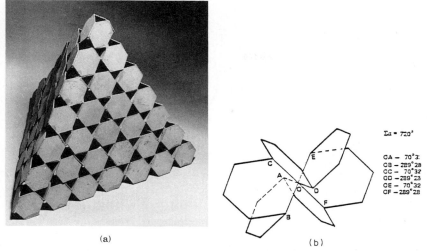

Figure 10.20   (a) Triangular faces removed from tetrahedra and truncated tetrahedra to form {6,6}; (b) vertex figure.

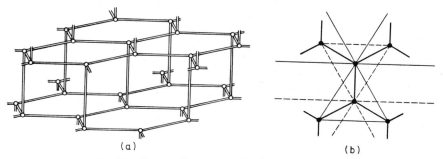

Figure 10.21   (a) Graphite lattice; (b) portion of a plane of carbon atoms in graphite.

diamond point complex turn out to be the space-filling combination of truncated tetrahedra with quarter tetrahedra attached to each triangular face as shown in Figure 10.22(b) [Loeb, 1986].

We see from the structure of its network that the structural weakness of graphite lies in its vulnerability to shearing forces lying within the plane of the hexagonally arranged carbon atoms, whereas the diamond network derives its strength from its ability to withstand forces equally in any direction.

Diamond crystal structures can also be derived from the close-packing of spheres (see Section 10.12). Carbons occupy both the ccp and the upturned tetrahedral interstice positions from Loeb's octet model of ccp crystals (just as we showed in Section 10.7.2 for

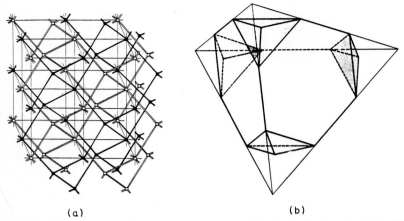

(a)   (b)

Figure 10.22 (*a*) Diamond point complex and self-dual; (*b*) D domain of the diamond point complex.

sphalerite). When the tetrahedron centers are connected to the vertices of the surrounding tetrahedra, they form a 4-connected network of edges. These edges are in the direction of the bonds between the carbons and meet at the Miraldi angle.

10.11   Soap Froths

The soap froth of Exercise 10.2 also forms along the diamond network, at least locally in the neighborhood of each corner of the froth. If you carry out this exercise, you will notice that, just as for the soap bubble in the tetrahedral and cubic frames of Section 8.9, four edges and six films meet at each corner of the froth at the Miraldi angle while three films meet at each edge with dihedral angles of 120 degrees. The edges and faces curve by just the right amount to satisfy these constraints on angle, and this curvature is regulated by pressure differences between the cells of the froth [Hilderbrant, 1984], [Stevens, 1974].

Lord Kelvin, a great British scientist of the nineteenth century, contemplated the precise sense of order represented by a froth of soap bubbles and wondered whether there existed a polyhedron that could stack to fill space meeting all the requirements of a froth. He noticed that 14-faced truncated octahedra stack to fill space in such a way that four edges meet at each vertex and three faces meet at each edge. Although the dihedral angles between the faces are not 120 degrees, and the angles between the edges are not 109.48 degrees, Kelvin proved that all the requirements of the froth could be met by transforming the usual truncated octahedron with six square and eight hexagon faces into one that had planar square faces and saddle-

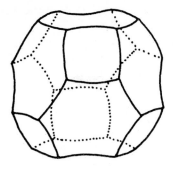

Figure 10.23   Kelvin's ideal soap film.

shaped hexagonal faces. His space-filling polyhedron is known as the *tetrakaidecahedron* and is shown in Figure 10.23.

But, alas, Kelvin's ideal soap bubble is rarely found in froths although it is closely approached in certain simple, homogeneous biological tissues [Williams, 1972]. In Exercise 10.2, you may have discovered, as did F. T. Lewis, that about 75 percent of the cells were either 12-, 13-, 14-, 15-, or 16-faced and that the average number of faces per cell was 13.96, an indication of the tendency toward Kelvin's ideal soap bubble. A froth subdivides three-dimensional space into random polyhedral cells with four edges incident to each vertex. The average number of faces $\langle F \rangle$ and the average number of edges per face $\langle p \rangle$ can be related by the formula [Rivier and Weaire, 1984]

$$\langle F \rangle = \frac{12}{6 - \langle p \rangle} \tag{10.1}$$

This formula is derived similarly to the two-dimensional version given by Equation (6.1) using only combinatorial properties such as Euler's formula and valency relationships. Although there is some question as to its generality in the serendipitous world of polyhedra, the value of $\langle p \rangle = 5.1$ does lead to the experimentally determined value of $\langle F \rangle = 13.96$. There is also strong evidence that this formula is valid with $\langle F \rangle$ in the range of 13.33 to 13.5 for the 24 Frank-Kasper metallurgical phases [Sadoc, 1983], [Shoemaker and Shoemaker, 1986].

10.12   A Unified Look at Nets Related to Cubic Lattices

Although scp, fcc, and bcc lattices and the diamond point complex were all derived from the geometry of a cube, Fuller sees them as all related to each other through the cubic close-packing of spheres. As we have seen, the ccp defines the edges of the octet network. Fuller refers to this network as the *isotropic vector matrix*, or IVM net, as shown in Figure 10.24 [Edmondson, 1987].

Fuller then defines what he calls a *dual net*, IVM′, by connecting the centroids of the cells of the IVM to its vertices. In this way, four edges meet at the centroid of the tetrahedron [Figure 10.25(*a*)] defining a portion of the diamond net, and six edges meet at the center of the octahedron in three mutually perpendicular directions defining the scp net [see Figure 10.25(*b*)]. Thus, at each vertex of the IVM there are six edges in the [1,0,0] directions and four edges in the [1,1,1] directions from the IVM′ system in addition to the twelve edges of the IVM system in the [1,1,0] directions.

Thus the IVM system with no right angles in sight defines a cartesian coordinate system as part of the IVM′ system. Also within the IVM system can be found some of the archimedean polyhedra (at least in a topological sense) from the cubic system. Within the IVM′ system other polyhedra related to the ones from the IVM system can be found. For example, when eight quarter-tetrahedra are added to each octahedron which they surround, an RD is formed by the construction of Figure 8.12.

Finally, if the vertices at the centroids of the tetrahedra are joined with the four octahedra centroids that surround it, as Burt did in defining his dual nets, the remaining [1,1,1] directions needed to define the bcc net are added to the IVM′ system. Now all basic point com-

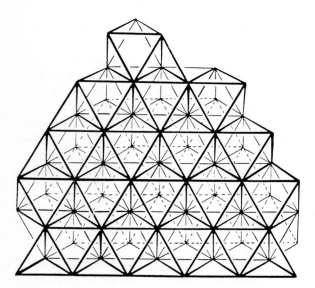

Figure 10.24   Isotropic vector matrix and octet truss.

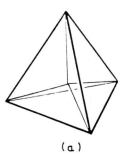
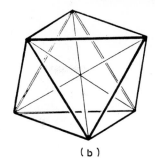

(a)   (b)

Figure 10.25   Central nodes of (*a*) tetrahedron and (*b*) octahedron define the edges of the IVM' net.

plexes, scp, fcc, bcc, and diamond, are defined by Fuller's system in a unified way.

10.13   Zonohedra

In Section 6.7 we showed that only parallelograms and certain kinds of hexagons tile the plane with the same orientation. In the same manner, Fedorov, a Russian crystallographer, gave a proof in 1879 in his book on geometry entitled *An Introduction to the Theory of Figures* that there are only five kinds of convex polyhedra that can be arranged to fill space with the same orientation. They are the parallellopiped, hexagonal prism, truncated octahedron, RD, and the elongated dodecahedron. All of these except the last have been previously discussed, and special representatives of all five kinds are shown in Figure 10.26

The five *Fedorov solids* are related to each other, as shown in Figure 10.26, where they are all generated from a cube by inserting a sequence of square or hexagonal prisms into the previous polyhedron of the sequence [Baracs et al., 1979], [Crapo, 1978*a*], [Coxeter, 1968]. Figure 10.27 shows another way to construct all five Fedorov solids, with the exception of the hexagonal prism, from a transformed parallelopiped whose vertices have been moved [Williams, 1972].

The Fedorov solids share with the parallelopiped from which they were generated the property that opposite faces are parallel and congruent. As a result, they belong to a family of polyhedra called *zonohedra*, the three-dimensional analogues of zonogons, which were discussed in Section 5.10.3. They also share with zonogons the property of being centrally symmetric. Just as zonogons were constructed to have all edges oriented in the direction of a set of vectors, referred to as a *vector star*, zonohedra may be constructed by specifying a star of vectors, not all of which lie in the same plane.

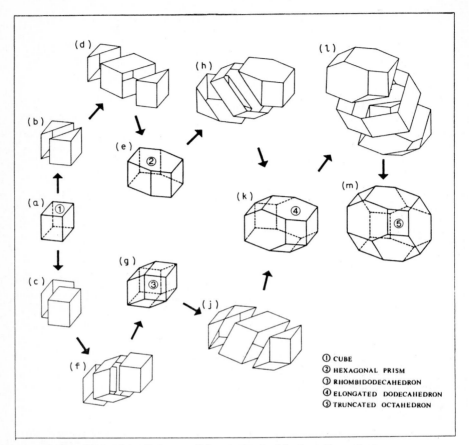

Figure 10.26 Each of the Fedorov solids are obtained by inserting prisms.

Figure 10.28 shows the images of several zonohedra projected onto the plane. In each case, the vector star produces a zonogon as the projected image of a zonohedron. As we saw in Section 5.10.3, each n-zonogon divides into $n(n - 1)/2$ parallelograms which are the projected images of the visible faces of the corresponding n-zonohedron. These faces are indicated in Figure 10.28 by solid lines. A second set of $n(n - 1)/2$ faces, drawn with dotted lines, are the hidden faces of the zonohedron under the projection. Therefore, an n-zonohedron, with all parallelogram faces, has [Coxeter, 1968]

$$F = n(n - 1) \qquad (10.2)$$

faces. Each vertex of an n-zonohedron has incident edges oriented according to some subset of the vector star while the edges divide into n

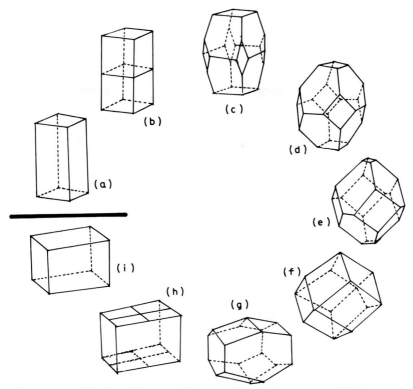

Figure 10.27 The five Fedorov solids generated by vertex motion transformations. A transformation from (*a*) a rectangular prism to (*d*, *e*) forms of 4 6^2; (*f*) rhombic dodecahedron; (*g*) elongated dodecahedron; (*h*, *i*) rectangular prism.

groups of parallel edges called *zones*, each group of which rings the zonohedron with *n* rhombic faces. A zonohedron with all parallelogram faces formed from *n* vectors can also be subdivided into

$$C_n = \frac{n(n-1)(n-2)}{6} \tag{10.3}$$

parallelopiped cells in a number of ways *N* given by the formula

$$N = 2^{n-i} \frac{\{^n_i\}}{C_n} \tag{10.4}$$

where $\{^n_i\} = n!/(n-i)!i!$ and *i* is the dimension of the space, i.e., two dimensions for zonogons, three dimensions for zonohedra, etc. Different sets of these parallelopipeds self-intersect. In this subdivision into intersecting parallelopipeds, all *n* vectors are incident to each vertex

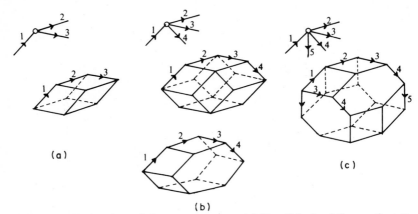

Figure 10.28 Zonohedra formed from vector stars. (a) Parallelopiped from a 3-vector star; (b) RD and hexagonal prism from a 4-star; (c) elongated dodecahedron from a 5-star.

and oriented the same from vertex to vertex. This constellation of parallelopipeds represents a projection of an n-dimensional cube into three-dimensional space (see Section 4.20).

The star of three vectors in Figure 10.28(a) generates a 3-zonohedron, or hexahedron. A star of four vectors gives rise to two possibilities. If no three of these four vectors lie in the same plane, an n-zonohedron in the form of an RD results and its projection is shown in Figure 10.28(b). Six parallelogram faces, represented by solid lines, are visible while the six dotted faces are hidden. Since the vector star contains four vectors, the RD decomposes into two self-intersecting sets of four parallelopipeds (i.e., $C_4 = 4$ and $N = 2$) according to Equations (10.3) and (10.4) with $n = 4$ and $i = 3$. With the RD subdivided by these eight parallelopipeds, the structure has the form of a four-dimensional cube or tesseract shown in Figure 4.68(e). Can you identify the eight intersecting parallelopipeds?

The other possibility for a four-vector star occurs if three of the vectors lie in the same plane. For example, if vectors 1, 2, and 4 of the four-vector star lie in the same plane, while vector 3 does not lie in this plane, the hexagonal prism of Figure 10.28(b) results.

Finally, in Figure 10.28(c) we show an elongated dodecahedron formed from a star of five vectors in which 1, 2, and 4 and 3, 4, and 5 are two sets of coplanar vectors resulting in two parallel pairs of hexagonal faces.

Problem 10.1 Construct the two-dimensional projections of a rhombic triacontahedron (all rhombic faces) and a truncated octahedron from a six-vector star. A six-dimensional cube, or 6-cube, whose outer surface is a rhombic triacontahedron is shown in Figure 10.29.

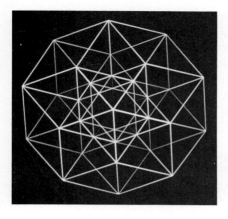

Figure 10.29 A 6-cube enclosed in a rhombic triacontahedron. (*Computer Graphics Labs, NYIT.*)

As for zonogons, what makes zonohedra so important to design is that they can be expanded, contracted, or distorted by altering the length and orientation of the vectors of the vector star. For example, if the lengths of the vectors are altered, the zonohedron is expanded or contracted along the directions of the vector star, i.e., zones, without altering the dihedral angles. On the other hand, edge lengths may be preserved while angles between the vectors change, in which case the dihedral angles of the zonohedra are altered and the zonohedra are accordingly deformed. The space-filling zonohedra or Fedorov solids remain space filling even after they are deformed [Lalvani, 1990].

This way of altering the size and shape of space-filling polyhedra gives zonohedra an advantage over geodesic domes as building structures. The shape of the geodesic dome is fixed by the polyhedron upon which it is based, and in order to change its size, all edges must be

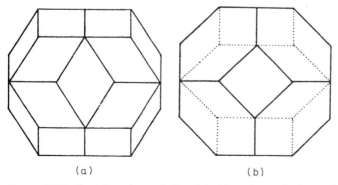

(a)   (b)

Figure 10.30 Certain edges of the rhombic triacontahedron of (*a*) collapse to form the truncated octahedron in (*b*).

altered. Thus the zonohedra has obvious implications for architectural design since it enables polyhedral enclosures to be built which
fit form to function. As a result, modules as versatile as the parallelopiped may be constructed, opening the way to new design possibilities. The structural topology group at the University of
Montreal has experimented along these lines with the RD as an alternative building form to the parallelopiped [Baracs, 1979]. H.
Lalvani has suggested that hypercubes be used as the possible basis
for building form [1987].

Another example of the transformability of zonohedra is shown in
Figure 10.30. By collapsing some of the edges of the rhombic faces of
the rhombic triacontahedron into a plane in a continuous manner, the
hexagonal faces of the resulting truncated octahedron are formed
[Lalvani, 1989]. Lalvani has also created sequences for a computer an-
imation of a continuous transformation of a 6-cube in its various
states. It starts from a portion of a simple cubic lattice through the
intermediate stage of a packed rhombic triacontahedron to the final
stage of a portion of an fcc lattice outlining a truncated octahedron
[1989]. The edges of the cubic lattice [see Figure 10.31(a)] are hinged
so that they open up into rhombic shapes until the triacontahedron
stage of the transformation [see Figure 10.31(b)], after which the
rhombuses close up until the fcc lattice stage is reached [see Figure
10.31(c)]. Notice that the outer shells of the lattice can be superimposed on the faces of Figure 10.30(b). It is interesting that, like in
Fuller's jitterbug, a figure of 5.3.2 symmetry (the triacontahedron in
place of the icosahedron) bridges the transformation from two structures with 4.3.2 symmetry.

There is a natural relationship between the 3-zonohedron (cube), 4-
zonohedron (RD), and the 6-zonohedron (rhombic triacontahedron)
and the three symmetry classes of the platonic solids: 3.3.2, 4.3.2, and
5.3.2, respectively (see Section 7.13.1) [Lalvani, 1989]. Recall from
Section 8.3 that each of these polyhedra has rhombic faces (square in
the case of the cube) whose diagonals are the paired edges of the pair
of intersecting polyhedra that are dual to each other. When the
rhombic faces of one of these zonohedra are divided along their diagonals into four triangles (each of the four sectors of the face can also be
projected to one of the spherical triangles on the inscribed sphere of
this polyhedron as described in Section 9.9 for the case of cubic symmetry 4.3.2), all of the archimedean polyhedra except the snubs can
then be generated by reflection in mirrors placed along these triangles
as we described in Section 9.9. The platonic and archimedean polyhedra can be derived from the 3-, 4-, and 6-zonohedra directly by placing
vertices in appropriate locations within the fundamental region de-

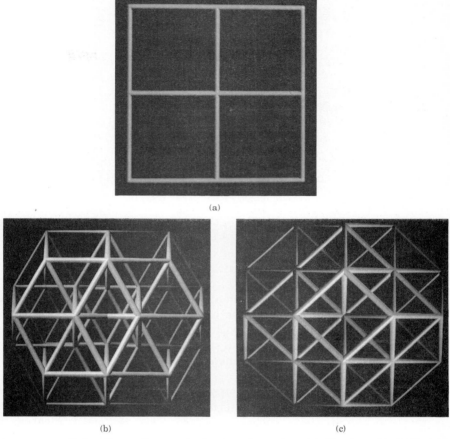

(a)

(b)   (c)

Figure 10.31 Lalvani's continuous 6-cube transformation. (a) Simple cubic lattice; (b) rhombic triacontahedron; (c) fcc lattice. (*Computer Graphics Lab, NYIT.*)

fined to be any one of the four sectors of the rhombic faces. Each sector is a 45 degree right triangle in higher space.

10.14 Golden Isozonohedra

A family of five zonohedra related to the golden mean, called *golden isozonohedra* (GIZ) by Coxeter [1968], have been studied by Baer [1970] and Miyazaki and Takada [1980]. The faces of these zonohedra are identical rhombuses whose diagonals are φ:1 (see Figure 10.32). We have already seen in Section 8.3 that the rhombic triacontahedron

Figure 10.32 Golden rhombus with spherical dodecahedron connectors.

(a six-vector star zonohedron) is a member of this family [see Figure 10.32(c)]. Two other hexahedra shown in Figure 10.32(a) and (b) are also derived from these golden diamonds as are a dodecahedron (four-vector star) and an icosahedron (five-vector star; not shown).

The two adjacent face angles of the golden diamond are equal to the two and only two angles under 180 degrees which are subtended at its body center by arbitrary pairs of vertices of the regular icosahedron. Golden diamonds can be made with Steve Baer's icosahedral system of sticks and connectors (available from [Biocrystal]).

Since they are zonohedra, each of the GIZ can be subdivided into the number of parallelopipeds given by Equation (10.3). In fact, each of the larger ones can be packed with the two golden hexahedra. For example, the dodecahedron can be subdivided into two of each kind of hexahedron while the triacontahedron can be packed with ten of each [see Figures 10.32(d) and (e)].

If a vertex of a triacontahedron packed with 20 golden hexahedra is truncated, the interesting pattern shown in Figure 10.33(a) is obtained [Lalvani, 1989]. Compare this with an identical truncation of

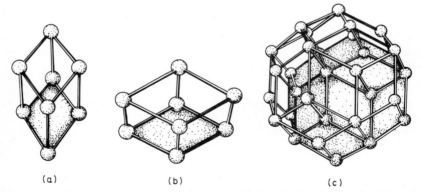

(a)   (b)   (c)

Figure 10.33   Golden isozonohedra. (a, b) Two golden hexahedra; (c) rhombic triaconta-hedron with golden parallelopipeds inserted.

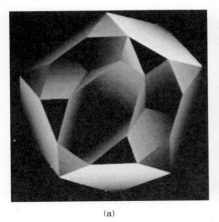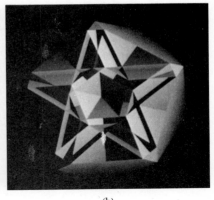

(a)   (b)

Figure 10.34 (*a*) A rhombic triacontahedron filled with 20 golden hexahedra truncated at a vertex; (*b*) a computer-generated picture of a rhombic triacontahedron with 160 self-intersecting hexahedra (6-cube) truncated at a vertex.

the triacontahedron in Figure 10.34(*b*) and on the cover of this book, packed with a full complement of 160 intersecting parallelopipeds predicted by Equations (10.3) and (10.4) (the three-dimensional projection of the 6-cube). Observe the symmetry of the second in contrast to the asymmetry of the first. Lalvani suggests a method of generating other hypercubes that leads to a large family of zonohedra which transform from one to another. Consider two basic examples: since the vector star of the golden triacontahedron is formed from the six directions defined by vectors from the center of the icosahedron to its 12 vertices, why not create a 10-cube and 15-cube from the vectors in the direc-

Figure 10.35 Two pairs of golden hexahedra forming a "left fist" and a "right fist" which result in nonperiodic tilings of three-dimensional space.

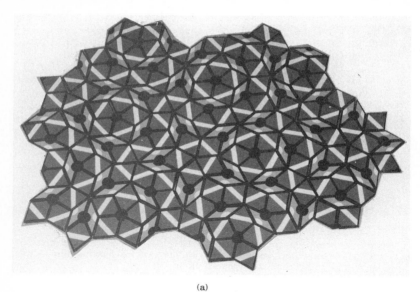

(a)

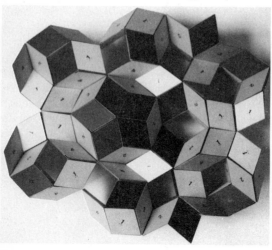

(b)

Figure 10.36 (a) A tiling with Penrose rhombuses superimposed with (b) a three-dimensional nonperiodic tiling with left and right fists.

tions of the icosahedron face centroids and edge bisectors, respectively? What would their truncations look like [Lalvani, 1990; 1986]?

Besides serving as the building blocks of some very interesting structures, these zonohedra may be useful in unlocking some of the mysteries behind quasicrystals (see Section 6.10). In fact, two mirror-image concave polyhedra are obtained by connecting the two different hexahedra across an arbitrary pair of faces. These polyhedra, shown in Figure 10.35, resemble the fists of a right and left hand. If the faces of these left- and right-handed units are properly color coded and numbered (see [Miyazaki and Takada, 1980]), they lead to nonperiodic tilings of three-dimensional space that are exact replicas of the two-dimensional Penrose tilings (see Sections 5.11 and 6.10). Figure 10.36 shows part of one such tiling that lies exactly over the corresponding Penrose tiling. Also, Lalvani's 6-cube, shown in Figure 10.31(b), can be visualized as a quasicrystalline lattice that is the intermediate state in a transformation between two crystalline states (simple cubic and fcc lattices).

Isometries and Mirrors

Is looking glass milk good milk to drink?
                    LEWIS CARROLL
            Through the Looking Glass

11.1   Introduction

Euclidean geometry governs the world of everyday experience. In this geometry, the properties of figures do not change if the figures are moved to a new position with a different orientation or if they are viewed in a mirror. For example, a cube sitting on a table can be moved to another room without the size or shape of either the cube or the table being changed by this process. The class of transformations that govern rigid-body motions and reflections in a mirror are called isometries. Mathematically, *isometries* are defined to be transformations that preserve distances between points. As a consequence of preserving distance, the size and shape of the object are also preserved.

If you take a string lying on a table in the shape of a circle, pick it up, and throw it back onto the table, its length is unchanged; however, it probably won't be a circle any longer, i.e., its shape has changed. So this operation is not an isometry. If you take a photograph and enlarge it, the shapes of the images in the photo are unchanged; however, all sizes are now larger. Hence, this transformation is not an isometry.

In this chapter we will first develop the mathematical concepts and language that will enable us to adequately describe transformations in general and isometries of the plane in particular and some of their properties. We will show that reflections play a fundamental role in describing isometries; in fact we shall show that any isometry of the plane can be carried out by reflections in no more than three mirrors. All this is preparation for the next chapter in which we shall show how symmetrical patterns that have decorated the structures and artifacts of mankind since the dawn of civilization are connected with isometries. We begin this chapter with a discussion of mirrors, and we suggest some experiments involving them [Crowe, 1986].

11.2 Mirrors

Mirrors present us with a world of strange and interesting illusions. Some animals never learn that mirror images are illusions and think that they are seeing another animal when they see themselves in a mirror [Gardner, 1964]. However, dogs and cats are more intelligent and lose interest with the mirror as soon as they realize that they are seeing a mere image of themselves. On the other hand, chimpanzees and young children find great satisfaction with the fact that, whereas the images they see in the mirror are themselves, there are certain subtle differences. They can spend hours exploring these differences. We would like you to go back in time and try to look again at mirrors with the curiosity of a young child. Martin Gardner, whose articles, books, mathematical games, and puzzles have entertained and intrigued millions over the years, suggests several things to do; however, you may add anything that you wish to this list. In response to each of these mirror experiments write a paragraph to describe what you see.

Exercise 11.1 Look in a mirror and wink your right eye. What does your image do?

Exercise 11.2 Write out the words of the following poem by looking at them in a mirror:

> ꙅɘvoꙵ ʏʜƚils ɘʜƚ bns ,gilliɿd ꙅswT
> :ɘdsw ɘʜƚ ni ɘldmig bns ɘɿʏg biᗡ ,
> ,ꙅɘvogoɿod ɘʜƚ ɘɿɘw ʏꙅmim llA
> .ɘdsɿgƚuo ꙅʜƚsɿ ɘmom ɘʜƚ bnA

Exercise 11.3 Place the following objects in various orientations before a mirror: a ball, a cube, an egg, a clock, a helical spring, a knot, your right hand, at least three other objects of your choosing.

Exercise 11.4 Look at a painting in a mirror. Does anything look strange about the painting or is it the mirror image of an equally valid painting?

Exercise 11.5 Take a pair of mirrors, as shown in Figure 11.1, and look at your face in the mirrors. Wink your left eye. What does your image do? How does your left hand look? Rotate the mirrors 90 degrees to each other. What do you see?

Exercise 11.6 Take a curved metal sheet like the one in Figure 11.2 and look at your reflected image in the sheet. What do you see? Change the orientation of

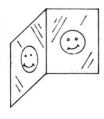 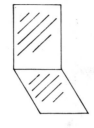

Figure 11.1 An image seen in two mirrors.

Figure 11.2 A curved mirror.

the sheet. What do you see? Try looking at objects in mirrors with different curvatures and record your observations [Thomas, 1980].

Exercise 11.7 The following sum is wrong:

$$
\begin{array}{r}
3414 \\
340 \\
74813 \\
\hline
4337813
\end{array}
$$

Look at the sum in a mirror and show that it is now correct.

Exercise 11.8 Look at the names in Figure 11.3 in a mirror. Why is TIMOTHY not reversed? Which of the letters of the alphabet look the same when seen in a mirror? Which will not look the same no matter how you orient them?

Exercise 11.9 Turn the example of Scott Kim's calligraphy in Figure 11.4 upside down. His book shows many other astounding examples [Kim, 1981].

Figure 11.3 Why is TIMOTHY unaffected by a mirror while REBECCA is altered?

Figure 11.4   An example of the symmetric calligraphy of Scott Kim.

Exercise 11.10   Why does an ordinary mirror appear to reverse right and left but not up and down?

When we look in a mirror and wink our right eye, we see our image wink its left eye so we say that mirrors reverse left and right. Actually, the mirror does not reverse left and right since it is really the eye on the right side of the mirror that winks when we wink our right eye. In fact it is front and back which the mirror reverses. We imagine that we can walk behind the mirror wherein the person in the mirror appears to have his or her left-right orientation reversed. Likewise, an asymmetric object such as a left glove becomes a right glove in the mirror in the sense that if the left glove were carried around to the other side of the mirror, it would not match up with its image, whereas a right glove would. Is there any way that a left glove can be turned into a right glove so as to match up with its mirror image? Strangely enough the answer is yes; however, the explanation is worthy of a science fiction story rather than a mathematics book and will be deferred to Section 11.9.

Physicists are particularly fond of symmetry. Recently, it was discovered that for every elementary particle there exists a mirror image particle (although it is not a strictly geometric mirror image). For example, corresponding to an electron there exists a positron with the same size but with opposite charge. Corresponding to protons there are antiprotons and for neutrons there are antineutrons. It has even been conjectured that there are mirror images of all forms of matter, known as antimatter. It is also thought that when these two mirror image forms of matter combine, they disintegrate with a large explosion which would seem to place Alice's "looking glass milk" in jeopardy.

11.3   Sets

Sets were introduced in Section 4.2 where they were defined as collections of objects along with rules enabling one to decide whether or not a given object belongs or does not belong to a particular set. The objects in a set are called its *elements*, or *members*. Two sets are equal if they have the same members.

If the set has finitely many (and not too many) members, we can display them. For example, {1,2,3,4,5} is the set consisting of the integers 1, 2, 3, 4, and 5. Even if a set is infinite, we can represent it by a display of this type if its elements fall into a sequential pattern; for example, {2,4,6,8,...} represents the set of positive even numbers. There is a convenient brace notation that can be used for any set, namely, $A = \{a \mid P(a)\}$. Here $\mid$ stands for "such that" and $P(a)$ is the rule of set membership. A is the set of all objects such that $P(a)$ is true. For example, $E = \{x \mid x$ is a positive and even number$\} = \{2,4,6,8,...\}$ while $\{x \mid x$ is a real number and $x^2 = 1\} = \{1, -1\}$.

We use captial letters for sets and lowercase for the members of a set. If an object x is a member of a set A, we write $x \in A$. The symbol $\in$ stands for "belongs to." If y does not belong to A, we denote this by $y \notin A$. For example, in the set of positive even numbers, $2 \in E$ but $1 \notin E$.

If $x \in A \Rightarrow x \in B$ (where $\Rightarrow$ denotes implies), we write $A \subset B$ and say set A is contained in or is a subset of set B.

11.4 Mappings

A mapping is a rule of correspondence from one set to another. Architects use mappings in drawing plans, constructing models, drawing pictures of buildings, and in many other ways.

> **Definition 11.1** A mapping (or map) consists of a pair of sets A and B and a rule of correspondence which associates each element of A to one and only one element of B. It is usually denoted by
>
> $$M: A \to B \qquad \text{or} \qquad A \to B$$
>
> where M can be thought of as the rule of correspondence. A is called the *domain* and B the *codomain* or *range* of the mapping.

We often denote the mapping $M: A \to B$ by M alone. If M associates $a \in A$ with $b \in B$, we call b the image of a or value of M at a and write $M(a) = b$ or simply $Ma = b$. By definition, a mapping has the property that $Ma = b$ for a unique $b \in B$. Thus $Ma = b$ and $Ma = c$ cannot both be true if $b \neq c$. Pictorially, $M: A \to B$ is represented by a set of arrows from A to B; one for each element $a \in A$ as shown as Figure 11.5.

Also, if $M: A \to B$ is a mapping and $C \subseteq A$, the image of C (under M) is defined by $M(C) = \{M(x) \mid x \in C\}$. See Figure 11.5.

The mathematical definition of a mapping is highly abstract, but it can be simply illustrated by placing two collections of objects in two

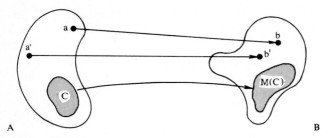

Figure 11.5   Diagram of a mapping from set A to set B.

cups, one labeled A and the other labeled B. For example, sets A and B might consist of

$$A = \{\text{pen, key, dime, chalk}\}$$

$$B = \{\text{nickel, quarter, eraser, handkerchief}\}$$

For convenience we use the abbreviations p = pen, k = key, d = dime, c = chalk, n = nickel, q = quarter, e = eraser, h = handkerchief.

The mapping $M_1 : A \rightarrow B$ is represented by emptying the objects from sets A and B onto the table. To each object of A we associate a unique element of B. For example, to the pen we associate the quarter which we set beside it. To the key we associate the nickel, to the dime we also associate the nickel, and to the chalk we associate the handkerchief. These associations are depicted schematically in Figure 11.6. Notice that while every element from set A is mapped to a unique element in set B, one element in set B, the nickel, is associated with (or mapped from) two elements in set A, the key and the dime, while another element of B, the eraser, is not associated with any element of A. With the above notation this mapping can be represented by

$$M_1 p = q \qquad M_1 k = n \qquad M_1 d = n \qquad M_1 c = h$$

It is thought that numbers originated from the need to keep track of such possessions as animals killed in the hunt or cattle out to pasture. At first, a stone was associated with each animal to be kept track of. Later the stones were replaced by counting numbers which served the

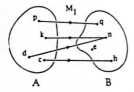

Figure 11.6   (a) A many-to-one map.

same purpose. The key to this procedure was a special mapping of animals to numbers. Such a mapping is an example of a one-to-one mapping.

Definition 11.2 A mapping $M: A \rightarrow B$ is said to be *one to one* if for each $b \in M(A)$ there is a unique $a \in A$ such that $Ma = b$.

Thus, mapping M_1 in Figure 11.6 is not one to one. In order to be one to one, arrows from different points in A cannot lead to the same point of B. For example, if d is associated with e instead of n, the resulting mapping, M_2,

$$M_2 p = q \qquad M_2 k = n \qquad M_2 d = e \qquad M_2 c = h$$

is one to one and is illustrated in Figure 11.7.

Since each element of B is mapped from some element of A, we say that M_2 is an onto map in contrast to M_1 which is not onto since e has no element of A to which it is associated.

If $M : A \rightarrow B$ is both *one to one* and *onto*, we say it is a *bijection* or one-to-one, onto correspondence. We shall use the word *transformation* to denote a bijection $M: A \rightarrow A$, where domain and range are the same sets.

If M is a one-to-one, onto mapping (bijection), it induces another mapping $M^{-1}: B \rightarrow A$ defined as follows:

$$x = M^{-1}(y) \qquad \text{if and only if} \qquad (\Leftrightarrow)M(x) = y$$

The mapping M^{-1} is called the *inverse* of M. Pictorially it is represented by reversing all the arrows in Figure 11.7. Note that for the inverse to exist, M must be onto so that $M(x) = y$ has a solution x for each $y \in B$, and it must also be one to one so that the solution is unique.

Example 11.1 Given any set A, the identity map $I: A \rightarrow A$ where I is defined by $I(a) = a$, for all $a \in A$ is a transformation of $A \rightarrow A$.

By the definition of the inverse map, we see that any element of A is mapped to itself, (i.e., is an identity transformation), by first trans-

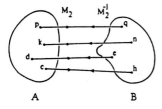

Figure 11.7 A one-to-one mapping.

forming it with M and then transforming the result of this mapping by M^{-1}. This is also true if the transformations are carried out in the reverse order. In other words,

$$M^{-1}(M(x)) = x = M(M^{-1}(x)) \qquad \text{or} \qquad M^{-1}M = I = MM^{-1}$$

for all x in either domain or range of M. The parentheses have been omitted in the second expression and the *product* of two transformations denoted by BA means "first carry out mapping A on an element, then carry out B on the result."

The remainder of this chapter is devoted to a discussion of the important class of transformations known as *isometries of the plane* which, as we described at the beginning of this chapter, transform sets of points in the the plane in such a way that distance between pairs of points is preserved. We will rely on the following two important theorems (given without proof) [Martin, 1982]:

Theorem 11.1 If a set of points in the plane is transformed by an isometry to another set of points in the plane, the transformation is either a translation, rotation, reflection, glide reflection, or the identity.

Theorem 11.2 The product of two isometries is another isometry.

The remainder of the chapter is devoted to describing these four basic isometries—translation, rotation, reflection, and glide reflection—and their properties.

11.5 Translations

Translations are familiar to us as the transformations that preserve the lattices in Section 6.7. A *translation T* of points in the plane is defined by specifying a vector **v** of translation. Points in the plane are transformed by translation when a typical image point p' is located at the tip of the vector whenever the object point p has its position at the tail of the vector as shown in Figure 11.8. If another point q is transformed to q', the distance d between object and image points is preserved as must be true for all isometries, i.e., $d\{p,q\} = d\{p',q'\}$. Also, it is clear that no point is mapped to itself under a translation; in other words, there are no fixed points under translation.

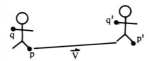

Figure 11.8 Translation of a figure in the plane.

Figure 11.9   Multiple translations. Kala nut box, Nigeria.

The translation is denoted, in the notation of Section 11.4, by T: $p \rightarrow p'$ or by $Tp = p'$. Since transformations are one to one onto mappings, according to Section 11.4, they have well-defined inverses. Thus there exists another translation, T^{-1}, with the property that $T^{-1}p' = p$ and therefore $T^{-1}Tp = p$ for all p. Also $TT^{-1}p' = p'$ or $T^{-1}T = I = TT^{-1}$ where I stands for the identity transformation. From Figure 11.8, it is clear that the inverse translation T^{-1} is defined by the vector $-\mathbf{v}$ directed opposite from $\mathbf{v}$.

A geometric pattern or motif and its translations are illustrated in Figure 11.9. If the motif is transformed by products of the translation T and its inverses,

$$\ldots T^{-3},\, T^{-2},\, T^{-1},\, I,\, T,\, T^2,\, T^3,\ldots,$$

the result is a linear train of reproductions of this set.

11.6   Rotations

A *rotation* S of points in the plane is defined by specifying the center of rotation O, the angle of rotation θ, and whether the rotation is clockwise or counterclockwise, with counterclockwise rotations being denoted by positive angles while clockwise rotations are denoted by negative angles. Figure 11.10 shows the result of rotating a typical pair of points p,q to image points p',q' counterclockwise through angle θ about the center O. Once again, distance is preserved by the rotation, i.e., $d\{p,q\} = d\{p'q'\}$. Also, O is the only fixed point of the rotation.

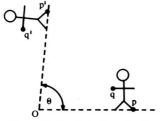

Figure 11.10   Rotation of a figure in the plane.

Figure 11.11   Multiple rotations.

In the notation of transformations,

$$S: p \rightarrow p' \quad \text{or} \quad Sp = p'$$

and the inverse rotation maps

$$S^{-1}p' = p \quad \text{or} \quad S^{-1}S = I = SS^{-1}$$

Such an inverse rotation must be the rotation about O through the same angle as S but in an opposite sense.

A geometric pattern or motif is transformed in Figure 11.11 by a sequence of rotations and inverse rotations:

$$I, S, S^2, \ldots, S^{n-1}$$

where $S^n = I$ and $S^{n-k} = S^{-k}$ to obtain a repeating circular pattern.

11.7   Reflections

A mirror of infinite length placed perpendicular to the plane defines a *reflection* R of the points of the plane in the mirror. The trace of the mirror on the plane is called a mirror line M, and the image p' of a typical point p lies on the perpendicular line to the mirror that contains p, the same distance d from the mirror as p but on the opposite side as shown in Figure 11.12. Again, distances between any pair of

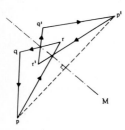

Figure 11.12   Reflection of a figure in the plane.

points p,q and their images p',q' are preserved, i.e., $d\{p,q\} = d\{p',q'\}$. All the points on the mirror line are fixed points of the reflection.

The transformation is denoted by

$$R : p \rightarrow p' \qquad \text{or} \qquad Rp = p'$$

and the inverse transformation is clearly identical to R, i.e., $R^{-1} = R$ since the reflection of a reflection maps any point to itself, i.e., $RR = I$.

In Figure 11.12, a set of points is mapped by the reflection to the image set on the opposite side of the mirror. Notice that the reflection reverses the direction of the boundary curve which appears counterclockwise on the object curve but clockwise on the corresponding image curve.

11.8   Glide Reflection

The *glide reflection G* is the least familiar of the isometries. One way of thinking about it is as a combination of a translation and a reflection. It too is defined by an axis of glide M, together with a vector $\mathbf{v}$ parallel to M. A typical point p is glide reflected to its image p' by translating it through vector $\mathbf{v}$ to p'' and then reflecting p'' in the *glide axis M* to p' as shown in Figure 11.13. It is easy to see that the axis of glide reflection bisects the line between any point and its image.

As was true for translations, a glide reflection has no fixed points under G, and as we saw for reflections, the sense of the boundary curves of geometric figures are reversed under G. Also, the inverse glide reflection has the same glide axis as G, but the translation is in the opposite direction $-\mathbf{v}$. It is also evident that the product of two glide reflections, i.e., a glide reflection of a point followed by another glide reflectlion of the image point, is a translation through $2\,\mathbf{v}$.

A pattern formed by subjecting a motif to a sequence of glides and their inverses is shown in Figure 11.14.

$$\ldots G^{-2}, G^{-1}, I, G, G^{2}, \ldots$$

As you see, glide reflections nicely describe the pattern of footprints in the snow. Figure 11.15 shows a more interesting pattern constructed from glide reflection of a motif.

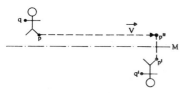

Figure 11.13   Glide reflection of a figure in the plane

Figure 11.14   Footprints along a line have a glide reflection symmetry.

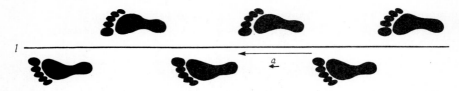

Figure 11.15   Multiple glide reflections.

11.9   Proper and Improper Transformations

E. A. Abbott's fable, *Flatland* [1952], is about how strange three-dimensional space would seem to creatures accustomed to living in a two-dimensional world. His creatures are all two-dimensional and reside in the plane. Each resembles the other [depicted in Figure 11.16(*a*)], with mouths on the right side of their faces and geometrically congruent. Each of these creatures can move in the plane with a rigid-body motion so that they match up with any of the others. One day a stranger moves into Flatland. Nothing like this individual has ever before been seen. As Figure 11.16(*b*) illustrates, his mouth is on the left side of his face although all other features appear normal. Try as they might, no creature from Flatland can match himself up with the stranger. That is, not until someone with the intuition of a Flatland's Einstein recognizes that by lifting the stranger out of the plane which comprises Flatland's universe, turning him over in an unknown higher-dimensional space (three-dimensional space), and replacing him, a match up can be achieved.

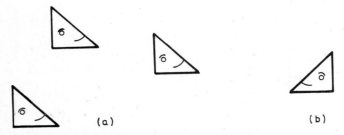

Figure 11.16   Three Flatland creatures encountering an alien.

Although all the Flatland creatures and the stranger are congruent by the standards of euclidean geometry, there is obviously something different about the two classes of images. Match ups between pairs of Flatlanders or pairs of strangers can be achieved by rigid-body movements (isometries) entirely in the plane; they are directly congruent, whereas superimposing a Flatlander over the stranger, or the opposite, requires a rigid motion out of the two-dimensional space that comprises Flatland: They are enantiomorphic (see Section 2.2).

The preceding discussion indicates why it is useful to subdivide isometries into two categories: proper isometries P which transform geometric figures by rigid-body motions entirely within the plane and improper isometries I which can be only carried out by rigid-body movements that remove the figure from the plane. Rotations and translations are examples of proper isometries while reflections and glide reflections are improper.

A little thought on this matter will lead you to the conclusion that the composite transformation of a proper transformation P followed by another proper transformation results in a proper transformation, i.e., $PP = P$. Likewise,

$$PI = I = IP \qquad \text{and} \qquad II = P \tag{11.1}$$

where improper isometries are denoted by I. This is analogous to the situation in which even and odd numbers are added together:

$$E + E = E \qquad E + O = O = O + E \qquad O + O = E$$

where E stands for even and O stands for odd numbers. For this reason proper isometries are sometimes called even while improper isometries are referred to as odd.

A similar situation exists for isometries in three-dimensional space. Translations and rotations have the effect of transforming figures by rigid-body motions in three-dimensional space. Reflections and glide reflections require the figure to be moved into a higher-dimensional space (this time four-dimensional space), turned over, and then replaced in three-dimensional space. In the same way a left glove can become a right glove if it is removed from three-dimensional space, inverted in a world of four dimensions, and returned to our three-dimensional world [Rucker, 1989], [Banchoff].

11.10   Isometries and Mirrors

Mirrors are objects of great fascination and have many nonintuitive properties as we showed in Section 11.2. They are also very much connected to the idea of symmetry. In fact, in Section 7.13.3, we mentioned that the reflection symmetry of a figure can be detected with the help of a mirror. The relation between mirrors and symmetry will be discussed in greater detail in the next chapter. In preparation for this discussion, we investigate the mathematics behind transformations of points by reflection in a mirror. We again limit ourselves to reflections in the plane, for which a mirror is represented by a line in the plane. We describe the transformations that result from points reflected first in a single mirror, then in two mirrors either intersecting or parallel to each other, and finally in three mirrors. It will turn out that no more than three mirrors are needed to generate any isometry. Before proceeding, we recommend that you carry out Exercise 11.11.

Exercise 11.11 [Martin, 1982]   The materials needed are a pencil and at least three sheets of waxed paper, each sheet about 30 centimeters square. A ruler and a protractor might also help. The first sheet of waxed paper is used to introduce the technique of using waxed paper to illustrate reflection. Fold the sheet in half. The crease represents the mirror line M. It is obvious how to find the image p' of any point p. We merely fold the sheet on M, trace the point p with the pencil (from either side of the sheet), unfold the sheet, and label the new point p'. Find the images of the integers 5, 6, 7 and the line N shown in Figure 11.17(a). You should get a result that looks like Figure 11.17(b).

Now trace the image of the image of 5, 6, and 7 under reflection in the line N. As Martin points out, this is not as easy to do as it sounds, and you should not be impatient with yourself when you find you have traced a wrong figure. Now ask yourself, "What do I conjecture is the result of first reflecting in line M and then in line N?" The protractor can be useful here in helping you make your conjecture. If you feel challenged, you might also carry out this experiment for triple reflections in three mirror lines that meet at a common point: reflect under M, then N, then L. Make another conjecture. Finally, try tracing the images of points under reflections in lines M and then N when these lines are parallel. Try reflections in three mirror lines M, then N, then L. What result do you get? Note that parallel lines can be constructed as those lines perpendicular to a given line U. When the paper is folded two or more times in such a way that U coincides with itself, the fold lines must be parallel.

Let's see how your conjectures bear up under the following analysis [Baglivo and Graver, 1983].

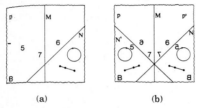

(a)   (b)

Figure 11.17   Figures reflected in mirror line M.

11.10.1   Reflection in a single mirror

Consider a mirror M and a point p located a distance d from M as shown in Figure 11.12. The transformed point p' is located distance d on the other side of the mirror.

Rule 1   A mirror M is always the perpendicular bisector of the line between a point and its reflected image.

If we use the letter R to stand for a reflection in mirror M, RR denotes a reflection followed by a reflection. Since a reflection followed by a reflection leaves all points p in the plane unchanged, we can write as we did in Section 11.7:

$$RR = I \quad \text{or} \quad R = R^{-1}$$

where I stands for the identity transformation.

11.10.2   Reflection in two mirrors

Consider mirrors M_1 and M_2 (assumed to be of infinite length). A reflection in M_1 followed by a reflection of the result in M_2, i.e., R_2R_1, by Theorem 11.2 is clearly an isometry. Since both R_1 and R_2 are improper transformations, according to Section 11.9, their product must be proper, i.e., either a rotation or a translation. But if the mirrors intersect, say at point O, then O must be a fixed point of R_2R_1 (why?). In this case, using Theorems 11.1 and 11.2, the product of the two reflections must be a rotation S since translations have no fixed points. On the other hand, if the two mirrors are parallel, the product of the reflections has no fixed point (why?) and the resulting proper isometry must be a translation. Let's now consider the geometry of these two cases in a little more detail.

Intersecting mirrors.   Let's say mirrors M_1 and M_2 intersect with angle θ at O and consider an arbitrary point p in the plane. Reflect p first in M_1 to p'', then in M_2 to p' as shown in Figure 11.18(a). Notice that $Op' = Op$ (why?). Can you prove that angle $pOp' = 2\theta$? (Do this!)

Let's now reflect p first in M_2 and then in M_1. Again, $Op' = Op$ and angle $pOp' = -2\theta$. (Prove this!) We can therefore state the following rule:

Rule 2   If any point in the plane is reflected successively in two intersecting mirrors, the transformed point is rotated about the point of intersection by twice the angle between the mirrors and with the same sense as the angle between the first and second mirror.

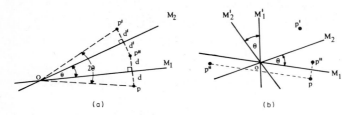

Figure 11.18　(a) Multiple reflection of a point in two intersecting mirrors produces a rotation; (b) the reflection is unaffected by the positioning of the mirror pairs.

If we use R_1 and R_2 to stand for reflections in mirror M_1 and M_2, respectively, and S for the rotation that results from successive reflections, rule 2 can be stated algebraically as

$$S = R_2 R_1 \qquad \text{and} \qquad S^{-1} = R_1 R_2 \tag{11.2}$$

where S^{-1} is, according to Section 11.6, a rotation with the same angle but opposite sense as S.

> **Remark 1**　Without proof we state that any two mirrors intersecting with angle θ at point O will have the same effect upon an arbitrary point of the plane regardless of the orientation of the mirrors. Thus, the pair of mirrors M_1' and M_2' of Figure 11.18(b) will have the same effect on point p as M_1 and M_2; namely, they will transform p to p'. (Prove this!)

Parallel mirrors.　Now let's consider two parallel mirrors, M_1 and M_2, a distance L apart. If p is reflected first in M_1 to get p'' and then p'' is reflected in M_2 to get p', the result of this multiple reflection is to translate p to p' by an amount $2L$ as Figure 11.19(a) shows. Likewise a reflection first in M_2 and then in M_1 translates p to p' by an amount $2L$ in the opposite direction. Thus we can state rule 3.

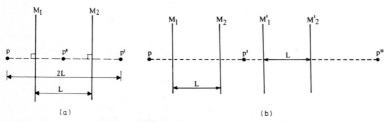

Figure 11.19　(a) Multiple reflections of a point in two parallel mirrors produces a translation; (b) the reflection is unaffected by the positioning of the mirror pairs.

Rule 3 Consecutive reflections in two parallel mirrors translate a point to its image point by an amount twice the distance between the mirrors in a direction from the first mirror to the second mirror.

If R_1 and R_2 stand for reflections in M_1 and M_2 and T refers to the translation corresponding to the multiple reflection, in accordance with Section 11.4, we can summarize rule 3 algebraically as follows:

$$T = R_2 R_1 \qquad \text{and} \qquad T^{-1} = R_1 R_2$$

Remark 2 If two parallel mirrors a distance L apart are oriented the same as M_1 and M_2 and placed anywhere, they have the same effect on points of the plane, i.e., they result in a translation T through twice the distance between the mirrors. For example, if the mirrors are placed so that M_1 is located on point p in Figure 11.19(b), it is directly evident that p' is translated a distance $2L$ in the direction from M_1 to M_2.

11.10.3 Reflections in three mirrors

Since reflections are odd transformations, according to Section 11.9, three consecutive transformations result in another odd transformation, i.e., a reflection or a glide reflection.

Now let's look at the different ways in which three mirrors can be oriented relative to each other. There are five distinct ways, and they are shown in Figure 11.20. (1) The mirrors can all intersect at a point, (2) they can all be parallel, (3) two can be parallel with the third perpendicular to them, (4) two can be parallel with the third cutting them obliquely, or (5) they can intersect each other but not all at the same point.

It might seem tedious to analyze each of these five cases. However, the first two cases have already been considered by rules 1, 2, and 3. The fact that we are now dealing with three mirrors presents no new problems.

Problem 11.1 Successive reflections in three mirrors intersecting at a point O are equivalent to a single reflection in a mirror through O. Where is the equivalent mirror located? In a similar vein, where is the equivalent mirror to three parallel mirrors located?

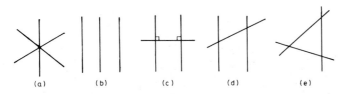

(a)　　　(b)　　　(c)　　　(d)　　　(e)

Figure 11.20 Five ways in which three mirrors can intersect.

Let's consider case (3). A point p is reflected first in parallel mirrors M_1 and M_2 a distance L apart and then in the mirror M_3 perpendicular to M_1 and M_2. The result of the first two reflections is to translate a typical point p through a distance $2L$ from M_1 to M_2 to p'' followed by a reflection in M_3 to p'. It is evident from Section 11.8 that such a transformation is a glide reflection and that M_3 is the axis of the glide reflection.

We are now ready to analyze case (4) illustrated in Figure 11.20(d). We shall show that it is also a glide reflection by reducing it to case (3). The succession of reflections in M_1, M_2, and then M_3 can be thought of first as a reflection of p in M_1 to p'' and then a rotation about O in Figure 11.21(a) through twice the angle between M_2 and M_3 to p', i.e., $R_3R_2R_1 = (R_3R_2)R_1 = S\,R_1$. Although this is fine, it is not evident that this transformation is a glide reflection, and we are left in the dark as to where the glide axis is located.

We can reapproach this transformation by using the fact that the effect of successive reflections in M_2 and M_3 does not depend on how these mirrors are oriented about O (see Remark 1 above). For example, M_2 and M_3 have been rotated about O until M_2' is perpendicular to M_1 as shown in Figure 11.21(b). Likewise, M_1 and M_2' are rotated about O' until M_2'' is perpendicular to M_3' as shown in Figure 11.21(c). This reduces case (4) to case (3) where M_2'' is the axis of the glide reflection.

By a similar argument, case (5) can also be reduced to case (3). We leave this to the reader.

11.10.4 A major theorem about isometries

We have described how reflections in one, two, or three mirrors give rise to all the isometries of the plane, namely, reflections, rotations, translations, and glide reflections. Now we summarize all the results of this section by an interesting theorem.

Theorem 11.3 Any isometry of the plane can be carried out by a series of no more than three reflections. (In other words, any two congruent geometric fig-

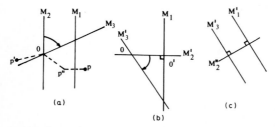

Figure 11.21 (a) Multiple reflections in three mirrors produces a glide reflection; (b), (c) geometric construction of the glide line.

ures can be made to coincide by subjecting one of them to a sequence of no more than three reflections.)

proof It is sufficient to match up the two congruent triangles ABC and $A'B'C'$ shown in Figure 11.22. The procedure is as follows:

1. Choose two corresponding vertices, for example A and A', and find the perpendicular bisector M_1 of AA'.
2. Reflect triangle ABC in a mirror at M_1 to triangle $A'B_1C_1$.
3. Reflect triangle $A'B_1C_1$ in the perpendicular bisector M_2 of two other corresponding vertices, say B' and B_1 to triangle $A'B'C_2$.
4. Reflect triangle $A'B'C_2$ to triangle $A'B'C'$ in a mirror M_3 placed on $A'B'$.

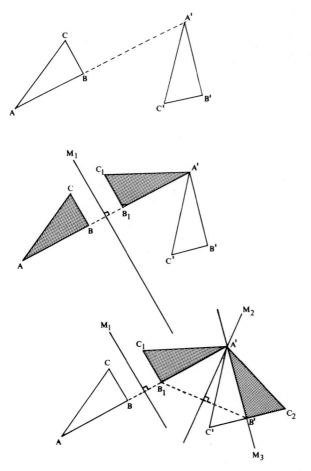

Figure 11.22 A triangle can be transformed to a congruent copy by no more than three reflections.

11.11   Some Reflection Exercises

Any pair of points p,q can be transformed to any other pair p',q' by both an even and an odd isometry using some combination of mirrors. The location of the mirrors can be found by a geometric construction using compass or dividers and a straightedge.

Example 11.2 [Baglivo and Graver, 1983] The four points p, q, p', q', where d $(p,q) = d(p',q')$, are located at the vertices of the parallelogram in Figure 11.23. Since p,q translate to p',q', the two mirrors needed to carry out this even isometry can be placed at p and at the midpoint of pp' and oriented perpendicular to pp'.

The odd isometry transforming p,q to p',q' could in theory be either a simple reflection or a glide reflection. However, it can't be a simple reflection since the perpendicular bisector of pp' is not, in general, the perpendicular bisector of qq', so the odd isometry must be a glide reflection. As we saw in Section 11.8, the axis of the glide reflection is located at the line joining the bisectors of pp' and qq'. The other two mirrors that generate this glide reflection are perpendicular to the glide axis and intersect this axis at A and B as shown in Figure 11.23(b). A is the point where the perpendicular line through p to the glide axis intersects this axis while B is the midpoint of AC where C is the point on the glide axis where the perpendicular through p' intersects the glide axis.

Example 11.3 Here q and q' are identical points in Figure 11.23(c), labeled O, while p and p' are equidistant from q,q'. It is obvious that the proper isometry is a rotation about O. The two mirror lines that carry out this rotation are line Op and the angle bisector of angle pOp'. The odd isometry is the reflection generated by the mirror line through O that bisects angle pOp'.

Problem 11.2 With compass and straightedge, construct the mirror lines that transform the pair p,q to p',q' which are illustrated in Figure 11.24 by both odd

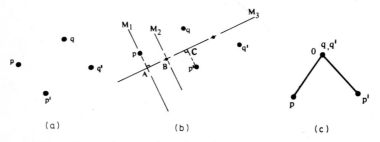

(a)   (b)   (c)

Figure 11.23 P,q can be transformed to p',q' by an even and odd isometry.

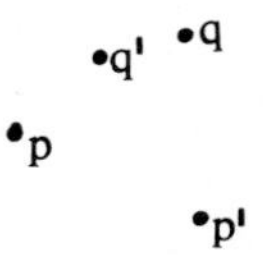

Figure 11.24 Find a rotation and glide reflection that maps p,q to p',q'.

(a glide reflection) and even (a rotation) isometries. The trick to finding the rotation is to first locate its center.

11.12 Some Additional Relations Involving Isometries

11.12.1 Half-turns

Among rotations, half-turns H deserve special attention. In the next chapter they will play a special role along with reflections in describing symmetric patterns along a line. First of all, they are generated, according to Section 11.10.2, by two perpendicular mirrors.

Next, a pair of half-turns generate a translation. For example, let H_A and H_B be half-turns about points A and B, respectively. If a typical point in the plane is transformed by a half-turn about A and the result is then transformed by another half-turn about B, the product of these transformations has the effect of translating any point in the plane in a direction from A to B through a distance twice the distance from A to B, i.e., $H_B H_A = T$. Satisfy yourself that this is true by choosing A and B to be any two points in the plane and transforming an arbitrary point p by successive half-turns about A and B.

Now carry out a sequence of three consecutive half-turns through points A, B, C not all lying on the same line. Show by experimentation that this has the same effect on a typical point p as rotating p through a half-turn about D, the fourth point on the parallelogram defined by A, B, and C.

11.12.2 Products of rotations in general

If consecutive rotations are carried out about point O through an angle θ followed by angle φ, and the result is clearly a composite rotation about O through angle $\theta + \varphi$, and the result is the same regardless of the order in which the rotations are carried out.

Let's see what happens if the successive rotations are carried out about two different centers [Martin, 1982]. If any product of rotations is carried out consecutively about A through angle θ and about B through angle φ, the result is a rotation through angle $\theta + \varphi$ about a third point C. The location of C is easily determined by looking at Figure 11.25(a). Mirrors a and c generate the rotation $S_{\theta,A}$ about A while c and b generate the rotation $S_{\varphi,B}$ about B. C lies at the intersection of a and b and corresponds to the rotation $S_{\theta+\varphi,C}$ about C. This follows from

$$S_{\varphi,B} S_{\theta,A} = R_b R_c R_c R_a = R_b R_a = S_{\theta+\varphi,C}$$

Check this result by constructing an example illustrating it. Show

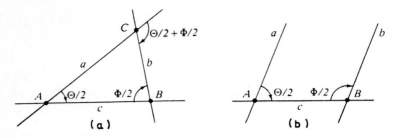

Figure 11.25   Multiple rotations about points A and B produce (a) a rotation about a third point C; (b) a translation when $\theta + \varphi = 360$ degrees.

that the result *is* affected by the order in which the rotations are carried out. If $\theta + \varphi = 360$ degrees, it can be shown that the product of rotations results in a translation.

Many additional relationships between isometries are described in *Transformation Geometry* by G. E. Martin [1982]. These relationships are often surprising and make excellent use of the concepts of euclidean geometry.

Symmetry of the Plane

Tyger! Tyger! burning bright
In the forests of the night,
What immortal hand or eye
Could frame thy fearful symmetry?
WILLIAM BLAKE

12.1   Introduction

To begin this book we questioned whether the beauty of a work of art
or architecture is due entirely to the craft of the designer or is intrinsic to its geometry. This book has tried to show that beautiful designs
must both exhibit a free flow of creative energy from the designer to
the work and obey the invisible hand restraining design due to the
geometric constraints of space. More often than not, the designer is
not conscious of these constraints; however, the success of a design depends to a large degree on how well the artist is attuned to the problems and possibilities presented by these constraints. Nowhere is this
tension between artists and their art more evident than with regard to
the issue of symmetry.

Symmetry is a concept that has inspired the creative works of artists and scientists; it is the common root of artistic and scientific endeavor. To an artist or architect symmetry conjures up feelings of order, balance, harmony, and an organic relation between the whole and
its parts. On the other hand, making these notions useful to a mathematician or scientist requires a precise definition. Although such a
definition may make the idea of symmetry seem less flexible than the
artist's intuitive feeling of it, that precision can actually help designers unravel the complexities of a design and see greater possibilities
for symmetry in their own work. It can also lead to practical tech-
niques for generating patterns.

The object of this chapter is to study several kinds of two-
dimensional symmetries: bilateral, point, line or frieze, planar or

wallpaper, and similarity symmetry. We will lay the mathematical foundation for the subject of symmetry and give some ideas of how to classify and generate symmetry patterns. This chapter is meant to serve as an introduction to several excellent books and articles that have been written on this subject [Shubnikov and Koptsik, 1974], [Martin, 1982], [Loeb, 1971], [Hargittai, 1986; 1987; 1988]. We begin with some introductory exercises.

The study of symmetry begins with observations of our own bodies. We share with all land animals the property of having approximate *bilateral symmetry* (see Figure 12.1). Since the force of gravity, in exerting a force upon us toward the center of the earth, distinguishes between up and down but not between left and right, the bodies of land animals are externally differentiated from head to feet but are symmetric from left to right. In other words, as Martin Gardner has described in his book by the same name, we live in an *ambidextrous universe* [1964].

Exercise 12.1   Your face appears to be symmetric. Let's see how symmetric it is. Place a mirror along the line of symmetry that divides the left side of a photograph of a human face from the right side and see whether the exposed portion of the face and its mirror image combine to give a realistic or distorted image of the entire face.

Symmetry around a point is familiar in patterns observed in the central spaces of buildings and in patterns such as the one in Figure 12.2. It is also the kind of symmetry that you see when you look into a kaleidoscope, which is why these patterns are sometimes called *kaleidoscope symmetries*.

Exercise 12.2   Place one corner of a small rectangular mirror at point O in Figure 12.3 and place the edge of the mirror on line M_1. Now place the corner of a second mirror at O and vary its angle with respect to M_1 until the image of the curve between the two mirrors repeats in the mirrors. You will notice that this occurs for a discrete set of angles. What are they?

Exercise 12.3   To discover how a kaleidoscope works, place two small rectangular mirrors perpendicular to a protractor and open them to a sequence of angles, $180/n$ degrees. Place a colored sequin between the mirrors and note the number of sequin images (including the original) that appear in the mirrors as a function of the angle of intersection between the mirrors. Record this information in a table. Also create a pattern from several sequins and look at the complete pattern that is generated by the two mirrors at each angle in the sequence.

All snowflakes exhibit hexagonal symmetry as illustrated by the pattern shown Figure 12.4. [Bentley and Humphreys, 1962]. In the next exercise, we invite you to create your own snowflake patterns.

Figure 12.1   Bilateral symmetry in a Senufo wooden mask, Ivory Coast/Mali/Upper Volta.

Figure 12.2   Circular symmetry pattern.

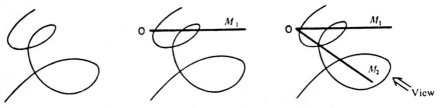

Figure 12.3   Place a smooth curve between two mirrors and move one mirror until a continuous curve is produced.

Figure 12.4   Snowflake pattern.

Exercise 12.4   Fold a piece of construction paper in half. Draw a line oblique to the fold of the paper at an angle of $180 / n$ degrees. Fold along this line. Next, cut away the excess paper (the portion of the paper not doubled up). Refold along the original fold, and again cut away the excess paper. Continue this process of folding and cutting until there is no longer any excess paper. Next, fold the resulting figure into a multilayered triangle and cut a pattern or motif into the triangle. Unfold to get a snowflake pattern around the central point consisting of several rotations and reflected images of your motif that are similar to an actual snowflake. Can you devise a way of creating snowflake patterns with rotations but no reflections of your motif?

12.2   The Mathematics of Symmetry

According to the examples of the last section, symmetry can be considered to be the *ordered repetition* of a basic pattern. The left half of the mask in Figure 12.1 is replicated by reflecting it onto the right half while one-sixth of the snowflake in Figure 12.4 is repeated in rotated and reflected positions to form the whole snowflake.

The mathematical definition of symmetry that we shall now present makes the notion of ordered repetition precise. If you do not wish to ponder the mathematical technicalities of this and the next two sections on your first reading, you can skip to Section 12.5 and return to these sections as needed.

Let A be a subset of the line L, plane P, or three-dimensional space S. A symmetry of A is any similarity T that leaves A invariant although the individual elements of A may be transformed; i.e., $T:A \to A$ or $T(A) = A$. (Note that this does not mean that individual points are invariant.) In most cases every similarity leaving A invariant turns out to be an isometry. For this reason, we prefer to use the following restricted definition for all but the last section of this chapter.

Definition 12.1 Let $X = L$, P, or S, and let $A \subset X$. A symmetry of A is an isometry $T:X \to X$ which leaves A invariant. The set

$$\text{Symm}(A) = \{T \mid T : X \to X \text{ is an isometry, and } T(A) = A\}$$

is called the symmetry group of A. If this group contains only the identity transformation I, A is said to have no symmetry. Why we use the word "group" will be explained shortly.

There is a natural product on $\text{Symm}(A)$ which is defined by the composition of mappings (see Section 11.2). If T_1, $T_2 \in \text{Symm}(A)$, the product T_2T_1 is just the result of operating on points first with T_1 and then following it by T_2. Suppose $T_1(p) = p'$ for points p and p'; then $T_2T_1(p) = T_2(p')$. This is shown graphically in Figure 12.5. Clearly, if $T_1(A) = A$ and $T_2(A) = A$, $T_2T_1(A) = A$, which means the product of symmetries is a symmetry. Recall that the identity map $I:X \to X$ is an isometry. Obviously $I(A) = A$, so I is a symmetry of any set A. An isometry T has an inverse T^{-1}, which is an isometry. If $T(A) = A$, $T^{-1}(A) = A$ as well. To see this, note that the definition of an inverse can be given in terms of products as $TT^{-1} = T^{-1}T = I$. Thus, if we multiply $A = T(A)$ on both sides by T^{-1}, we obtain $T^{-1}(A) = T^{-1}T(A) = I(A) = A$. Paraphrasing, we say that the inverse of a symmetry is a symmetry. To summarize these facts (and others) we state the following; the proof is left to the reader:

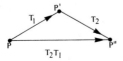

Figure 12.5 The product of translations is a translation.

Theorem 12.1 Let T_1, T_2, T_3,... $\in$ Symm(A). Then the identity I, the inverse T_i^{-1}, and all products T_iT_j belong to Symm(A). Furthermore, you can verify that the following properties are satisfied:

(i) $(T_iT_j)T_k = T_i(T_jT_k)$ for any i, j, and k (associativity)

(ii) $IT_i = T_iI = T_i$ for every i (identity)

(iii) $T_iT_i^{-1} = T_i^{-1}T_i = I$ for every i (inverse)

12.3   Symmetry Groups

Theorem 12.1 shows that Symm(A) is an example of the important mathematical structure called a *group* [Budden, 1972].

Definition 12.2 Let G be a set together with a composition law $*$, which associates to each pair $g, h \in G$ another element $g*h \in G$ called the product of g and h. Furthermore, suppose that $*$ satisfies the following properties:

(G1) $(g_1 * g_2) * g_3 = g_1 * (g_2 * g_3)$ for all g_1, g_2, $g_3 \in G$ (*associative law*).

(G2) There exists an identity element $e \in G$ such that $e * g = g * e = g$ for all $g \in G$ (*identity*).

(G3) For each $g \in G$, there exists an element $g^{-1} \in G$ called the inverse of g such that $g * g^{-1} = g^{-1} * g = $ e (*inverse*).

Then G is said to be a group. If, in addition, G satisfies

(G4) $g * h = h * g$ for all $g, h \in G$ (commutative law), we say G is a commutative, or abelian, group.

The number of elements in the set is called the order of the group.

Example 12.1 Let Z denote the set of integers. Define $m * n = m + n$. It is easy to see that Z is an (additive) abelian group with identity 0 and inverse $-m$ for each m.

Example 12.2 Let $G = \{ x \mid x \in R$ and $x \neq 0\}$ where R stands for the set of real numbers. Define $x * y = xy$ where xy is the product of real numbers. Then G is a commutative group with identity 1 and inverse $1/x$ for each x.

Example 12.3 The groups in the preceding two examples have infinitely many elements. A group of finite order can be constructed as follows: Let $G = \{0,1\}$ and let $*$ be defined by the following multiplication table which denotes $0 * 0 = 0$, $0 * 1 = 1$, and $1 * 1 = 0$. Then G is an abelian group with identity 0 and inverses $0^{-1} = 0$, $1^{-1} = 1$:

*	0	1
0	0	1
1	1	0

Example 12.4 Consider the symmetry group of the letter E shown in Figure 12.6. Clearly Symm(E) consists of two elements: the identity I and a reflection R in the line AA'. Observe that $RR = R^2 = I$.

A — - — E — - —A'

Figure 12.6 Symmetry of the letter E.

Although the groups G and Symm(E) in Examples 12.3 and 12.4 appear quite different, they are actually equivalent in a certain sense. To see this, we define a mapping ϕ: Symm(E) $\to G$ by $\phi(I) = 0$ and $\phi(R) = 1$. Then ϕ is a bijection (see Section 11.4) which, furthermore, preserves the law of composition. Thus, $\phi(IR) = \phi(I) * \phi(R)$, $\phi(RI) = \phi(R) * \phi(I)$, $\phi(II) = \phi(I) * \phi(I)$, and $\phi(RR) = \phi(R) * \phi(R)$. Such a mapping is called an *isomorphism*.

Definition 12.3 Let G and H be groups with composition laws $*$ and $\circ$, respectively. A one-to-one correspondence ϕ: $G \to H$ such that $\phi (g * h) = \phi (g) \circ \phi(h)$ for every $g, h \in G$ is said to be an isomorphism, and the groups G and H are said to be isomorphic. We denote this by $G \simeq H$.

Although the two isomorphic groups appear as different as apples and aardvarks, they are identical as abstract groups, i.e., they have the same mathematical structure. This is the same situation that we encountered when we studied isomorphic graphs in Section 4.2.

12.4   Subsets of a Group

Definition 12.4 The *generators* of a group are the smallest subset of elements of a group, say f_1, f_2, whose products $f_1^n f_2^m$ for m and n integers generate all the elements of the group, i.e., every element g of the group is of the form $g = f_1^m f_2^n$.

Definition 12.5 A *subgroup* H of a group G is a subset of elements from the group that itself constitutes a group. (Every subset of elements of G generates either G or some subgroup of G.)

Theorem 12.2 (*Fundamental theorem of group theory*). The order of any subgroup of G is a divisor of G, i.e., if G has n elements and a subgroup H has k

elements, $n / k = m$ for n, k, and m positive integers; m is called the *index* of the subgroup.

Definition 12.6   If $a \in G$ and H is a subgroup of G, the set $aH = \{ah \mid h \in H\}$, gotten by multiplying each element of H by a, is called a *left coset* of H. The set $Ha = \{ha \mid h \in H\}$ is called the *right coset* of H.

Now, if $a \notin H$, aH has no elements in common with H (why?). As a result of this and the fundamental theorem, if H is a subgroup of G of index m and order k, G is partitioned into exactly m disjoint sets (no elements in common), each of order k. H is one of the sets of the partition.

Definition 12.7   If H is a subgroup of G and $aH = Ha$ for all $a \in G$, H is called a *normal subgroup* of G. This defining relationship can also be written $H = aHa^{-1}$.

Theorem 12.3   If H is a normal subgroup of G, the cosets of H form a group in which the product is defined by $aH \circ bH = (a * b)H$. This group is called the *quotient group* of G and is symbolized by G/H. The identity element of this group is $H = eH$. H is called the kernel of the mapping $\phi \colon G \to G/H$, i.e., $\phi(a) = aH$.

It is a little difficult to get a geometric handle on the meaning of a normal subgroup in this abstract context; however, we will attempt to do this. In Figure 12.7, the group G is schematically partitioned into its cosets aH where the cosets are the elements of the quotient group G/H while the normal subgroup H functions as the identity element of the quotient group. Each coset may be thought of as a "universe" unto itself. This universe can be transformed to the base universe H by multiplying each of its elements by a^{-1}. What is most notably true about these universes is that they are in some sense self-similar. What we mean by this is that any two elements in coset aH are related to each other through multiplication by an element of H just as in the base universe and so they share the same structure as H. This relationship is shown schematically in Figure 12.7 and is proven in Problem 12.1.

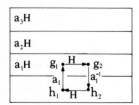

Figure 12.7   Diagram showing self-similarity of a normal subgroup.

Problem 12.1   Show that if g_1 and g_2 are two elements of coset aH of a normal subgroup H, $g_2 = hg_1$ for some $h \in H$. Hint: The solution uses the fact that $aHa^{-1} = H$.

Example 12.5   The even numbers are a normal subgroup of index 2 of the group of integers under the operation of addition. The other coset consists of the odd numbers. Access from one universe (coset) to the other is by addition of an odd number.

Example 12.6   All the isometries of the plane form a group of infinite order. The even isometries (rotations and translations) form a subgroup of index 2. The odd isometries (reflections and glide reflections) form the other coset. As we saw in Section 11.9, the even and odd cosets can be imagined to be separate universes as Abbott did in his Flatland allegory (Section 11.9). Access from one universe to the other is via either a reflection in the plane or a rotation into a higher-dimensional space.

In Section 12.7, we will discover that the quotient group is the key to understanding the nature of colored symmetry where the number of colors corresponds to the index of the underlying normal subgroup.

We now look at the structure of the symmetry group of the equilateral triangle and see how it relates to point or kaleidoscope symmetry.

12.5   Kaleidoscope Groups

12.5.1   The group of the equilateral triangle

In this section we elaborate on Exercises 12.2, 12.3, and 12.4 in order to study the mathematics of kaleidoscope symmetry groups. First, consider the group that keeps an equilateral triangle $\{3\}$ invariant. It is obvious from Figure 12.8 that $\{3\}$ is mapped to itself by the following isometries: the reflections R_i in mirrors M_i and the rotations S and S^2 of 120 and 240 degrees about the centroid of the triangle O, i.e.,

$$\text{Symm}\,(\{3\}) = \{I, R_1, R_2, R_3, S, S^2\}$$

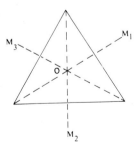

Figure 12.8 The equilateral triangle is invariant under the elements of the dihedral (kaleidoscope) group D_3.

Symm ($\{3\}$) is usually known as the *dihedral*, or *kaleidoscope*, group of order 3 (it has three reflections). It is usually abbreviated D_3. Even though D_3 was constructed for an equilateral triangle, it is also the symmetry group of many other sets of points. That D_3 is a group follows from the fact that each element has its own inverse within the set and that the product of any two elements lies in the set. For example,

$$R_iR_i = I \qquad SS^2 = I \qquad SR_1 = R_2$$

The last relation follows from the fact that according to Section 11.7, $S = R_2R_1$ and thus $SR_1 = (R_2R_1)R_1 = R_2(R_1R_1) = R_2I = R_2$.

Problem 12.2 Construct a multiplication table similar to the table in Example 12.3 for the elements of D_3.

According to Definition 12.4, any two reflections of D_3, say R_1 and R_2, generate all the elements of the group. To see this, we showed in the last section that $S = R_2R_1$. It also follows from Section 11.10.2 that $S^2 = R_1R_2$. What about R_3? We leave it to the reader to show that $R_3 = R_1R_2R_1$. It is this ability to generate all the transformations of D_3 by two mirrors which lies at the basis of the kaleidoscope (see Exercise 12.3).

If the three reflections are removed from D_3, another group called the *cyclic group* (abbreviated C_3) results, i.e.,

$$C_3 = \{I, S, S^2\}$$

Since every isometry of C_3 is also in D_3, C_3 is a subgroup of D_3. Any figure invariant under D_3 is also invariant under C_3 but not the other way around. For example, Figure 12.9 is a symmetry of C_3 but not of D_3. The reflections do not leave the set invariant because of the flags on the spikes. C_3 has a single generator, namely S.

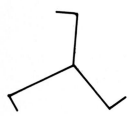

Figure 12.9 A figure invariant under the cyclic group C_3.

12.5.2   Other kaleidoscope groups

D_n for $n \geq 3$ is the symmetry group of an n-sided regular polygon as illustrated for the dihedral group of the triangle. D_n symmetry patterns contain n lines of mirror reflection and $2n$ isometries. D_n is generated by reflections in two mirrors intersecting at an angle of $180/n$ degrees and serves as the symmetry group of the kaleidoscope patterns whose generating region is the segment of the plane between the two mirrors. D_1 and D_2, with one and two mirror lines, respectively, are special cases of dihedral groups. D_1 is the symmetry group corresponding to bilateral symmetry (generated by one mirror), i.e., the group in Example 12.4. D_2 is the symmetry group of a rectangle (generated by two perpendicular mirrors).

As the number of mirrors increases, the angle between them decreases. In the limiting case of D_x, the kaleidoscope group contains an infinity of parallel (zero angle of intersection) lines of reflection symmetry. This corresponds to the doubly infinite train of images observed in the parallel mirrors of a barber shop. D_x can also be viewed as the symmetry group of a circle in which an infinite number of mirror lines intersect at the center of the circle. Although the symmetry of a circle may appear trivial, it plays an important role in electromagnetic theory and other areas of science [Shubuikov, 1988(a)].

If the mirror lines are removed from D_n, the cyclic group C_n remains. Again, C_n is a group of order n generated by a rotation through $360/n$ degrees; it is the group that generates rosette patterns such as the one in Figure 12.12. C_x corresponds to a doubly infinite series of images translated along a line as in certain molding patterns to be illustrated in Section 12.10.

12.6   Pattern Generation and the Kaleidoscope

We are now in a position to understand the relationship between pattern generation and group structure. Starting with a simple region of the plane or tile called a *fundamental region*, or *domain*, determined by the nature of the symmetry group, one can obtain a *symmetry tiling* by constructing the collection of images obtained by operating on the fundamental domain with members of a group of isometries. This technique has been employed extensively in art and architecture. In this section we show how it works for kaleidoscope patterns.

The shaded area in Figure 12.10 is the fundamental domain of the symmetry group of {3}. It is transformed by each isometry of the group to form a tiling of the equilateral triangle. Five replicas are generated

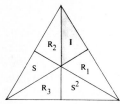

Figure 12.10 Each fundamental domain of the equilateral triangle is labeled according to the element of D_3 that maps it from the fundamental domain marked I (identity).

Figure 12.11 A fundamental pattern mapped by the elements of D_3 to form a symmetry pattern.

from the fundamental domain which, itself, corresponds to the identity element of D_3. Thus, the number of replicas (including the original tile) equals the order of the group. This is also what you discovered in Exercises 12.2 and 12.3.

Fundamental domains of D_3 with other shapes may also be created in the 60-degree sector formed by the rays corresponding to two mirror lines of D_3 and replicated by the elements of the group to form a kaleidoscope tiling. A *fundamental pattern*, or *motif* as it is called, may also be emblazoned on the fundamental domain to form a symmetry pattern or design such as the snowflake patterns of Exercise 12.4 or the circular symmetry pattern of Figure 12.2. In Figure 12.11 the fundamental domain is taken to be the entire sector, and a comma is taken as the motif of another D_3 symmetry pattern. In general, a motif placed between two mirrors intersecting at an angle of 180/n degrees will create a snowflake pattern with the symmetry of D_n. Each pattern will consist of $2n$ replicas of the motif. It is important to note that the motif can be created only in the fundamental region. Any part of the motif drawn outside the fundamental region would be superimposed by other parts of the design during the pattern generation process.

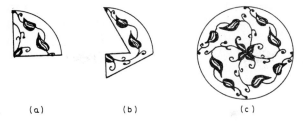

(a)   (b)   (c)

Figure 12.12 A rosette design with fourfold rotational symmetry has symmetry group C_4. The wedge in (a) is a generating region for the design; the altered wedge in (b) is another generating region for the same design [note that the generating region in (b) fits the motif better].

The fundamental region corresponding to C_3 is made up of a 120-degree sector about the center of rotation (see Figure 12.9). Again, any motif drawn in the fundamental region is replicated three times including itself by C_3 to form a rosette pattern. Unlike the fundamental region of D_3, which is bounded by mirror lines, the fundamental region of C_3 can take many forms. Two fundamental regions of a C_4 rosette pattern are shown in Figure 12.12. [Schattschneider, 1986] describes how to create a variety of fundamental domains for the symmetry patterns of the plane.

12.7 A Colored Kaleidoscope Symmetry

Color may be used to enhance the aesthetic effect of a symmetry pattern [Schattschneider, 1986]. For the case of tilings that use only two colors, say black and white, the symmetries of the tiling are most enhanced whenever each symmetry of the uncolored tiling either (1) transforms all white tiles to white tiles and all black tiles to black tiles or (2) transforms all black tiles to white tiles and all white tiles to black tiles.

Every symmetry of the colored tiling which satisfies (1) or (2) is called a *two-color symmetry* of the tiling. When every symmetry of the uncolored tiling is also a two-color tiling of the colored tiling, we say that the coloring is *perfect*, or compatible. (Each symmetry of the uncolored tiling induces a permutation of the colors in the colored tiling.)

Perfect colorings can be described by the language of group theory introduced in Section 12.4. Let's see how this works for the kaleidoscope pattern D_3. According to Section 12.4, C_3 is a subgroup of D_3 with order 3 and index 2. Therefore, according to Theorem 12.2, the quotient group D_3/C_3 has two elements. Beginning with a tile of one color, say black, color all elements black that are generated by the action of C_3 on this tile. Color any uncolored tile white. The action of C_3 on this element results in coloring the remaining congruent images of the tile white. Thus, in a sense, we have colored the two elements of the quotient group black and white, as Figure 12.13 illustrates for {3}. The only perfect two-coloring of the C_4 rosette tiling is shown in Figure 12.14. You will also notice that conditions (1) and (2) are also satisfied.

In a sense, the tiling has been decomposed into two distinct but oppositely congruent (self-similar) universes accessible to each other by mirror reflection as we described in Section 12.4 in an abstract sense. The coloring brings out this fact. In general, a perfect coloring with m colors is defined by a normal subgroup of the pattern of index m and separates the pattern into m congruent subsets, each marked by a different color.

A colored symmetry can also be generated by a motif emblazoned on the fundamental domain or tile. In this case the motif is usually called the *figure* and the remainder of the fundamental domain is called the *ground*. The colored symmetry proceeds to permute the colors of the motif, leaving the ground the same color, or it may permute both figure and ground as described above for colored tilings.

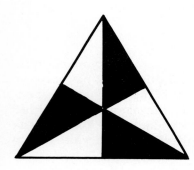

Figure 12.13 A perfect coloring of D_3.

Figure 12.14 A perfect coloring of C_4.

12.8 Some Other Examples of Pattern Generation

In general, given a set of isometries subject to certain constraints (to be described in Sections 12.13 to 12.16) and a starting pattern or motif, a symmetry pattern can be created with these isometries as the generators.

Example 12.7 Start with a right triangle represented by the identity I in Figure 12.15. Let R_1 and R_2 be reflections in mirrors M_1 and M_2, respectively. Let S be a 90-degree rotation (quarter-turn) in the center c. Every member of the group can be written in the form $f_1 f_2 \ldots, f_k$, where each is one of R_1, R_2, S or R_1^{-1}, R_2^{-1}, S^{-1}. A pattern formed by some of the images of the triangle under this group is shown in Figure 12.15.

Example 12.8 Perhaps the most dramatic example of pattern generation is illustrated by the dihedral kaleidoscope (see Section 7.13.4). Here the image of a cube is obtained by combining the 47 images of an irregular tetrahedron (orthoscheme) generated by three mirrors. In other words, each of the 48 elements of the symmetry group of the cube contributes one transformed replica of the fundamental domain (orthoscheme) of the cube to recreate the entire cube.

Example 12.9 In Section 9.9 the sphere was divided into 48 congruent curvilinear triangles (fundamental domains). The 48 images of a single vertex recreated the vertices of one or another of the archimedean polyhedra with the symmetry of a subgroup of the cube, depending on the position of the starting vertex.

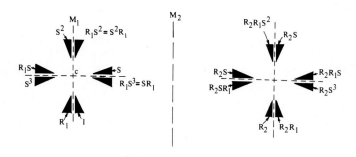

Figure 12.15   An example of pattern generation. A right triangle is repeatedly acted upon by reflections in two mirrors and fourfold rotations around a center.

12.9   Pattern Generation in Hyperbolic Geometry

A group of isometries generates a symmetry pattern consisting of congruent images of some motif in euclidean space. A repeating pattern composed of hyperbolically congruent copies of a motif can also be generated (see Appendix 2.B). M. C. Escher has created beautiful patterns with *hyperbolic symmetry*, and Douglas Dunham [1986] has recreated some of Escher's patterns in addition to many of his own with the aid of a computer. Escher's and Dunham's patterns are created in the Poincaré circle model of the hyperbolic plane (see Appendix 2.B) in which lines are represented by circular arcs in the disc which are perpendicular to the boundary circle.

A {6,4} tiling of the Poincaré circle is shown in Figure 12.16. The edges of the tessellation are the solid lines while the dual tiling is represented by the dotted lines. Both of these sets of lines represent lines of mirror symmetry of the tiling. The dashed lines are additional lines of mirror symmetry. In hyperbolic geometry, a point and its reflected

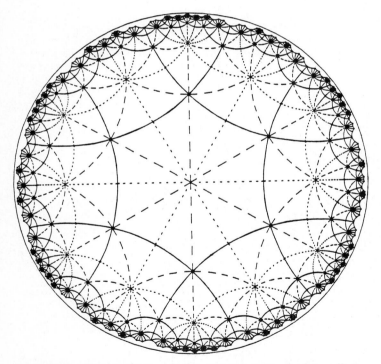

Figure 12.16 The tessellation {6,4} (solid lines), the dual tessellation {4,6} (dotted lines), and other mirror lines (dashed) of the two tessellations.

image with respect to an arc (line) is represented by the point and its inverse with respect to the arc (see Appendix 2.B). The vertices of this tiling are also fourfold rotocenters while the vertices of the dual are sixfold rotocenters.

The fundamental domain of this tiling is the curved right triangle made up of solid, dotted, and dashed lines. In general, a regular $\{p,q\}$ tiling with congruent tiles exists in hyperbolic geometry whenever $(p - 2)(q - 2) > 4$. The fundamental domain of these tilings also consists of right triangles with acute angles π/p and π/q. Any design created in the curvaceaous triangle would be multiply reflected in the convex mirrors that make up its sides to produce the entire pattern, which would also possess p- and q-fold rotocenters.

In Figure 12.17, Dunham creates flounder-like fishes arranged in a repeating hyperbolic pattern in the style of Escher's famous picture "Circle Limit I." Either the right or the left side of any one of the fishes serves as the motif. (If color is taken into account, a motif may be formed from half of a black fish together with half of an adjoining white fish.) The backbones of the fish all lie on hyperbolic lines (arcs) which are also lines of reflection symmetry. The mirrors are removed from the fourfold rotocenters but retained at the sites of the sixfold rotocenters to produce a symmetry pattern corresponding to a subgroup of the symmetry group of the $\{4,6\}$ tiling. Further information

Figure 12.17　A hyperbolic pattern.

about these tilings, including a computer program to generate them, can be found in [Dunham, 1986].

12.10   Line Symmetry

A *frieze group* is the symmetry group of a repeated pattern on a strip which is invariant under a translation along the strip, although it can also be the symmetry group of a pattern on the whole plane (or half-plane) provided all its translations are in one direction. The finite width of the strip permits a variety of patterns which may be invariant under reflections in horizontal or vertical mirrors, half-turns, and glide reflections. Many beautiful examples of these symmetries, such as the ones in Figures 11.9 and 11.15, are to be found in the work of the ancient Egyptians, Greeks, Romans, and the Moors and of modern artists and designers.

There are seven essentially different frieze groups, and starting with some figure having no symmetry (invariant only under I), say a P, we can enumerate them. The seven frieze groups are listed in Table 12.1, along with their generators and commonly accepted symbols.

In this table, --- denotes the mirror line of a reflection or glide reflection and ∘ denotes the center of a half-turn. The group symbolized in the right column is isomorphic, in the sense of Section 12.3, to the frieze groups in the left column. Notice that C_x is identical with F_1 and F_2 while D_x is identical with F_1^2. Also C_x is isomorphic to F_1^3 while D_x is isomorphic to F_1^2 and F_2^2. The direct products $C_x \times D_1$ and $D_x \times D_1$ means that every element of these groups is the product of one element from D_1 and one element from either C_x or D_x. We say that these groups have a reflection in a center.

TABLE 12.1

F_1	p p p p p					1 translation	C_∞
F_1^3	‾p‾ ‾ ‾$\frac{b}{p}$‾ ‾ ‾$\frac{b}{p}$‾					1 glide reflection	
F_1^2	q¦p ¦ q p ¦ q p					2 reflections	D_∞
F_2	d∙p ∘ d p ∘ d p					2 half-turns	
F_2^2	q p ∘ d b ∘ q p					1 reflection and 1 half-turn	
F_1^1	b b b b b ‾‾‾‾‾‾‾‾‾‾‾ p p p p p					1 translation and 1 reflection	$C_\infty \times D_1$
F_2^1	$\frac{d¦b}{q¦p}$ ¦ $\frac{d\ b}{q\ p}$ ‾ $\frac{d\ b}{q\ p}$ ‾ $\frac{d\ b}{q\ p}$					3 reflections	$D_\infty \times D_1$

Rather than analyzing the structure of these groups, our primary interest in presenting them is to enable the reader to recognize them on sight. Since the overwhelming majority of patterns use either F_1, F_1^2, or F_1^1, perhaps the ability to correctly read patterns will result in an appreciation of the wider range of possibilities. Figure 12.18 gives a flowchart for determining frieze groups [Martin, 1974].

Problem 12.3   Use Figure 12.18 to determine the frieze groups corresponding to the patterns in Figure 12.19 collected by D. W. Crowe [1986].

In the next exercise we invite you to try your hand at creating seven tilings representing each of these classes of line symmetry. To tile a frieze means to fill a band in the plane with identical tiles, without overlaps or voids and without turning the tiles over (with the tiles having distinct tops and bottoms).

Exercise 12.5 [Findeli, 1986]   A practical method of obtaining tiles for the seven frieze patterns starts with pasted paper cut according to the following procedures. The name of the symmetry class is specified, and illustrations of the construction are shown in Figure 12.20.

F_1. Make a cylinder by gluing two opposite edges of a rectangle. Join the two edges of the cylinder by an arbitrary line and cut along it; the designer has complete liberty as to the choice of the shape of this line (motif) and of course also for the decoration of the tile itself (color, etc.).

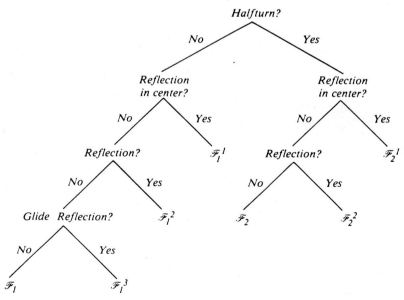

Figure 12.18   A flowchart for classifying frieze patterns.

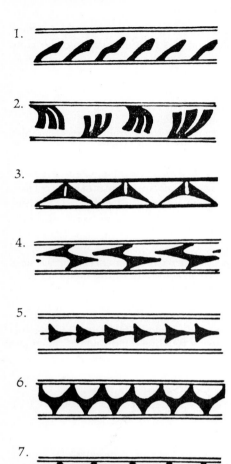

Figure 12.19 Examples of the seven frieze patterns appearing on the pottery of San Ildefonso pueblo, New Mexico.

F_2. Make an envelope by gluing three edges of two superimposed paper rectangles, leaving the fourth open. Join each of the corners of the envelope to the open boundary of an arbitrary line; cut along these lines and unfold the tile thus obtained.

$F\,{}^1_1$. Fold a rectangle in two along a median and make a cylinder from the rectangle obtained by gluing the edges. Proceed as for F_1.

$F\,{}^2_1$, $F\,{}^1_2$. No liberty is permitted for these groups by reason of the presence of mirrors. The elementary tiles are mere rectangles.

$F\,{}^3_1$. Make a Möbius band (see Section 4.12) from a rectangle; cut the band along an arbitrary line and unfold the tiling obtained.

$F\,{}^2_2$. Take a half-envelope such as that constructed for F_2 and join the corner to the upper boundary by an arbitrary line; cut and unfold.

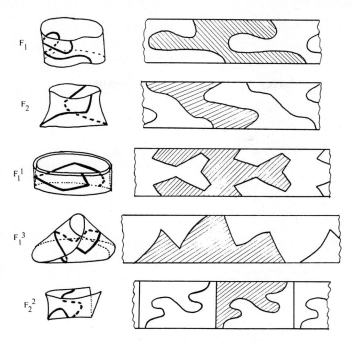

Figure 12.20   Construction of frieze tiles.

The frieze groups can also be perfectly colored by identifying normal subgroups. In fact, there are 23 two-colorings of the frieze groups. Examples of some of them are found in [Schattschneider, 1986].

12.11   The Two-Dimensional Ornamental Symmetry Groups

For thousands of years, elaborate patterns of ornamentation along lines and planes have been created for aesthetic purposes. Even so, the mathematical limitations of these patterns were not fully understood until the end of the nineteenth century, when a deeper analysis was inspired by their application in crystallography. In 1881, Fedorov proved that there are 17 essentially different wallpaper groups. Patterns representing all of these groups were known to the ancient Egyptians, who exhibited them in the decoration of their temples and tombs.

Perhaps the most dazzling display of these groups was rendered by the Moors in the ornamentation of the Alhambra. Following the dictates of the Koran forbidding the likeness of living things in decorative art, the Moors covered the walls of the Alhambra with myriad geometric patterns of breathtaking beauty and variety [Grünbaum, and Shephard, 1986]. Escher, who was unfettered by any such theological restraints, spent a great deal of time creating patterns for the 17 ornamental groups consisting of strange looking creatures, an example of which is shown in Figure 12.21 [McGillavry, 1976], [Escher and Locher, 1971], [Coxeter, 1986]. In this section we give examples of the 17 distinct classes of symmetry patterns that can cover the plane. The rest of this chapter shows how a study of the geometry of these patterns can lead to an understanding of how to generate them. Our aim is to provide a creative spark akin to that experienced by Frank Lloyd Wright and Le Corbusier after reading Jones' *Grammar of Ornament* [1928] to fire the intellect and the imagination.

An example of a two-dimensional ornamentation is shown in Figure 12.22. The pattern is potentially infinite in two nonparallel directions A and B. Translations of appropriate lengths in these directions leave the pattern invariant. In other words, a typical point within the pattern is replicated in the whole pattern as a two-dimensional lattice. Reflections in CC' and a series of mirror lines perpendicular to CC'

Figure 12.21   Escher pattern. (© *M. C. Escher Heirs/Cordon Art - Baarn - Holland*)

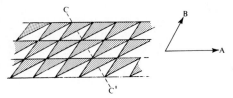

Figure 12.22   A wallpaper symmetry pattern.

also leave it invariant. Crystallographers symbolize the group of this pattern by *pm m*.

The two-dimensional ornamental groups (or wallpaper groups) are the symmetry groups of repeated patterns in the plane which are left invariant by a pair of independent translations; in other words they are patterns with underlying lattice structures. Each point of the pattern is replicated as a two-dimensional lattice. The fundamental domains, group generators, and group structures of the 17 distinct wallpaper groups have been nicely described by other references [Schattschneider, 1978], [Martin, 1982]. Many examples of patterns from ancient cultures illustrating these groups can be found in the classic book *Pattern Design* by Christie [1969] and the more recent books *Symmetries of Culture* by Donald W. Crowe and Dorothy K. Washburn [1989] and *Handbook of Regular Patterns* by Peter Stevens [1981].

Problem 12.4   Some beautiful wallpaper patterns from a variety of ancient sources are illustrated in Figure 12.23 [Stevens, 1981]. Use the flowchart in Figure 12.24 to identify the symbol of each crystallographic group. The crystallographic symbols are given so that you can identify them with the aid of the flowchart.

Once again, perfect colorings of the wallpaper patterns can enhance their interest. For wallpaper patterns with two opposing colors there is always the question of what is to be considered the *figure* and what the *ground*. Designers have brought off feats of incredible ingenuity in order to play with the idea of figure and ground. An example of this is shown in Figure 12.25 where the word 'Allah' in Arabic letters is set out in diagonal bands across the surface. The black letters dovetail the white so that no ground appears. Another example is shown in Figure 12.26(*b*) in which the symmetry of the pattern of Figure 12.26(*a*) is broken by eliminating every alternate crossing with the result that a new element appears, completely changing its character. Illustrations of the 46 possible two-colorings along with listings of their groups are given by Crowe in [1986*a*] [Senechal, 1975; 1979] and [Grünbaum and Shephard, 1987].

Figure 12.23 Examples of the 17 wallpaper symmetries from different cultures. (*a*) p6mm; (*b*) pg; (*c*) p2; (*d*) p3; (*e*) p4; (*f*) p4g; (*g*) p6; (*h*) c2mm; (*i*) p2mm; (*j*) p31m; (*k*) p1; (*l*) p2gg; (*m*) pm; (*n*) p3m1; (*o*) cm; (*p*) p2mg; (*q*) p4mm.

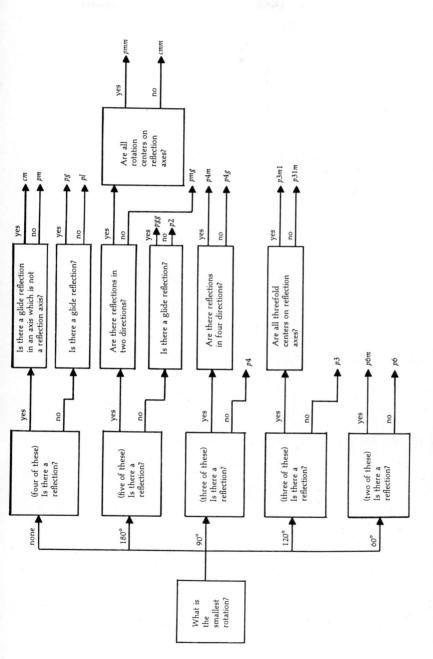

Figure 12.24 A flowchart for classifying the 17 wallpaper symmetry patterns.

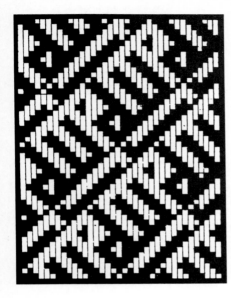

Figure 12.25 Panel of enamelled earthenware tiles form the collegiate Mosque of el-Ayny, Cairo, fifteenth century.

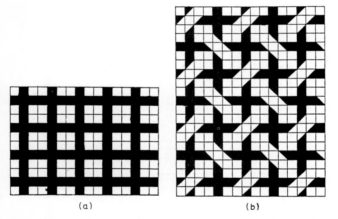

(a)   (b)

Figure 12.26 The basic pattern in (a) is transformed to the design in (b) from a drawing in the Mirza Akbar collection.

12.12 Symmetry and Design

What are the ingredients of an aesthetically pleasing ornament or design? This remains one of the great mysteries. After all, the smallest unit of a pattern or motif bears little resemblance to the entire pattern. The quest for the essence of beautiful ornamentation led to the following debate between two designers [Gombrich, 1979]:

My friend had been maintaining that the essence of ornament consisted in three things: contrast, series and symmetry. I replied that none of them together would produce ornament. Here, [making a ragged blot with the back of my pen on the paper; see Figure 12.27(a)] you have contrast; but it isn't ornament: and here—1,2,3,4,5,6—you have series; but it isn't ornament: and here [sketching the figure in Figure 12.27(a)] you have symmetry; but it isn't ornament. My friend replied, "Your materials were not ornament, because you did not apply them. I send them back to you, made up into a choice sporting neckerchief [see Figure 12.27(b)].

Exercise 12.6   Gombrich places the mathematics of ornamentation into perspective with his comment that "there is no danger that the resources of the pattern-maker will ever be exhausted by the constraints of geometry because any one of the groups or devices described by [mathematicians] can be combined with others in an infinity of combinations and permutations." By rotating the divided two-colored square in Figure 12.28(a), four variants are produced. Sixteen different motifs are created by pairing these four units, and 256 possibilities are created from combining four of them. Try your hand at creating a motif from these units and repeating it to form a planar ornamentation such as the ones shown in Figure 12.28(b).

It is a good exercise to look at a design and to characterize it according to the group to which it belongs. The anthropologist Washburn in collaboration with the mathematician Crowe has even used some of these assignments as clues to help trace the origins of designs left behind by primitive cultures [1989]. Nevertheless, we are left with a feeling of incompleteness. On the one hand, the group categories appear to be rather coarse. Designs which appear to bear no visual re-

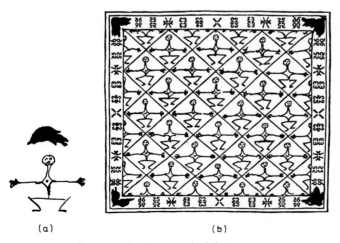

(a)   (b)

Figure 12.27   An example of ornamental design.

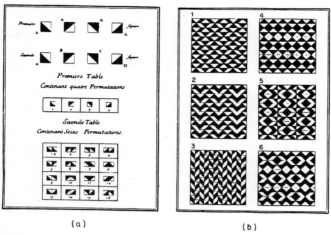

Figure 12.28   Designs based on divided two-color squares.

semblance to each other may belong to the same group. On the other hand, the fundamental domain of a wallpaper pattern gives a limited view of how the entire finished pattern will appear and how to create the whole pattern from its smallest part or fundamental pattern [Grünbaum, 1990].

In his book *Color and Symmetry* [1971] A. L. Loeb presents a way of both generating and reading patterns in a more global manner by first studying their underlying geometric structure: location and order of points of rotational symmetry or *rotocenters* and location of mirror lines and glide lines. For example, rotocenters lie at the vertices of meshes of a semiregular tiling and that mirror lines may constitute the edges of this mesh or lines run through the mesh according to precise geometric laws. H. Lalvani has shown how this mesh can be used to generate the motifs of a great many of the classical Islamic tilings as we shall see in Section 12.18. In the next five sections of this chapter, we will reproduce some of Loeb's analysis and description of the structure of mosaics. First we will see how the nature of space places a severe constraint on the kind of rotocenters that are possible in a wallpaper symmetry.

12.13   A Fundamental Postulate

Let's consider the interaction between two rotocenters of a pattern, one of order k located at A_k and one of order l located at B_l. We mean by an nth-order rotocenter that n successive rotations of any point about the rotocenter results in the identity transformation. By the na-

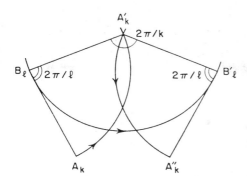

Figure 12.29 Interaction of rotocenters. (*From Loeb, 1971.* ©
1971, Wiley.)

ture of rotational symmetry, there will be other directly congruent k-
fold rotocenters located on a circle of radius B_lA_k about B_l as shown in
Figure 12.29. A_k' is the nearest of these k-fold rotocenters to A_k, and it
is rotated counterclockwise from A_k through an angle of $2\pi/l$. Likewise, other l-fold rotocenters lie on a circle of radius $A_k'B_l$ about A_k'.
B_l' is the nearest of these l-fold centers to B_l. It is rotated counterclockwise from B_l through an angle of $2\pi/k$.

An example of how this interaction works for a sixfold and a threefold rotocenter is shown in Figure 12.30. As a consequence of the interaction between these centers, another threefold and sixfold
rotocenter appears at the vertices of a parallelogram. This interaction
can also be applied to generating patterns. Beginning with a single
sixfold rotocenter and a motif having threefold rotational symmetry,
Figure 12.30 shows that another motif and sixfold center is a necessary consequence of the interaction.

Let's see what happens if we consider the interaction between a
threefold and a fourfold rotocenter. Starting with rotocenters A_3, B_4,
we trace a sequence of rotocenters generated by each other similarly,
as in Figure 12.31: $A_3B_4A_3'B_4'A_3''B_4''\ldots$. But before continuing, let's

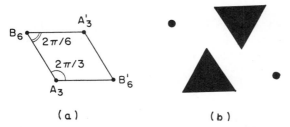

(a)   (b)

Figure 12.30 (*a*) Interaction of a threefold and a sixfold
rotocenter; (*b*) interaction of a motif having threefold
symmetry with a sixfold rotocenter. (*From Loeb, 1971.*
© *1971, Wiley.*)

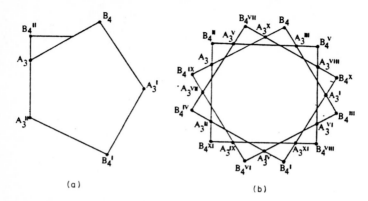

Figure 12.31 (*a*) The interaction of a threefold and a fourfold rotocenter. Note in (*b*) that after five cycles, the polygonal line closes upon itself. (*From Loeb, 1971. © 1971, Wiley.*)

stop to examine the point B_4''. Observe that it lies closer to the original A_3 than B_4 does. Let's call the ratio of these distances

$$\frac{A_3B_4'}{A_3B_4} = r \qquad \text{for } r < 1$$

Instead of continuing the polygonal line beyond B_4'', as shown in Figure 12.31(*b*), we could have started all over again with the pair of rotocenters A_3, B_4''. If we do this, it is clear that we will generate another rotocenter distant from A_3 by $r^2A_3B_4$. If we repeat this procedure n times, there will be a fourfold rotocenter at a distance $r^nA_3B_4$ from A_3. Since there is no limit to the number of times this procedure can be repeated and since $r < 1$, this implies that there are rotocenters arbitrarily close to each other.

Continuous distribution of rotocenters is not a situation that occurs in crystallography or in wallpaper patterns, and so this possibility must be excluded. To do this, Loeb first considers the set of all rotocenters congruent to each other (e.g., A_3 and A_3' are congruent while A_3 and B_4 are not), which he calls a *rotosimplex*. Thus all rotocenters congruent to A_3 form one rotosimplex while all rotocenters congruent to B_4 form another. Now we can state the fundamental postulate of crystallography.

Postulate of Closest Approach For every pair of rotosimplexes in a plane, there exists a finite distance such that no point in one simplex is closer to any point of the other simplex than that given distance.

It follows from this postulate that even within a rotosimplex, rotocenters are not arbitrarily close (why?) so that no pair of rotocenters in the entire pattern can be arbitrarily close. This postulate imposes a severe constraint on space and limits the kinds of wallpaper patterns that are possible. At every step along the way this postulate enters into the geometrical arguments that determine the structure of mosaics. We will now summarize some of this structure and encourage the interested reader to look at *Color and Symmetry* for the details.

12.14 Interaction of Two Rotocenters Implies a Third

Returning to the polygonal sequence of k-fold and l-fold rotocenters generated in Figures 12.29 and 12.31, in order to satisfy the fundamental postulate, the polygon must close. Loeb finds that the rotocenters congruent to A_k lying at the vertices of this polygon also lie on a circle while the rotocenters congruent to B_l lie on another circle concentric with the first as shown in Figure 12.32. The center C of these circles lies at the intersection of the angle bisectors to $A_k B_l A_k'$ and $B_l A_k' B_l'$ and is itself an m-fold rotocenter. Referring to the quadrilateral $B_l A_k' B_l' C$, it follows, upon summing its angles, that

$$\frac{2\pi}{l} + \frac{2\pi}{k} + \frac{2\pi}{m} = 2\pi$$

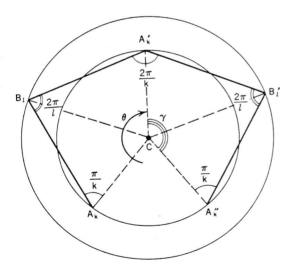

Figure 12.32 Generation of a polygonal line by the interaction of two rotocenters. (*From Loeb, 1971.* © *1971, Wiley.*)

or

$$\frac{1}{k} + \frac{1}{l} + \frac{1}{m} = 1 \qquad (12.1)$$

for k, l, m integers or ∞ ($1/\infty = 0$). This latter equation was discovered jointly by Loeb and Le Corbeiller and is the key to unlocking the structure of wallpaper patterns. Let's see how it works.

Reconsider the interaction between the sixfold rotocenter and the motif with threefold symmetry. This interaction results in the pattern shown in Figure 12.33. Since $k = 3$ and $l = 6$, Equation (12.1) predicts that $m = 2$, which is borne out. The pattern also shows that the interaction of threefold and twofold rotocenters equally well produces sixfold centers and that twofold and sixfold centers result in threefold centers.

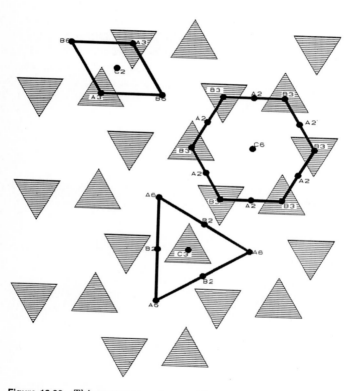

Figure 12.33   This pattern can be considered equally as the result of the interaction between an A_2 and a B_3, an A_3 and a B_6, or an A_6 and a B_2. (*Computer generated by David Henig-Elona based on figure in Loeb, 1971. © 1971, Wiley.*)

TABLE 12.2

k	l	m
1	∞	∞
2	2	∞
2	3	6
2	4	4
3	3	3

In fact, there are only five solutions to Equation (12.1) involving positive integers or ∞, and they are listed in Table 12.2. Rotocenters of order ∞ signify that the corresponding rotations degenerate to translations as they would if the center of the rotation were located at infinity. For example, the patterns for which $k = 1$, $l = \infty$, $m = \infty$ are the ones generated by William J. Gilbert's method in Section 6.8. Loeb refers to the symmetry groups of the plane with rotational subgroups of these orders by the symbols 1 $\infty\infty$, 22 ∞, 236, 244, 333 where the subgroup corresponding to the symbol ∞ is understood to be a group of translations.

12.15 Nets

If lines are drawn so that every rotocenter is joined to its nearest neighbor in each of the other two rotosimplexes, a net is formed whose nodes are the rotocenters. For example, the net for another pattern, with $k = 2$, $l = 3$, $m = 6$, is shown in Figure 12.34.

The net is subdivided into directly congruent and enantiomorphic regions (see Section 12.16) called *meshes*. The number of meshes meeting at any rotocenter equals twice the symmetry value of that rotocenter. But the number of directly congruent points at a given distance from a rotocenter equals the symmetry value of that rotocenter. Therefore two points in adjacent meshes (meshes that share an edge) cannot be directly congruent, though they may be enantiomorphic.

When a pattern does not contain enantiomorphy (no mirror or glide lines), the fundamental region can be taken to be the content of any two adjacent meshes. But in this case, the fundamental region is not unique, and Figure 12.34 shows that such a pairing of meshes can be accomplished in three different ways. If the pattern possesses enantiomorphy, the pattern contains mirror lines and glide lines. As we shall show in the next section, if all rotocenters lie on mirror lines, each mesh constitutes a fundamental domain and is bounded by mirror lines; in this case the shape as well as the area of the fundamental region is uniquely defined.

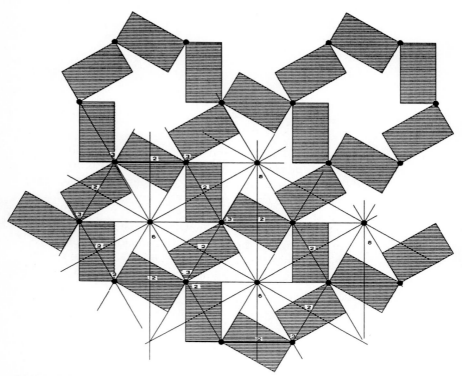

Figure 12.34   Pattern and net representing the solution $k = 2$, $l = 3$, $m = 6$. (*Computer generated by David Henig-Elona based on figure in Loeb, 1971. © 1971, Wiley.*)

12.16   Enantiomorphy

By definition, two figures are enantiomorphic if each can be transformed to the other by an improper isometry (see Section 11.9).

12.16.1   Mirror lines

Two points P and P' of a symmetry pattern are said to be directly congruent if they have identical environments. In other words, any region of the pattern that includes P can be transformed to some region of the pattern including P' by a proper isometry. For example, in Figure 12.35(a) triangles PQR and $P'Q'R'$ are identical environments of points P and P'. On the other hand, P and P' are said to be enantiomorphic if any region including P can be mapped by an improper isometry to a corresponding region including P'. Thus, triangles PQR and $P'Q'R'$ are enantiomorphic environments of points P and P' [see Figure 12.35(b)].

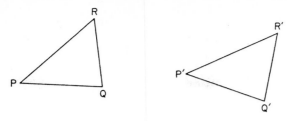

(a) P and P′ are congruent

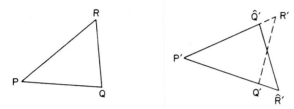

(b) P and P′ are enantiomorphic

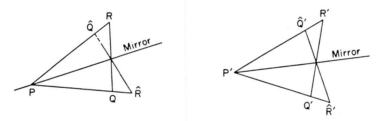

(c) P and P′ are congruent as well as enantiomorphic

Figure 12.35 Two points being simultaneously congruent and enantiomorphic. (*a*) *P* and *P′* are congruent; (*b*) *P* and *P′* are enantiomorphic; (*c*) *P* and *P′* are congruent as well as enantiomorphic. (*From Loeb, 1971.* © *1971, Wiley.*)

It is also possible for two points to be both directly congruent and enantiomorphic. In this case, both proper and improper isometries result in transforming the environment of P into the corresponding environment of P'. Figure 12.35(*c*) illustrates such a situation in which the pair of triangles PQR and $P\hat{Q}\hat{R}$ are both directly congruent and enantiomorphic to triangles $P'Q'R'$ and $P'\hat{Q}'\hat{R}'$. Notice that for this to take place, it is necessary for both P and P' to lie on mirror lines. Thus,

Theorem 12.4 Two points that are both directly congruent and enantiomorphic necessarily lie on mirror lines.

This theorem refers in particular to enantiomorphic rotocenters since these must also have the same orders. Therefore, enantiomorphy between twofold rotocenters from the 236 system implies that all twofold centers lie on mirror lines. The same holds for enantiomorphy between threefold and sixfold rotocenters, which must also lie on mirror lines. Figure 12.36 illustrates a 236 pattern in which all rotocenters are enantiomorphic.

What about 244 patterns? By the same argument, the twofold rotocenters must lie on mirror lines. However, the two fourfold rotosimplexes either may or may not be directly congruent. If they are directly congruent, they must, once again, all lie on mirror lines as shown in Figure 12.37. If they are not directly congruent, they need not lie on mirror lines as Figure 12.38 illustrates. (Note that here mirrors bisect the meshes.)

12.16.2 Glide lines

As a final look at the structure of wallpaper symmetries, we ask where the glide lines are located. Consider two k-fold enantiomorphic rotocenters A_k and $\hat{A}_k$ (see Figure 12.39). Choose an arbitrary point P_0; the rotocenter A_k implies the existence of k points $P_0, \ldots, P_{k-1}$ equidistant from A_k. Enantiomorphy implies the existence of k corresponding points $\hat{P}_0 \cdots \hat{P}_{k-1}$ equidistant from $\hat{A}_k$. From a property of

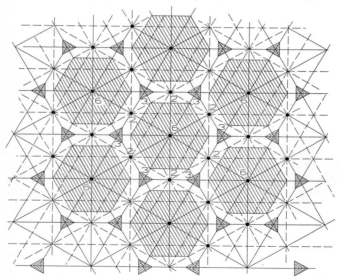

Figure 12.36 Enantiomorphy in the 236 system: all rotocenters located on mirror lines. Solid lines represent mirrors, dashed lines represent glide lines. (*Computer generated by David Henig-Elona based on figure in Loeb, 1971.* © *1971, Wiley.*)

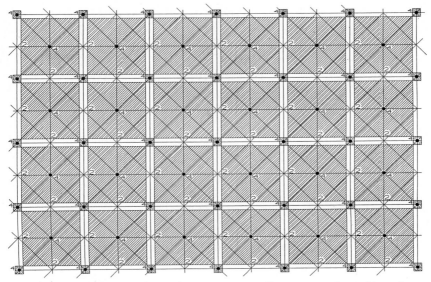

Figure 12.37 Enantiomorphy in the 244′ system: all rotocenters located on mirror lines. (*Computer generated by David Henig-Elona based on figure in Loeb, 1971. © 1971, Wiley.*)

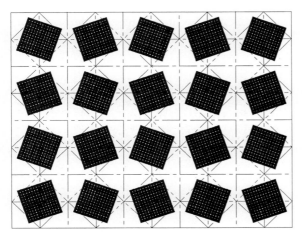

Figure 12.38 Enantiomorphy in the 244′ system: non-congruent fourfold rotocenters enantiomorphically paired. The 4′ indicates enantiomorph of 4. (*Computer generated by David Henig-Elona based on figure in Loeb, 1971. © 1971, Wiley.*)

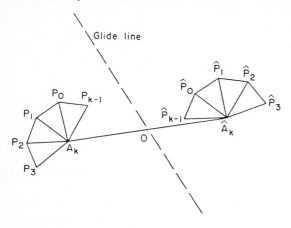

Figure 12.39 Two enantiomorphically noncongruent k-fold rotocenters. (*From Loeb, 1971. © 1971, Wiley.*)

glide reflections (see Section 11.8), a line bisecting the line segments $A_k\hat{A}_k$ and $P_0\hat{P}_0$ joining equivalent pairs of points is a glide line. This glide line also bisects the line segments $P_i\hat{P}_i$ where i is an integer joining the other equivalent pairs of points. Since A_kP_0 and $\hat{A}_k\hat{P}_1$ are also equivalent, there is another glide line that bisects the line segments $A_k\hat{A}_k$ and $P_0\hat{P}_1$ and others that bisect $A_k\hat{A}_k$ and $P_0\hat{P}_2$, etc. In total, therefore, k reflection lines intersect at the point halfway between A_k and $\hat{A}_k$. In general, these are all glide lines, but if one of them perpendicularly bisects $A_k\hat{A}_k$, it is a mirror. Thus,

Theorem 12.5 Two enantiomorphic k-fold rotocenters imply the intersection, at a point midway between them, of either k glide lines or $k - 1$ glide lines and a single mirror line.

What's more, Loeb goes on to prove:

Theorem 12.6 Two glide lines can only intersect at angles of 0, 30, 45, 60, and 90 degrees, and this can occur only in patterns of the type: 1 ∞∞, 244, 333, 2222 (not mentioned before but it has rotocenters of order 2 at the vertices of a parallelogram).

and,

Theorem 12.7 A rotocenter located on a glide line is limited to the symmetry numbers 2, 4, and ∞ and implies its enantiomorph on the same glide line. The shortest distance between enantiomorphs equals the translation component of the glide lines.

12.16.3   Some enantiomorphic patterns

Enantiomorphy in the 236 system is well represented in Figure 12.36. All rotocenters lie on mirror lines as required by Theorem 12.4. Mirror lines are represented by solid lines and glide lines are dotted. By Theorems 12.5 and 12.7, a glide line connects enantiomorphic twofold rotocenters and intersects mirror lines midway between these rotocenters and at right angles to them.

Enantiomorphy in the 244 system takes two forms. The first is illustrated in Figure 12.38. Here, the fourfold rotocenters are not directly congruent and do not lie on mirror lines. Consistent with Theorem 12.7, the line between the enantiomorphic fourfold rotocenters is a glide line and is intersected midway between the fourfold centers by perpendicular mirror lines that bisect the mesh. The second 244 symmetry pattern is shown in Figure 12.37. Here, all fourfold rotocenters are directly congruent and as a result of Theorem 12.4, all lie on mirror lines. There are also glide lines connecting twofold rotocenters as permitted by Theorem 12.7.

Loeb has developed his own notation for the 17 classes of wallpaper symmetries. Since his notation is based on the underlying structure of these classes, it helps you to read the pattern much better than with the more standard crystallographic symbols. The symbols also convey all the information needed to construct the underlying geometric scaffolding for the patterns, i.e., the rotosimplexes and mirror and glide lines. Table 12.3 lists all 17 wallpaper symmetries along with the usual notation of crystallographers. In Loeb's notation, rotocenters completely define the symmetry of a pattern in most cases. The symbols ' and ˆ denote distinctness and enantiomorphy of a rotocenter, respectively. The symmetry number of a rotosimplex is underlined when the rotocenters lie on mirror lines. The use of a diagonal, e.g., m/g and g/g' indicates mutually perpendicular glide and mirror lines; two symbols beside each other, e.g., mm', mg, and gg', indicate parallel mirror and glide lines.

12.17   Aesthetics of Wallpaper Patterns

Loeb commented on the aesthetics of wallpaper patterns as follows:

> Patterns may have rotational symmetry only, may have enantiomorphically paired noncongruent rotocenters, or may have all rotocenters located on mirror lines....These kinds of patterns have decided characteristics of their own: the patterns without enantiomorphy are wildly dynamic, whereas the patterns whose rotocenters are located on mirror lines are excessively static....[Loeb] has found the frameworks with enantiomorphically paired (but not directly congruent) rotocenters the most satisfactory from an aesthetic point of view because they give a fine balance between the static and the dynamic.

Exercise 12.7   Figure 12.23 shows some lovely wallpaper symmetry patterns reproduced from ancient sources. We leave it to the reader to explore their underlying structures in the light of Loeb's analysis. Try to relate the aesthetic qual-

TABLE 12.3

Number of distinct rotocenters with finite symmetry value	Combination of symmetry values	Configurations	I.T. notation
0	1	1 $1m$	
	∞	∞ $\infty mm'$ ∞m ∞g	
	$1\infty\infty$	$1\infty\infty'$ $1\infty\infty'mm'$ $1\infty\infty'mg$ $1\infty\infty'gg'$	$p1$ pm cm pg
1	k	k km	C_k D_k
2	22∞	$22'\infty$ $\underline{22}'\infty$ $2\hat{2}\infty$	
3	236	236 $\underline{236}$	$p6$ $p6m$
	244	$244'$ $\underline{244}'$ $2\underline{4}\hat{4}$	$p4$ $p4m$ $p4g$
	333	$33'3''$ $\underline{33}'3''$ $3\hat{3}\underline{3}'$	$p3$ $p3m1$ $p31m$
4	2222	$22'2''2''''$ $\underline{22}'2''\underline{2}'''$ $2\hat{2}\underline{2}'2''$ $2\hat{2}\hat{2}'2'g/g'$ $2\hat{2}\hat{2}'2'm/g$	$p2$ pmm cmm pgg pmg

ities of these patterns to their geometry. After all, it has been the aim of this book to show that through the microscope of geometry, a great many areas of artistic and scientific endeavor can find their common ground.

12.18   The Symmetry of Islamic Tilings

Although Islamic tilings exhibit a breathtaking variety of patterns, they have a common underlying structure [Lalvani, 1982, 1989]. First, Islamic patterns all have wallpaper symmetry. Next, consider the fundamental domain of the tiling. For example, the tiling shown in Figure 12.40 has symmetry 236, which means that its fundamental domain is a typical right triangle in the net shown in Figure 12.34.

The motifs of many of the Islamic tilings and other patterns such as Chinese lattices are directly related to the symmetry of the overall pattern. For example, in Figure 12.41 Lalvani replicates the 236 net shown in Figure 12.34 at a smaller scale within the fundamental domain itself. The sixfold, threefold, and twofold rotocenters in the fundamental domain are denoted by Pi, Qj, Rk where i, j, k are indices to distinguish one net point from the other. The motif of the tiling is traced along the lines of this replicated net. For example, $Q1.P2.Q3.P3.Q6$ traces the fundamental pattern of Figure 12.40. Interestingly, Lalvani is able to manipulate the sequence of net points within the fundamental region to transform one pattern to another so

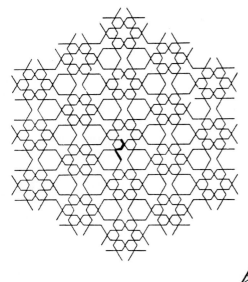

Q 1. P 2. Q 3. P 3. Q 6.

Figure 12.40   An Islamic pattern with 236 symmetry. Fundamental pattern outlined in boldface and reproduced in a 236 net superimposed on the fundamental domain.

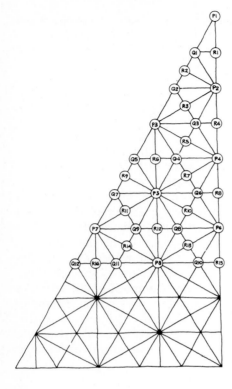

Figure 12.41 A magnification of the fundamental domain with 236 net superimposed. The fundamental pattern of Figure 12.40 is outlined by $Q1.P2.Q3.P3.Q6$.

that there is a dynamic between all tilings within a symmetry group. It is also remarkable that, once again, the basic structure of a pattern is repeated at different scales.

12.19 Symmetry of Similarity

Throughout this book, we have tried to show how geometry mediates between the harmony and unity of forms in the natural world and our human capabilities to grasp these forms with our senses. The similarity of natural forms has been one of the keys to understanding this unity and harmony and it has been a recurring theme throughout this book. The first three chapters showed how the concept of similarity is related through geometric series, spiral growth in nature, fractals and their application to the geometry of natural forms from mountains ranges to coastlines, proportion in architecture, the musical scale, and Fibonnaci series. In Chapters 5 and 6 the Penrose tilings could each be inflated to tilings with kites and darts of arbitrarily large scale. All origami patterns were shown to be made up of self-similar elements, as were the polygons within the Islamic tiling discovered by

Chorbachi and described in Section 5.13.3 and the ones studied by Lalvani and illustrated in the previous section. In Chapter 6, two-dimensional patterns of soap bubbles, biological tissues, rural market networks, and the patterns of plant growth were shown to be made up of self-similar cells. Also, in Chapter 8 the close packing of spheres led to a series of similar high-frequency polyhedra where the frequency depends on the number of layers in the packing. It is therefore fitting to conclude this book with a brief discussion of the symmetry of similarity.

12.19.1   The mathematics of similarity

In Section 2.9, the self-similarity of the right triangle led to the construction of the logarithmic spiral which arises in nature in the form of the Nautilus shell and other patterns of growth. In polar coordinates, the logarithmic spiral was described by $r = a^\theta$ for some $a > 0$. For each real number t, the transformation $T^t = (r, \theta) \rightarrow (ra^t, \theta + t)$ is a similarity which leaves the logarithmic spiral invariant. In other words, there is a whole continuum of similarities that leave the logarithmic spiral invariant. We now examine this similarity transformation and some of its variants.

12.19.2   The three similarity-symmetry operations in the plane

Shubnikov has made a complete study of the *symmetry of similarity* [1988b] and described the three similarity transformations that keep a plane similarity pattern invariant, which he refers to as Operations K, L, and M.

Operation K.   Operation K is the simplest of the three operations. It translates a figure parallel to itself so that all lines joining corresponding points converge to a common point, O, the center of similitude (see Appendix 2.B). Any pair of corresponding points A,B and A',B' are enough to establish both the center of similitude and the growth factor k of the similarity where

$$k = \frac{AB}{A'B'}$$

as shown in Figure 12.42. Repeated applications lead to a symmetry pattern such as the one shown in Figure 12.43.

Operation N.   Corresponding points within two directly similar figures related by Operation N are rotated with respect to each other so that they simultaneously define and lie on a logarithmic spiral. For exam-

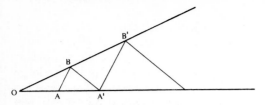

Figure 12.42 Operation K translates a figure parallel to itself. Any pair of points A,B and corresponding points A',B' establish a center of similitude O.

Figure 12.43 Example of a figure having similarity symmetry K.

ple, corresponding points A,B and A',B' spanning the sides of two triangles in Figure 12.44 define the growth factor k, where $k = AB \,/\, A'B'$ of the similarity. The two triangles are rotated with respect to each other by angle φ. The two quantities k and φ uniquely define a logarithmic spiral upon which can lie any sequence of corresponding points under this transformation as shown in Figure 12.44. For this spiral, $k = r_2/r_1$ where r_1, r_2 are the two radii separated by angle φ that joins A and A' to the center O of the spiral. In fact, if

$$r_2 = a^{\theta_2} \quad \text{and} \quad r_1 = a^{\theta_1}$$

then

$$\frac{r_2}{r_1} = a^{\varphi} \quad \text{where } \varphi = \theta_2 - \theta_1$$

or, taking logarithms,

$$\ln \frac{r_2}{r_1} = \varphi \ln a$$

Figure 12.44 Operation L rotates a figure through angle φ.

Setting $r_2/r_1 = k$ and using Equation (2.10) yields

$$\tan \psi = \frac{\varphi}{\ln k}$$

where ψ is the angle between the tangent and radial directions to the spiral, which completely defines the spiral as described in Section 2.9.

Repeated applications of Operation N are shown in Figure 12.44. Figure 12.45 represents another pattern with symmetry L ($\theta = -\pi / 5$). In other words, successive segments of the pattern are displaced by one-tenth of a revolution, with the negative number signifying that the pattern grows in a clockwise direction. You will also notice that the pattern subdivides into 10 congruent segments with each segment representing similarity symmetry by Operation K. This segmenting of an L symmetry into K symmetries will always be possible when the angle of rotation θ is a rational fraction of 2π. It is this Operation L that lies behind the symmetry of the Nautilus shell shown in Section 2.10.

Operation M.   Operation M is the analogue of a glide reflection. Corresponding enantiomorphically similar figures such as the two right triangles with hypotenuses spanning AB and $A'B'$ are shown in Figure 12.46. They have a growth factor $k = AB/A'B'$, and they are related to each other by a combination of translation by the K-operation and a kind of similarity-reflection glide axis G. To be more specific, notice in Figure 12.46 that the line AA' joining corresponding points A and A' is oriented at an angle φ with respect to any characteristic line of the original pattern such as AB. Thus, the similar image $A'B'$ is oppositely oriented by the same angle φ to the line $A'A''$ joining A' with the corresponding point A'' of the hypotenuse $A''B''$ where $AA'/A'A'' = AB/A'B' = k$. $A''B''$ can be seen to be related to AB by Operation K. The center of similitude O defined by Operation K is also the center for a K-operation relating $A'B'$ and $A'''B'''$. The similarity glide line G is the angle bisector of $\angle A'OA$.

Figure 12.45 Example of a figure having the similarity symmetry L ($\theta = -\pi/5$).

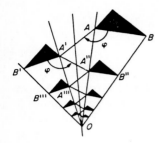

Figure 12.46 General method of constructing a figure having the similarity symmetry M from two given parts AB and $A'B'$ having mirror-similarity relation. The similarity plane bisects angle AOA'.

The result of repeated applications of the M-operation to a series of right triangles is shown in Figure 12.46.

12.19.3 Similarity symmetry groups

Operations K, L, and M can be combined with the usual point symmetry transformations to generate patterns that have point similarity symmetry. For example, the pattern in Figure 12.47 is called $6.L$ ($\varphi = -\pi/8$) because it is invariant under sixfold rotations about the center. The black polygons transform to the black and the white to the white. The angle $\varphi = -\pi/8$ signifies the angular intervals through which all the main spirals, which are readily visible, run along the short diagonals of the black (or white) quadrilaterals.

Figure 12.48 illustrates a pattern with symmetry $8.M$ since it combines the M-operation with eightfold rotational symmetry. Figure 12.49, demonstrating similarity symmetry $12.mL$ ($\varphi = \pi/12$), is made up of all similar hexagons as in two-dimensional space-filling patterns of soap bubble and biological tissue of Section 6.3. It can be seen to have twelvefold dihedral symmetry about the center and similarity invariance under the L-operation through a central angle of $\pi/12$.

As a final example, we return in Figure 12.50 to another geometric diagram illustrating the archetype 12 contained in the sacred geome-

Figure 12.47 Figure having the similarity symmetry $6.\,L$ ($\varphi = -\pi/8$).

Figure 12.48 Figure having the similarity symmetry 8.*M*.

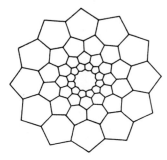

Figure 12.49 Figure having the similarity symmetry 12.*mL*.

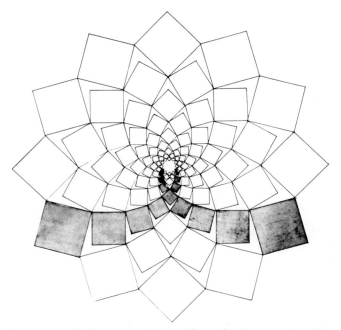

Figure 12.50 A 12-square pattern with similarity symmetry either 12.*mL* ($\varphi = \pi/24$) or 12.*mK*. Growth factor between successive squares is $\sqrt{2}$.

try of the New Jerusalem (see Section 1.2). Keith Critchlow states that this pattern was taken as a proportioning diagram used by Islamic craft schools to illustrate controlled proportional increase or decrease [1976]. He also describes this diagram's relevance to the design of the Chartres Cathedral in the film *Reflections* [1977]. The similarity symmetry is either $12.mL$ ($\varphi = \pi/24$) or $12.mK$, with a growth factor of $\sqrt{2}$ between successive squares within each of the 12 spirals. In other words, the diagonal of each square is the side of the next larger square in the spiral. A sequence of nine stages is shown from the outermost to the innermost square. These spirals of growth are models for the growth of plants while also serving as a proportional guide for the design of an entire building or a single tile.

Epilogue

When I undertook the writing of this book 4 years ago, I had a dim awareness of how similarity, proportion, the theory of graphs, two- and three-dimensional geometry and tiling, and symmetry were interrelated and formed a common language for the arts, architecture, the sciences, and engineering. At the conclusion of this project, I stand in awe of the degree to which the knowledge of these areas connect. These various disciplines, rather than being nations at the Tower of Babel, all speaking in different tongues, have much in common.

I have also discovered that mathematics provides the sinew and bone that knits these diverse areas of knowledge together. However, this is not enough. In order to gain life, ideas must travel from their roots in abstraction to the sights, sounds, smells, and textures of the world of experience. Here is where designers enter the picture as coequals. In a sense, the mathematician tears at the heart of a problem and reduces its essence while, paradoxically, gaining a deeper and more general understanding of it. On the other hand, the designer sees a problem as a whole and offers up a personal solution and through this special involvement, sheds light for all to see. Each needs the other's insights. It is only through this duality of approaches to problems of all kinds that lasting and useful solutions can come about.

It is my feeling that the most worthwhile achievements of humankind come about not merely for utilitarian purposes but to fulfill certain spiritual yearnings in common to us all. It is here that the work of fine artists, composers, the great works of sacred architecture, and some of the writings of ancient civilizations have something to tell us. I have tried as best as I can in this book to present some of this material and point to the need to take a more serious account of it.

Although two- and three-dimensional euclidean geometry has traditionally nourished the roots of mathematical thought, today it is a much neglected subject studied only by a few specialists. It is our hope that educators will be stimulated by the material that we have presented to introduce geometry to their students through constructions and applications.

It is out of the need to rediscover geometry as the language of the arts and sciences that design science had its origins. This book only touches the periphery of this vast discipline of design science, yet we hope that it will provide readers with a compass to find their own way. The reward will be to put the reader not only in touch with colleagues from other areas of thought but also to help recreate the linkages that bind us to the work of the centuries that have preceded us.

References

Abbott, E. A., *Flatland*. New York: Dover (1952).

Ackland, J. H., *Medieval Structure: The Gothic Vault*. Toronto: Univ. of Toronto Press (1972).

Alexander, C., *Notes on the Synthesis of Form*. Cambridge, MA: Harvard Univ. Press (1964).

———, "Perception and Modular Coordination." *RIBA Journal*, 425-9 (October 1959).

Andrews, W. S., *Magic Squares and Cubes*. New York: Dover (1960).

Ash, P., et al., "Convex Polyhedra, Dirichlet Tessellations, and Spider Webs." (In Senechal, 1988, following.)

Bachmann, T., and Bachmann P., "An Analysis of Béla Bartók's Music through Fibonaccian Numbers and the Golden Mean." *Music Quarterly*, 72–82 (1979).

Baer, S., *Zome Primer*. Alberqueque, NM: Zomeworks (1970). (See Biocrystal below.)

Baglivo, J. A., and Graver, J., *Incidence and Symmetry in Design and Architecture*. New York: Cambridge Univ. Press (1983).

Ball, N. R., *Mathematical Recreations and Essays*. Revised by H. S. M. Coxeter. New York: Macmillan (1967).

Banchoff, R., *Visualizing Dimensions,* New York: Scientific Am. Books (in press).

Baracs, J., Private communication (1989).

———, et al., "Polyhedral Habitat." *Structural Topology*, no. 2 (1979).

Barnette, D. W., and Grunbaum, B., "On Steinitz's Theorem Concerning Convex 3-Polytopes and on Some Properties of Planar Graphs." Lecture Notes in *Mathematics*, vol. 110. New York: Springer-Verlag (1969).

Barnsley, M., *Fractals Everywhere*. New York: Academic Press (1988).

Beck, A., Bleicher, M. N., and Crowe, D. W., *Excursions into Mathematics*. New York: Worth (1969).

Benade, A. A., *Fundamentals of Musical Acoustics*. New York: Oxford Univ. Press (1976).

Bentley, W. A., and Humphreys, W. J., *Snow Crystals*. New York: Dover (1962).

Bern, M. W., and Graham, R. L., "The Shortest-Network Problem." *Sci. Am.*, vol. 260, no. 1: 84–89 (Jan. 1989).

Biocrystal created by P. Hilderbrandt and M. Peletier (based on design by S. Baer), 185 Hungry Hollow Rd., Spring Valley, NY 10977. (See Baer, S., above.)

Blake, R. N., "The Spider and the Fly: A Geometric Encounter in Three Dimensions." *Math. Teacher*. (Feb. 1985).

Bloss, F. D., *Crystallography and Crystal Chemistry*. New York: Holt (1971).

Bolker, E. D., and Crapo, H., "How to Brace a One-Story Building." *Environment and Planning*, B, 4, 125–52 (1977).

Bonner, J. T., *Morphogenesis*. Princeton, NJ: Princeton Press (1952).

Bouleau, *The Painter's Secret Geometry: A Study of Composition in Art*. New York: Hacker (1963).

Bourgoin, J., *Arabic Geometrical Pattern and Design*. New York: Dover (1973).

Boys, C. V., *Soap Bubbles*. New York: Dover (1959).

Budden, F. J., *The Fascination of Groups*. New York: Cambridge Univ. Press (1972).

Burckhardt, T., *Art of Islam: Language and Meaning.* London: World of Islam Festival Publishers (1976).

——, *Mirror of the Intellect: Essays on Traditional Science and Sacred Art,* W. Stoddart (trans. and ed.).. Albany, NY: SUNY Press (1987).

Burt, M., "Saddle Polyhedra and Close-Packing." *Zodiac 22* (1973)

——, "The Wandering Vertex Method." *Structural Topology,* no. 6 (1982).

——, Kleinmann, M., and Wachman, A., *Infinite Regular Polyhedra.* Haifa: Technion (1974).

Chorbachi, W., "In the Tower of Babel: Beyond Symmetry in Islamic Design." (In Hargittai, 1988, following.)

Christie, A. H., *Pattern Design.* New York: Dover (1969).

Chu, R., "Octet Truss Expansion System." *Synergetica,* vol. 1, no. 4 (Nov. 1986).

Cole, R. V., *Perspective for Artists.* New York: Dover (1976).

Cook, T., *The Curves of Life.* New York: Dover (1979).

Courant, R., and Robbins, H., *What Is Mathematics?* New York: Oxford Univ. Press (1941).

Coxeter, H. S. M., *Contributions of Geometry to the Main Stream of Mathematics.* Notes by W. O. J. Moser. NSF Summer Institute, Oklahoma Agricultural and Mechanical College (1955).

——, "Golden Mean Phyllotaxis and Wythoff's Game." *Scripta Mathematica,* vol. XIX, no. 2-3 (1953).

——, *Introduction to Geometry.* New York: Wiley (1961).

——, "Music and Mathematics." *Math. Teacher,* 312–320 (1968).

——, "Regular and Semiregular Polyhedra." (In Senechal, 1988, following.)

——, *Regular Polytopes.* New York: Dover (1973).

——, *Twelve Geometrical Essays.* Carbondale, IL: Univ. of Southern Illinois Press, (1968).

Coxeter, H. S. M., et al. (eds.), M. C. Escher: *Art and Science,* New York: Elsevier Sci. Pub. B.B. (1986).

Crapo, H., "Mathematical Questions Concerning Zonahedral Spacefilling." *Structural Topology,* no. 2. (1978).

——, "Transpolyhedra: A Review." *Structural Topology,* no. 6 (1982).

—— (ed.), *Structural Topology: A Journal of Design and Structure (1978–)* Univ. of Montreal.

Critchlow, K., *Islamic Patterns: An Analytical Approach.* London: Thames Hudson (1984).

——, *Order in Space: A Design Sourcebook.* London: Thames Hudson (1987).

——, (consultant), *Reflections.* L. Moore (producer/director). Arts Council of Great Britain. Distributed by American Federation of the Arts (1977).

——, *Time Stands Still.* New York: St. Martin's Press (1982).

——, *Zodiac 22.* (1971).

Crowe, D. W., *Albrecht Dürer and the Regular Pentagon.* (In Hargittai, following.)

——, *Symmetry, Rigid Motions, and Patterns.* Arlington, VA: COMAP Inc. (1986).

Cundy, H. M., and Rollett, A. P., *Mathematical Models.* New York: Oxford Univ. Press (1961).

Davis, P. A., and Chinn, W. G., *3.1416 and All That.* New York: Simon and Schuster (1969).

Dewdney, A. K., "Imagination Meets Geometry in the Crystalline Realm of Latticeworks." *Sci. Am.,* vol. 258, no. 6: 120–123 (June 1988).

Doczi, G., *Power of Limits: Proportional Harmonies in Nature, Art and Architecture.* Shambhala (1981).

Dormer, K. J., *Fundamental Tissue Geometry for Biologists.* New York: Cambridge Univ. Press (1980).

Dunham, D., "Hyperbolic Symmetry." (In Hargittai, 1986, following.)

Dunstone, P. H., *Combinations of Numbers in Building.* London: Estates Gazette (1965).

Edmondson, A., *A Fuller Explanation: The Synergetic Geometry of R. Buckminster Fuller.* Boston: Birkhauser (1987).

Edwards, L., *Projective Geometry.* Phoenixville, PA: Rudolph Steiner Institute (1985).

Engel, P., *Folding the Universe: Origami from Angelfish to Zen*. New York: Vintage Books (1989).

———, "Origami Gallery." *Discover*, vol. 9, no. 6, 54–61 (June 1988).

Erickson, R. O., "The Geometry of Phyllotaxis." In *The Growth and Functioning of Leaves*, J. E. Dale and F. L. Milthrope (eds.). New York: Cambridge Univ. Press (1983).

Ernst, B., *The Magic Mirror of M. C. Escher*. New York: Ballantine (1976).

Escher M. C., and Locher, J. L., *The World of M. C. Escher*. New York: Abrams (1971).

Euler, L., "The Koenigsberg Bridges." In *Mathematics: An Introduction to its Spirit and Use*, J. R. Newman (ed.). New York: W. H. Freeman (1979).

Falconer, K., *Fractal Geometry*. New York: Wiley (1990).

Findeli, A., *Structural Topology*, no. 12 (1986).

Firby, P. A., and Gardiner, C. F., *Surface Topology*. Chichester: Ellis Horwood, Ltd.; Halstead Press, New York, distributors (1982).

Fomenko, A. T., "Symmetries of Soap Films." (In Hargittai, 1986, following.)

Franceschelli, A., Private communication.

Francis, G. K., *A Topological Picturebook*. New York: Springer-Verlag (1987).

Frank, A., Private communication.

Frederico, P. J., *Descartes on Polyhedra*. New York: Springer-Verlag (1982).

Fuller, R. B., *Synergetics* (appendix by Arthur Loeb). New York: Macmillan (1975).

———, and Marks, R., *The Dymaxion World of Buckminster Fuller*. New York: Doubleday Anchor Books (1973).

Gabriel, F., "The Space Within. A Guided Tour." *J. of Space Struct.*, vol. 1, no. 1, 3–12 (1985).

Gadol, J., *Leon Battista Alberti: Universal Man of the Early Renaissance*. Chicago: Univ. of Chicago Press (1969).

Galileo, G., *Two New Sciences*. H. Crew and A. DeSalvio (trans.). New York: Dover (1954).

Gardner, M., *Aha! Insight*. New York: W. H. Freeman (1978).

———, *The Ambidextrous Universe*. New York: Basic Books (1964).

———, "Extraordinary Nonperiodic Tiling." *Sci. Am.*, 110–121 (1978)

———, *Mathematical Circus*. New York: Knopf (1979).

———, "On the Remarkable Czaszar Polyhedron and Its Application to Problem Solving." *Sci. Am.*, 102–106 (May 1975).

———, *Penrose Tiles: Trapdoor Ciphers*. New York: W. H. Freeman (1989).

———, "Szilassi Polyhedron." *Sci. Am.*, vol. 239, no. 5, 22–32 (1978).

———, "White and Brown Music, Fractal Curves and One-Over-F Fluctuations." *Sci. Am.*, vol. 238, no. 4 (1978).

Ghyka, M., *A Practical Handbook of Geometrical Composition and Design*. London: Tiranti (1952).

———, *The Geometry of Art and Life*. New York: Dover (1978).

Gilbert, W. J., "An Easy Way to Construct Spacefillings." *Structural Topology*, no. 8 (1983).

Gombrich, E. H., *The Sense of Order: A Study in the Psychology of Decorative Art*. Ithaca, NY: Cornell Univ. Press (1979).

Gordon, J. E., *Structures*. New York: Plenum (1978).

Gorman, P., *Pythagoras—A Life*. Boston: Rutledge and Keegan (1979).

Gray, J. C., "Design Science 25: Celebrating 25 Years of Design Science at Harvard." April (1988).

Grünbaum, B. Private communication, (1989).

———, "Periodic Ornamentation of the Fabric Plane: Lessons from Peruvian Fabrics," *Symmetry*, vol. 1 (1990).

———, "Regular Polyhedra—Old and New." *Aequationes Mathematicae 16*, 1–20 (1977).

———, "Shouldn't We Teach Geometry?" *The Two-Year College Math. J.*, vol. 12, no. 4, (Sept. 1981).

———, Grünbaum, Z., and Shephard, G. C., "Symmetry in Moorish and Other Ornament." (In Hargittai, 1986, following.)

——— and Shephard, G. C., "Duality of Polyhedra." (In Senechal, 1986, following.)

—— and ——, "Perfect Colorings of Transitive Tiles." *Discrete Math.*, 20, 235–247 (1977).

—— and ——, *Introduction to Tilings and Patterns.* New York: W. H. Freeman (1989).

—— and ——, *Tilings and Patterns.* New York: W. H. Freeman (1987).

—— and ——, "Tilings by Regular Polygons." *Math. Mag.*, vol. 50, no. 5, 227–247 (1977).

Guidoni, E., *Primitive Architecture.* New York: Abrams (1978).

Hambridge, J., *The Fundamental Principles of Dynamic Symmetry as They Are Expressed in Nature and Art.* Alberqueque: Gloucester Art (1979).

Hargittai, I., *Five-Fold Symmetry in a Cultural Context.* New York: VCH Publishers (in press).

——, *Symmetry: Unifying Human Understanding.* Elmsford, NY: Pergamon (1986). Also appeared in *Comp. and Math. with Appl.*, vol. 12B, nos. 1–4, 1–1046 (1986).

—— (ed.), *Symmetry: An International and Interdisciplinary Journal.* VCH Publishers (1989–)

——, *Symmetry II.* Elmsford, NY: Pergamon (1988). Also published in *Comp. and Math. with Appl.*, vol. 16, nos. 5–8 (1988).

—— and Hargittai, M., *Symmetry Through the Eyes of a Chemist.* New York: VCH Publishers (1987).

Haughton, E., and Loeb, A. L., "Symmetry, the Case History of a Program," *J. Res. in Science Teaching*, 2, 132–145 (1964).

Hazen, R. M., "Perovskites." *Sci. Am.*, 74–81 (June 1988).

Hecker, Z., "Architectural Works." *Zodiac 19* (1970).

Hilderbrandt, S., and Tromba, A., *Mathematics and Optimal Form.* New York: Sci. Am. Books (1984).

Hillier, B., and Hanson, J. W., *The Social Logic of Space.* New York: Cambridge Univ. Press (1984).

Hillis, W. D., "The Connection Machine." *Sci. Am.* (June 1987).

Hofstadter, D., "Parquet Deformations: Patterns that Shift Gradually in One Dimension." *Sci. Am.*, 14–20 (July 1983).

Holden, A., *Orderly Tangles.* New York: Columbia Univ. Press (1983).

——, *Shapes, Space, and Symmetry.* New York: Columbia Univ. Press (1971).

Huntley, H. E., *The Divine Proportion: A Study in Mathematical Beauty.* New York: Dover (1970).

Jacobs, H., *Geometry.* New York: W. H. Freeman (1987).

Jean, R. V., "A Synergetic Approach to Plant Pattern Generation." *Math. Biosci* (in press).

——, *Mathematical Approach to Pattern Form in Plant Growth.* New York: Wiley (1984).

Jones, O., *Grammar of Ornament.* London: Quartich (1856); reprinted 1910 and 1928.

Kappraff, J., "A Course in the Mathematics of Design." (In Hargittai, 1986, preceding.)

——, "The Geometry of Coastlines." (In Hargittai, 1986, preceding.)

——, "The Relationship Between Mathematics and Mysticism of the Golden Mean Through History." (In Hargittai, 1990, preceding.)

——, "The Spiral in Nature, Myth, and Mathematics." (In Pickover and Hargittai, 1990, following.)

Kapusta, J. Private communication.

Kenner, H., *Geodesic Math and How to Use It.* Berkeley: Univ. of Calif. Press (1976).

Khinchin, I. A., *Continued Fractions.* Chicago: Univ. of Chicago Press (1964).

Kim, S., *Inversions.* New York: W. H. Freeman (1989).

——, and Samuelson, R. F., *Letterforms and Illusions: A Software Collection.* New York: W. H. Freeman (1989).

Klee, P., *The Nature of Nature.* J. Spiller (ed.). New York: Wittenborn (1961).

Knuth, D. E., *The Art of Computer Programming*, vol. 3, p. 543. Palo Alto: Addison Wesley (1980).

Lalvani, H., *Coding and Generating Islamic Patterns.* Ahmedabad: National Institute of Design (1982).

——, "Coding and Generating Complex Periodic Patterns." *Visual Computer*, vol. 5, 180–202 (1989).

———, Continuous Transformations of Nonperiodic Tilings and Space-Fillings. Preprint (1990).

———, "Morphological Aspects of Space Structures." Preprint (1987) to appear in *Space Structures: Theory and Practice,* H. Nooshin (ed.).

———, "Non-periodic Space Structures." *Space Structures 2,* 93–108 (1986/87).

———, *Structures on Hyper-structures: Multi-dimensional Periodic Arrangements of Transforming Space Structures."* New York: H. Lalvani (1982). Also Ph.D. thesis, Ann Arbor, Mich.: Univ. Microfilms Int. (1981).

———, "Transpolyhedra: Dual Transformations by Explosion-Implosion." Papers in theoretical morphology. New York: Lalvani (1977).

———, unpublished (1989).

———, Personal communication (1989).

Lawlor, R., *Sacred Geometry.* New York: Crossroad (1982).

Laycock, M., *Bucky for Beginners.* Hayward, CA: Activity Resources Co. (1984).

Le Corbusier. *Modulor.* Cambridge: M.I.T. Press (1968).

———, *Modulor 2.* Cambridge: M.I.T. Press (1968).

Lendvai, E., "Duality and Synthesis in the Music of Béla Bartók." In *Module, Proportion, Symmetry, Rhythm,* G. Kepes (ed.). New York: George Braziller (1966).

Loeb, A. L., Preface and Contributions (In Fuller, 1975, preceding.)

———, "Algorithms, Structure and Models," in *Hypergraphics,* AAAS Selected Symposium Series, 49–68 (1978).

———, "The Architecture of Crystals." In *Module, Proportion, Symmetry, Rhythm,* G. Kepes (ed.). New York: George Braziller (1966).

———, "A Binary Algebra Describing Crystal Structures with Closely Packed Anions." *Acta Cryst.,* vol. 11, 469–476 (1958).

———, *Color and Symmetry.* New York: Wiley (1971).

———, "Color Symetry and Its Significance for Science," in *Patterns of Symetry,* G. Fleck and M. Senechal (eds.). Amherst: U. of Mass. Press (1977).

———, Private communication (1988).

———, "The Magic of the Pentangle: Dynamic Symmetry from Merlin to Penrose," *J. Computers & Mathematics with Applications,* 17, 33–48 (1989).

———, "A Modular Algebra for the Description of Crystal Structures," *Acta Cryst.,* 15, 219–226 (1962).

———, "Polyhedra in the Work of M. C. Escher." [In H. S. M. Coxeter et al. (eds.), 1986, preceding.]

———, "Remarks on Some Elementary Volume Relations Between Familiar Solids." *The Math Teacher,* LVIII, 417–419 (May 1965).

———, "Sculptural Models, Modular Sculptures," in *Structures of Matter and Patterns in Science.* M. Senechal and G. Fleck, (eds.). Cambridge, MA: Schenkman (1979).

———, *Space Structures: Their Harmony and Counterpoint.* Reading: Addison Wesley (1976).

———, "Structure and Patterns in Science and Art," *Leonardo,* 4, 339–345 (1971) reprinted in *Visual Art, Mathematics and Computers,* Frank J. Malina (ed.), Elmsford, NY: Pergamon Press (1979)

———, "A Studio for Spatial Order," *Proc. Intl. Conference on Descriptive Geometry and Engineering Graphics, Fiftieth Anniversary Symposium of the Engineering Graphics Division of the American Society for Engineering Education,* 13–20 (1979).

———, "The Subdivision of the Hexagonal Net and the Systematic Generation of Crystal Structures." *Acta Cryst.,* 17, 179–182 (1964).

———, "Symmetry and Modularity. (In Hargittai, 1986, preceding.)

———, "A Systematic Survey of Cubic Crystal Structures." *J. of Solid State Chem.,* 1, 237–267 (1970).

———, "Vector Equilibrium Synergy," *J. Space Structures,* 1, 99–103 (1985).

——— and Gray, J., "The Rhombic Dodecahedron and Its Relation to the Cube and the Octahedron." (In Senechal and Fleck, 1988, following.)

———, and Haughton, E., "The Programmed Use of Physical Models," *J. Progr. Instr.,* III 9–18 (1965).

———, and Pearsall, G. W., "Moduledra Crystal Models." *Amer. J. Phys.,* 31, 190–196 (1963).

MacGillavry, C. H., *Symmetry Aspects of M. C. Escher's Periodic Drawings.* Utrecht: Oosthoeck (1965). Reprinted as *Fantasy and Symmetry,* New York: Abrams (1976).

Malkevitch, J., "Milestones in the History of Polyhedra." (In Senechal, 1988, following.)

Mandelbrot, B., *The Fractal Geometry of Nature.* New York: W. H. Freeman (1982).

March, L., and Steadman, P., *The Geometry of Environment.* Cambridge, MA: M.I.T. Press (1974).

Martin, G. E., *Transformation Geometry: An Introduction to Symmetry.* New York: Springer-Verlag (1982).

Marzec, C., and Kappraff, J., "Properties of Maximal Spacing on a Circle Relating to Phyllotaxis and to the Golden Mean." *J. Theor. Biol.* 103, 201–226 (1983).

Michell, J., *The Dimensions of Paradise.* San Fransisco: Harper and Row (1988).

———, *The New View Over Atlantis.* San Fransisco: Harper and Row (1983).

Minke, G., "Tensegrity." *Zodiac 21* (1971)

Miyazaki, K., and Takada, I., "Uniform Ant-hills in the World of Golden Isozonahedra." *Structural Topology,* no. 4 (1980).

Morris, I. L., and Loeb, A. L., "A Binary Algebra Describing Crystal Structures with Closely Packed Anions, Part II: A Common System of Reference for Cubic and Hexagonal Structures," *Acta Cryst.* 13, 434–443 (1960).

Nelson, D. R., "Quasicrystals." *Sci. Am.,* 43–51 (August 1987).

Niven, I., *Numbers: Rational and Irrational.* New York: Random House (1961).

Nooshin, H., and Makowski, Z. S. (eds.), *Intl. J. of Space Structures* (1985–).

Olds, C. D., *Continued Fractions.* Washington, DC: Math. Assoc. of Amer. Books (1963).

Ore, O., *Graphs and Their Uses.* New York: Random House (1963).

Panofsky, E., *Meaning in the Visual Arts.* Chicago: Univ. of Chicago Press (1955).

Pauling, L., and Hayward, R., *The Architecture of Molecules.* New York: W. H. Freeman (1964).

Pearce, P., *Structure in Nature: A Strategy for Design.* Cambridge: M.I.T. Press (1978).

Penrose, R., "Pentaplexity." *Eureka 39* (1978) 16–22. Reprinted in *Math. Intelligencer,* 2, 32–37 (1979).

Pickover, C. A., and Hargittai, I. (eds.), *Spiral Symmetry* (special issue of *Symmetry,* in press).

Plato, *Timaeus,* D. Lee (trans). New York: Penguin (1977).

Plattner, S., "Rural Market Networks." *Sci. Am.,* vol. 232, no. 5 (1975).

Pugh, A., *An Introduction to Tensegrity.* Berkeley: Univ. of Calif. Press (1976).

———, *Polyhedra: A Visual Approach.* Berkeley: Univ. of Calif. Press (1976).

Purce, J., *The Mystic Spiral.* New York: Thames and Hudson (1980).

Rivier, N., Occelli, J., and Lissowski, A., "Structure of Bénard Convection Cells, Phyllotaxis and Crystallography in Cylindrical Symmetry." *J. Physique,* 45, 49–63 (1984).

——— and Weaire, D., "Soap, Cells and Statistics—Random Patterns in Two Dimensions." *Contemp. Phys.,* vol. 25, no. 1, 59–99 (1984).

Rhombics, Inc. Watertown, MA 02172.

Rotgé, J., "Rotating Polyhedra with Congruent Plane Pentagonal Faces." *Structural Topology,* no. 9 (1984).

Rowe, C., *The Mathematics of the Ideal Villa and Other Essays.* Cambridge, MA: M.I.T. Press (1976).

Rucker, R., *The Fourth Dimension: A Guided Tour of the Higher Universe.* Providence: Janson (1989).

Sadoc, J. F., "Period Network of Disclination Lines: Applications to Metal Structures." *J. of Phys. Lettre,* vol. 44 (1983) L-707.

Safdie, M. "New Environmental Requirements for Urban Buildings." *Zodiac 19* (1969).

Salvadori, M., *Architecture and Engineering: Why Buildings Stand Up.* New York: N.Y. Academy of Science (1983).

Schattschneider, D., "In Black and White: How to Create Perfectly Colored Symmetric Patterns." *Comp. and Math. with Appl.,* vol 12B, no. 3/4 (1986).

———, "The Plane Symmetry Groups: Their Recognition and Notation." *M.A.A Monthly,* 85, 439–450 (1978).

Schechtman, D., et al., Letters in *Phys. Rev.,* vol. 53, no. 20, 1951–3 (1984).

Scholfield, P. H., *The Theory of Proportion in Architecture.* New York: Cambridge Univ. Press (1958).

Sedlak, V., "Paper Shelters." Structural Plastics Research Unit, Univ. Guildford, England (1973).

Senechal, M., "Color Groups." *Discrete Appl. Math.*, 1, 51–73 (1979).

———, "Point Groups and Colour Symmetry." *Z. Kristall.*, 142, 1–23 (1975).

——— and Fleck, G., *Shaping Space: A Polyhedral Approach*. Boston: Birkhauser (1988).

Shoemaker, C. B., and Shoemaker, D. P., *Acta Crysl*. B, 42, 3 (1986).

Shubnikov, A. V., "On the Works of Pierre Curie on Symmetry." (In Hargittai, 1988, preceding.)

———, "Symmetry of Similarity." Reprinted from *Soviet Physics Crystallography*, vol. 5, no. 4, 469–476 (1961) by American Institute of Physics in *Crystal Symmetries*, I. Hargittai and B. K. Vainshtein (eds.). Oxford: Pergamon (1988).

——— and Koptsik, V. A., *Symmetry in Science and Art*. New York: Plenum Press (1974).

Slawsky, N., "The Artist as Mathematician." *Math. Teacher*, vol. 70, no. 4 (Apr. 1977).

Smith, C. S., "Structure Substructure Superstructure." In *Structure in Art and Science*, G. Kepes (ed.). New York: George Braziller (1965).

———, "The Shape of Things." *Sci. Am.*, vol 190, no. 1, 58–65 (1954).

Sonea, S., "The Global Organism." *The Sciences* (June 1988)

Steinhaus, H., *Mathematical Snapshots*. New York: Oxford Univ. Press (1969).

Stevens, P., *A Handbook of Regular Patterns: An Introduction to Symmetry in Two Dimensions*. Cambridge: M.I.T. Press (1981).

———, *Patterns in Nature*. Boston: Little, Brown (1974).

Stewart, A., Private communication (1986).

Stewart, B. M., *Adventures among the Toroids*. Author: 4494 Wausau Road, Okemos, MI 48864 (1980).

Struble, M., *Stretching a Point*. Philadelphia: Westminster (1971).

Szilassi, L., "Regular Toroids." *Structural Topology*, no. 13, 69–80 (1986).

Tensegrity Systems, Octa-jitterbug. RD 1, Box 270. Tivoli, NY 12583.

Thomas, D. E., "Mirror Images." *Sci. Am.*, 206–228 (Dec. 1980).

Thompson, D'Arcy, *On Growth and Form*. New York: Cambridge Univ. Press (1966).

Tietze, H., *Famous Problems in Mathematics*. Baltimore: Graylock (1965).

Trudeau, R., *Dots and Lines*. Kent, OH: Kent State Univ. Press (1976).

Turnbull, H. W., *The Great Mathematicians*. New York: New York Univ. Press (1961).

Tyng, A. G., "Geometric Extensions of Consciousness." *Zodiac 19* (1969)

———, "Simultaneous Randomness and Order: The Fibonacci Divine Proportion as a Universal Forming Principle." Ph.D. thesis. Ann Arbor, MI: Univ. Microfilms Int. (1975).

Varney, W., Private communication (1988).

Verheyen, H. F., *The Icosahedral Design of the Great Pyramids*. Preprint (1989).

Vitruvius. *Ten Books of Architecture*. M. H. Morgan (trans.). New York: Dover (1960).

Walker, J., "Methods for Going Through a Maze without Becoming Confused." *Sci. Am.* (Dec. 1986).

Washburn, D. K., and Crowe, D. W., *Symmetries of Culture: Theory and Practice of Plane Pattern Analysis*. Seattle: Univ. of Washington Press (1989).

Watts, D. J., and Watts, C., "A Roman Apartment Complex." *Sci. Am.*, vol. 255, no. 6, 132–140 (Dec. 1986).

Wells, A. F., *The Third Dimension in Chemistry*. New York: Oxford (1956).

Wenninger, M., *Dual Models*. New York: Cambridge Univ. Press (1983).

———, *Polyhedral Models*. New York: Cambridge Univ. Press (1971).

Weyl, H., *Symmetry*. Princeton, NJ: Princeton Univ. Press (1952).

Williams, R. W., *The Geometrical Foundation of Natural Structures*. New York: Dover (1972).

Winter, D., 9411 Sandrock Rd., Eden, NY 14057.

———, *Alphabet of the Heart*, Eden, NY: Blazing Star Press (1989).

Wittkower, R., *Architectural Principles in the Age of Humanism*. New York: Norton (1971).

Wright, V., Private communication (1989).

Yanagi, S., *Unknown Craftsman*. Han Shan Tang, Ltd. (1978).

Index

ABOUT THE AUTHOR

Jay Kappraff is an associate professor of mathematics at the New Jersey Institute of Technology where he has developed a course in the mathematics of design for architects and computer scientists. Prior to that, he taught at Cooper Union College and held the position of aerospace engineer at NASA. He has published numerous articles on such diverse subjects as fractals, phyllotaxy, design science, plasma physics, passive solar heating, and aerospace engineering,. He has also lectured widely on the relationship between art and science. Dr. Kappraff holds a Ph.D. from the Courant Institute of Mathematical Science at New York University.